Graph Theory

Application to the Calculation of Electrical Networks

STUDIES IN ELECTRICAL AND ELECTRONIC ENGINEERING

Vol. 1 Solar Energy Conversion: The Solar Cell (Neville)
Vol. 2 Charge Storage, Charge Transport and Electrostatics with their Applications (edited by Wada, Perlman and Kokado)
Vol. 3 Integrated Functional Blocks (Novák)
Vol. 4 Operational Amplifiers (Dostál)
Vol. 5 High-Frequency Application of Semiconductor Devices (Kovács)
Vol. 6 Electromagnetic Compatibility in Radio Engineering (Rotkiewicz)
Vol. 7 Design of High-Performance Negative-Feedback Amplifiers (Nordholt)
Vol. 8 Discrete Fourier Transformation and Its Applications to Power Spectra Estimation (Geçkinli and Yavuz)
Vol. 9 Transient Phenomena in Electrical Machines (Kovács)
Vol. 10 Theory of Static Converter Systems: Mathematical Analysis and Interpretation Part A: Steady-State Processes (Slonim with Delson)
Vol. 11 Power Sources for Electric Vehicles (McNicol and Rand, Editors)
Vol. 12 Classical Electrodynamics (Ingarden and Jamiołkowski)
Vol. 13 Reliability of Analogue Electronic Systems (Klaassen)
Vol. 14 Electrets (Hilczer and Małecki)
Vol. 15 Graph Theory: Application to the Calculation of Electrical Networks (Vágó)

STUDIES IN ELECTRICAL AND
ELECTRONIC ENGINEERING 15

Graph Theory

Application to the Calculation of Electrical Networks

ISTVÁN VÁGÓ

Department of Electromagnetic Theory
Technical University of Budapest
Hungary

ELSEVIER
Amsterdam–Oxford–New York–Tokyo 1985

The original:
A gráfelmélet alkalmazása villamos hálózatok számításában
was published by Műszaki Könyvkiadó, Budapest

Translated by
O. Bíró

The distribution of this book is being handled by the following publishers
for the U.S.A. and Canada
Elsevier Science Publishing Company, Inc.
52 Vanderbilt Avenue
New York, New York 10017, U.S.A.

for the East European Countries, Democratic People's Republic of Korea, People's Republic of Mongolia, Republic of Cuba and Socialist Republic of Vietnam
Kultura Hungarian Foreign Trading Co., P.O.Box 149, H-1389 Budapest, Hungary

for all remaining areas
Elsevier Science Publishers
P.O.Box 211, 1000 AE Amsterdam, The Netherlands

Library of Congress Cataloging in Publication Data

Vágó, István.
Graph theory.

(Studies in electrical and electronic engineering; 15)
Translation of: A gráfelmélet alkalmazása villamos hálózatok számításában.
Bibliography: p.
Includes index.
1. Electric network analysis. 2. Graph theory.
I. Title. II. Series.
TK454.2.V3413 1985   621.319'2'015115   84-21177
ISBN 0-444-99589-7

ISBN 0-444-99589-7 (Vol. 15)
ISBN 0-444-41713-3 (Series)

Joint edition published by
Elsevier Science Publishers, Amsterdam, The Netherlands
Akadémiai Kiadó, The Publishing House of the Hungarian Academy of Sciences, Budapest, Hungary

Printed in Hungary

PREFACE

A means of describing the connections of electrical networks is provided by graph theory. Its application yields a method for solving network analysis problems, by means of a systematic derivation of an appropriate number of linearly independent equations. Digital computers are readily utilized for writing the necessary relationships and solving them. It is for this reason that the application of graph theory to the calculation of electrical networks has gained widespread use in recent decades.

After the earliest work on graph theory (that of L. EULER published in 1736 [17]) it was G. KIRCHHOFF, as far as we know, who was the first to deal with this subject in his treatise issued in 1847 [30], examining primarily the laws of electrical networks. The first comprehensive book to discuss graph theory was that by D. KÖNIG [31] and it was published in 1936. A detailed study of the application of graph theory to electrical networks is presented in the book of S. SESHU and M. B. REED [42], published in 1961.

Lectures on electrical networks at the Faculty of Electrical Engineering of the Technical University of Budapest have included the applications of graph theory for more than a decade. On the basis of the experience gained in teaching this subject, a book in the Hungarian language was published by the Technical Publishing House, Budapest, in 1976. The present volume is a revised and enlarged, English language edition of this book.

I wish first of all to express my thanks to Academician Prof. OTTÓ P. GESZTI, who did pioneer work by introducing graph theory methods into education. The suggestions made by Prof. GESZTI during the writing of the first four chapters, by Prof. ANDRÁS AMBRÓZY on the completion of chapters 5 and 6, and by Assistant Professor MIKLÓS BOHUS during the writing of chapters 7 and 8 were invaluable to me. My discussions with Prof. GYÖRGY FODOR on the subject were extremely useful. In writing the theoretical parts and in the collection of examples I was helped by my immediate colleagues: assistants ISTVÁN BÁRDI and OSZKÁR BIRÓ and principal assistants EDIT HOLLÓS and IMRE SEBESTYÉN. I herewith express my thanks to all of them.

István Vágó

NOTATIONS

$\mathbf{X}$	column matrix
$\boldsymbol{X}$	matrix
$\mathbf{X}^+$	transpose of $\mathbf{X}$, row matrix
$\boldsymbol{X}^+$	transpose of $\boldsymbol{X}$
$\boldsymbol{X}^*$	conjugate of $\boldsymbol{X}$
$\dot{x}$	time derivative of x
$\langle x_1 x_2 \ldots x_n \rangle$	diagonal matrix, $x_1, x_2, \ldots, x_n$ being the elements on the main diagonal
$\boldsymbol{A}$	basis incidence matrix
$\mathbf{A}^+$	row matrix of vertex
$\boldsymbol{A}_t$	non-basis incidence matrix
$\boldsymbol{A}_0$	basis incidence matrix of undirected graph
$\mathbf{B}^+$	row matrix of loop
$\boldsymbol{B}_t$	non-basis loop matrix
$\boldsymbol{B}$	basis loop matrix
C	capacitance
$\mathbf{C}^+$	row matrix of path
$\boldsymbol{F}$	a block in the normal form of the loop matrix
G	conductance
$\boldsymbol{G}$	matrix of conductances
I	current
$\mathbf{I}$	column matrix of currents
I_g	source-current
J	loop current
$\mathbf{J}$	column matrix of loop currents
L	inductance
$\boldsymbol{L}$	incidence matrix of self-loops
$\boldsymbol{L}_i$	matrix Lagrange-polynomials
$\mathscr{L}$	symbol for Laplace transformation
$\mathscr{L}^{-1}$	symbol for inverse Laplace transformation
$\boldsymbol{P}$	matrix formed by the parameters p_i or $\boldsymbol{p}_i$ of a transmission line network
$\mathbf{Q}^+$	row matrix of cutset

$\boldsymbol{Q}_t$	non-basis cutset matrix
$\boldsymbol{Q}$	basis cutset matrix
R	resistance
$\mathbf{R}$	column matrix of excitation signals
$\boldsymbol{R}$	matrix of resistances
$\boldsymbol{R}$	matrix formed by the parameters r_i or $\boldsymbol{r}_i$ of a transmission line network
$\boldsymbol{S}$	reflection matrix
T	period
U	voltage
$\mathbf{U}$	column matrix of voltages
U_g	source-voltage
V	cutset-voltage
$\mathbf{V}_Q$	column matrix of cutset-voltages
$\mathbf{V}$	column matrix of internal signals
$\boldsymbol{W}_0$	transfer matrix
$\boldsymbol{W}_t$	vertex transfer matrix
Z	impedance
$\boldsymbol{Z}$	impedance matrix
$\boldsymbol{Z}_B$	loop-impedance matrix
$\mathscr{Z}$	symbol for z-transformation
$\mathscr{Z}^{-1}$	symbol for inverse z-transformation
Y	admittance
Y_0	characteristic admittance
$\mathbf{Y}$	column matrix of response signals
$\boldsymbol{Y}$	admittance matrix
$\boldsymbol{Y}_A$	vertex-admittance matrix
$\boldsymbol{Y}_Q$	cutset-admittance matrix
$\boldsymbol{Y}_0$	characteristic admittance matrix
b	number of edges
c	number of components
d_{ik}	inverse hybrid parameter
$f^*(t)$	sampling of $f(t)$
h_{ik}	hybrid parameter
i	current
$\mathbf{i}$	column matrix of currents
i_g, i_0	source-current
m	number of chords
n	number of vertices, nullity
$p_i, \boldsymbol{p}_i$	an admittance parameter of the i-th transmission line section
r	rank
$r(t)$	time function of excitation signal
$r_i, \boldsymbol{r}_i$	an admittance parameter of the i-th transmission line section
s	variable of Laplace transform
t	time

u	voltage
$\mathbf{u}$	column matrix of voltages
u_g, u_0	source-voltage
$v(t)$	time function of internal signal
$y(t)$	time function of response signal
$\delta(t)$	Dirac delta
γ	propagation coefficient
$\boldsymbol{\Gamma}$	matrix of propagation coefficients
Φ	node-potential
$\mathbf{\Phi}$	column matrix of node-potentials
ω	angular frequency
$1(t)$	unit step function
$\boldsymbol{1}$	unit matrix
$\boldsymbol{0}$	zero matrix

Contents

CHAPTER 1

BASIC CONCEPTS OF GRAPH THEORY

Graphs and subgraphs

The currents and voltages in electrical networks and the relationships between them depend upon the characteristics of the network elements constituting the network and on the way these elements are interconnected. The relation between the current and voltage of a two-terminal element as well as between the currents and voltages of *multi-terminal* elements is expressed with the aid of the element characteristics. Such characteristics are the resistance R, inductance L, capacitance C, source-voltage U_g, source-current I_g of linear *two-terminal* elements, the impedance and admittance parameters of linear two-ports, the characteristics of nonlinear elements, etc. The network elements form one or more branches of the network, which connect at nodes. The *graph* of the network represents the manner in which the branches and nodes are interconnected in the network independently of the network elements forming the branches. Thus, graph theory is applicable to network analysis [26, 28, 33, 36, 42, 44].

Instead of the rigorous mathematical definition of graphs the following illustrative description suffices for our discussion [9, 27].

Graphs are formed from two types of element: *edges* and *vertices*. The terms *branch* instead of *edge* and *node* instead of *vertex* are often used, particularly in an engineering, as opposed to mathematical context. The graph is a union of sets of edges and vertices with two vertices associated with each edge. One vertex may be associated with several edges. The graph can be illustrated as follows. An edge is indicated by a line or curve, and a vertex by a small circle. Each edge has two distinct endpoints, called vertices, which belong to the edge. Any two edges in the graph may only have vertices as common points. Graphs consisting of a finite number of edges and vertices are called *finite graphs*. For the solution of electrical network problems we shall only consider the use of finite graphs.

Let us examine networks made up of coupled and non-coupled two-terminal elements only. A graph of the network is drawn by representing the two-terminal elements of the network by lines, with the intersections of the lines corresponding to the connection points between the elements, these lines having no other common points. In the case of a planar or two-dimensional representation of the graph, any other intersections of the lines are not considered common points of the edges (as is customary for connection diagrams of networks). The edges of the graph

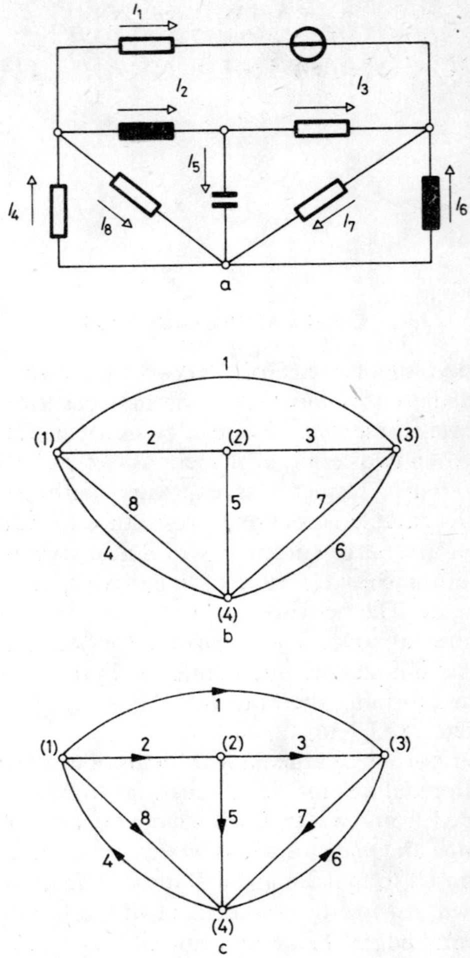

Fig. 1.1

correspond to the branches of the network, and the vertices of the graph to the joining points of edges, i.e. to the nodes. A graph of the network shown in Fig. 1.1, a is drawn in Fig. 1.1, b. The edges and vertices of the graph are numbered, using parentheses for the vertex numbers.

A direction may be assigned to the edges of the graph (Fig. 1.1, c). The direction of edge k between vertices (i) and (j) may point from (i) to (j) or from (j) to (i). If every edge of the graph is assigned a direction*, a *directed graph* is obtained. As is well

* The terms *orientation* and *oriented graph* are also commonly used.

known, a reference direction for branch-currents and for branch-voltages of the network must be chosen in order to write the basic Kirchhoff equations of the network. When applying directed graphs to network analysis it is expedient to choose the directions of the edges to correspond to the reference directions of branch-currents or branch-voltages.

The number of edges incident with a vertex gives the *degree* of the vertex. E.g. vertex (1) in Fig. 1.1 is incident with edges 1, 2, 4, 8 and so its degree is 4.

A vertex incident with one edge only is called an *end-vertex*, and the edge incident to it is an *end-edge*. The degree of an end-vertex is 1.

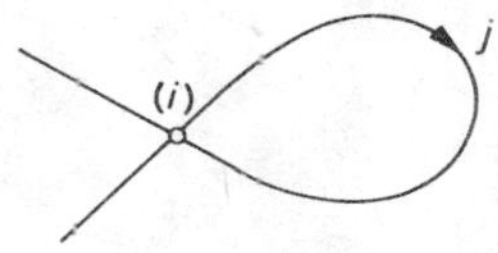

Fig. 1.2

A vertex not incident with any edge is called an *isolated vertex*, which has a degree of zero.

As has been mentioned, the two vertices incident with an edge are considered to belong to the edge. The edge without the vertices is called an open edge to make a distinction.

It should be noted that edges with coinciding endpoints, so called *self-loops*, are also discussed in graph theory (Fig. 1.2). To admit them would necessitate a generalization of the above definition of graphs. The discussion which follows primarily concerns graphs and theorems about graphs without self-loops. In those cases where self-loops may be permitted attention will be drawn to the fact.

A subset of the elements of the graph is called a *subgraph* of the graph. Certain types of subgraphs have great importance in network analysis, and will be introduced in the following sections. Particular degenerate subsets, such as sets of open edges or isolated vertices are also considered subgraphs. A few subgraphs of the graph in Fig. 1.3, a are shown in Figs 1.3, b . . . f.

If two subgraphs of the graph together contain every edge and vertex of the graph, the subgraphs being edge-disjoint, the two are called *complementary subgraphs*. Two complementary subgraphs of the graph in Fig. 1.3, a are shown in Figs 1.3, c and d.

Only certain edges of the graph are elements of any subgraph under consideration. Their membership of the subgraph may be characterized by a row matrix. If the graph contains b edges, this row matrix has b elements:

$$\mathbf{X}^+ = [x_1 \, x_2 \, \dots \, x_j \, \dots \, x_b], \tag{1.1}$$

where $x_j = 1$ if branch j is an element of the subgraph, and $x_j = 0$ otherwise.*

* $\mathbf{X}^+$ denotes a row matrix, the transpose of column matrix $\mathbf{X}$.

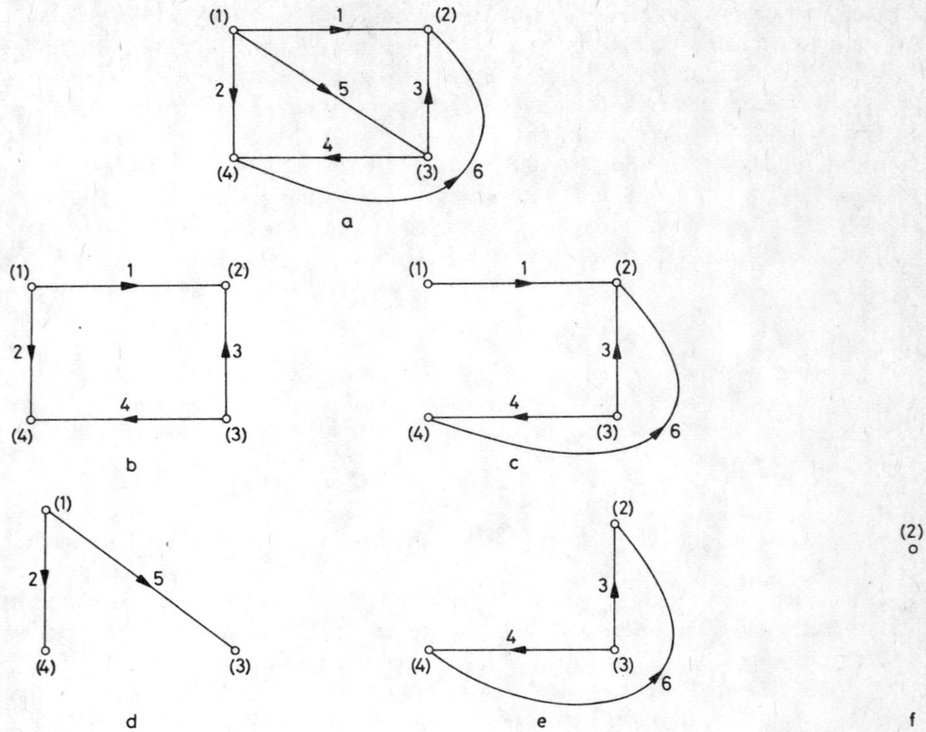

Fig. 1.3

An orientation will be found to be associated with subgraphs of directed graphs in several cases. The directions of the branches in comparison with this can then also be indicated by the elements of the row matrix, with x_j equalling $+1$, -1 or 0 according to the following:

$x_j = 1$, if branch j is an element of the subgraph, with its direction coinciding with the orientation of the subgraph;

$x_j = -1$, if branch j is an element of the subgraph, with its direction opposite to that of the subgraph;

$x_j = 0$, if branch j is not an element of the subgraph.

Path

A subgraph with two endpoints and all further vertices of degree two in the subgraph, is called a *path*. This implies that there is a unique route from one of the endpoints to the other along all the edges belonging to the path. A path can traverse

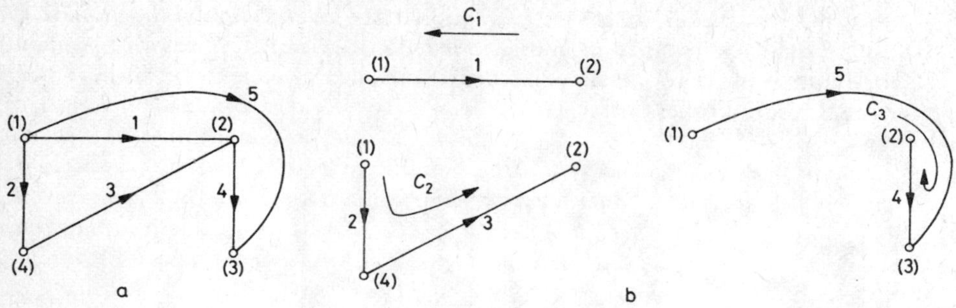

Fig. 1.4

an edge at most once, e.g. between vertices (1) and (2) in Fig. 1.4, a the subgraphs shown in Fig. 1.4, b form paths. The path may also be directed. Its direction points from one of its endpoints to the other along its edges. The direction of the path is independent of the direction of its edges: it may coincide with the direction of some edges and be opposite to that of others. In the paths shown in Fig. 1.4, b directions have been chosen. Since it is a subgraph, any path can be characterized by a row matrix with b elements (b is the number of edges in the graph):

$$\mathbf{C}^{+}=[x_1\, x_2\, \ldots\, x_j\, \ldots\, x_b]. \tag{1.2}$$

Thus the row matrices of the paths in Fig. 1.4, b as subgraphs of the graph in Fig. 1.4, a are:

$$\mathbf{C}_{10}^{+}=[1\ 0\ 0\ 0\ 0],$$
$$\mathbf{C}_{20}^{+}=[0\ 1\ 1\ 0\ 0],$$
$$\mathbf{C}_{30}^{+}=[0\ 0\ 0\ 1\ 1],$$

or indicating directions:

$$\mathbf{C}_{1}^{+}=[-1\ 0\ 0\ \ \ 0\ 0],$$
$$\mathbf{C}_{2}^{+}=[\ \ 0\ 1\ 1\ \ \ 0\ 0],$$
$$\mathbf{C}_{3}^{+}=[\ \ 0\ 0\ 0\ -1\ 1].$$

If there exists a path between any two vertices of the graph, the graph is *connected*. In general, a graph consists of one or more connected subgraphs (components).

Loop

A *loop* (circuit) is a connected subgraph with the degree of all its vertices in the subgraph equalling two. A path is obtained by deleting any open edge from a loop. The numbers of vertices and edges in a loop are equal. Among the subgraphs of the graph in Fig. 1.3, a, the ones shown in Figs 1.3, b and e are loops.

A loop may also be directed. The orientation of the loop points away from any of its vertices and back towards this initial vertex along the edges of the loop (Fig. 1.5). The orientation of the loop may coincide with the direction of some edges and be opposite to that of others.

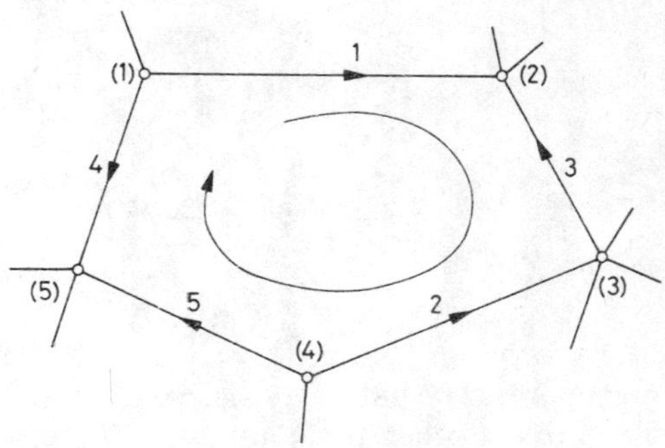

Fig. 1.5

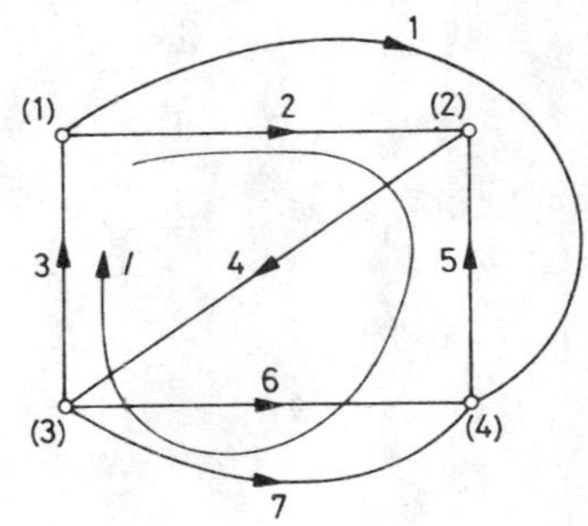

Fig. 1.6

The loop can be characterized by a row matrix:

$$\mathbf{B}_h^+ = [x_1\ x_2\ \ldots\ x_j\ \ldots\ x_b]. \tag{1.3}$$

E.g. for loop I of the graph in Fig. 1.6:

$$\mathbf{B}_{I0}^+ = [0\ \ 1\ \ 1\ \ 0\ \ 1\ \ 0\ \ 1],$$

or indicating orientations:

$$\mathbf{B}_I^+ = [0\ \ 1\ \ 1\ \ 0\ \ -1\ \ 0\ \ -1].$$

The row matrix representing the loop will be employed for the application of one of the basic laws of electrical networks, Kirchhoff's voltage law.

Block

A graph is said to be non-separable if it either consists of one edge and the two vertices incident with it, or there are two paths forming a loop between any two vertices. An example of a graph which is separable is shown in Fig. 1.7. Here, no two paths forming a loop can be given between vertices (1) and (6), since all paths between them contain vertex (3).

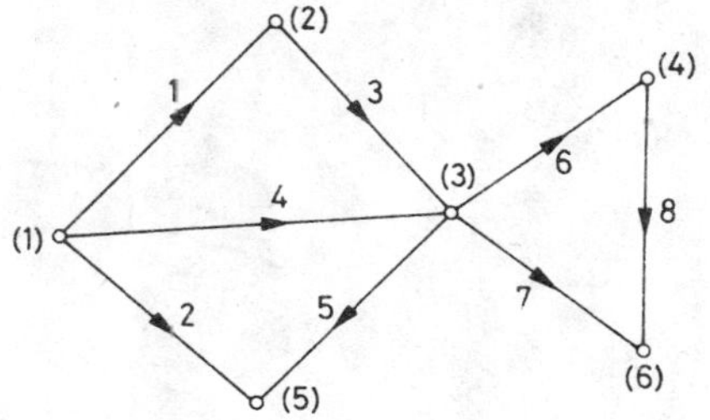

Fig. 1.7

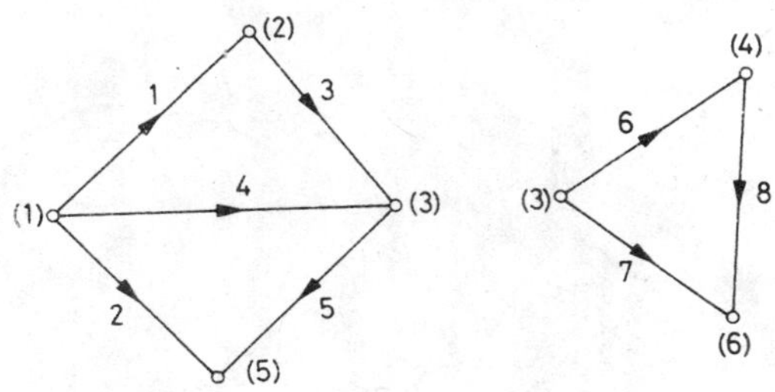

Fig. 1.8

If a graph is separable, two vertices can always be found with any paths between them containing a third vertex, called a *cutpoint* of the graph (e.g. vertex (3) in the graph shown in Fig. 1.7). If the graph is separated at all cutpoints, without creating open edges, i.e. attributing the cutpoint to both subgraphs formed by each separation (Fig. 1.8), non-separable subgraphs called *blocks* of the graph are obtained.

Tree and forest

One of the important types of subgraphs is the tree. A connected subgraph of a connected graph, containing all vertices of the graph, and not including any loop is called a *tree*. A few trees of the graph shown in Fig. 1.9, a have been drawn in Fig. 1.9, b. It follows from the definition of a tree that, ignoring any path orientations, exactly one path can be found between any two vertices in a tree of a connected graph. Otherwise the tree would contain a loop.

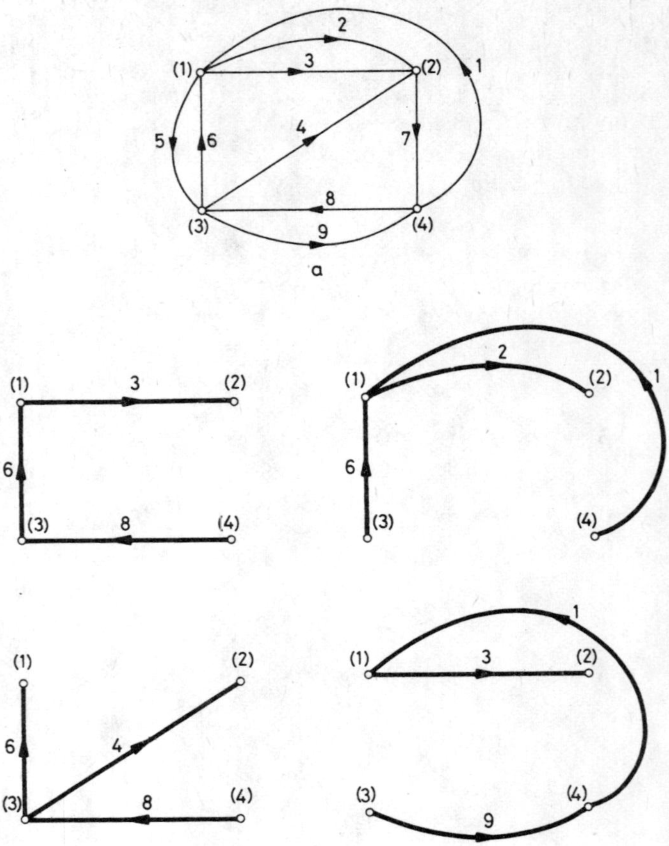

Fig. 1.9

If the number of vertices in a connected graph is n, naturally all of its trees include n vertices. The number of edges in such a tree is therefore $n-1$. This can be shown as follows. Let us construct the tree starting from any one edge of the graph. This subgraph contains two vertices. Let us join to this edge another edge of the tree incident with one of these two vertices. There are two edges and three vertices in this subgraph. If subsequent edges of the tree are always chosen to be incident with one vertex of the subgraph already constructed, the numbers of vertices and edges both increase by one at the connection of each new edge. Thus for a tree with n vertices the number of edges is $n-1$. If different trees of a graph with n vertices are constructed, every tree will have $n-1$ edges. The edges of the tree are called *tree-branches* (tree-edges).

Let the path pointing from vertex (*i*) to (*j*) in the tree of a directed graph be represented by row matrix $\mathbf{C}^+_{ij}$, and the path pointing from vertex (*j*) to (*m*) by row matrix $\mathbf{C}^+_{jm}$. Since there is only one path in the tree between vertices (*i*) and (*m*), this can only be the one characterized by row matrix

$$\mathbf{C}^+_{im} = \mathbf{C}^+_{ij} + \mathbf{C}^+_{jm}, \qquad (1.4)$$

called the sum of the two former paths.*

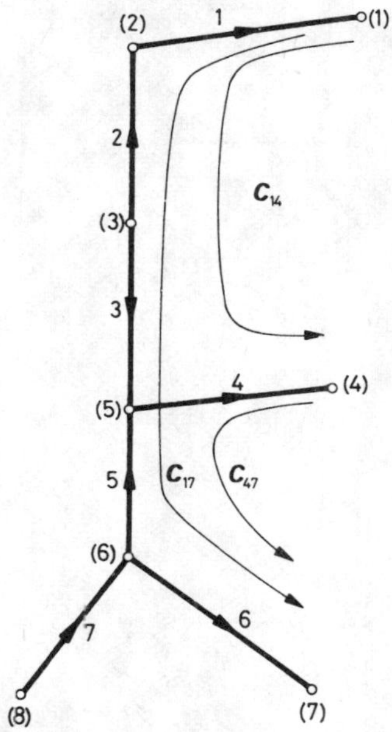

Fig. 1.10

For example, the row matrix of the directed path between vertices (1) and (4) of the tree in Fig. 1.10 is

$$\mathbf{C}^+_{14} = [-1 \quad -1 \quad 1 \quad 1 \quad 0 \quad 0 \quad 0]$$

and that of the path between (4) and (7) is:

$$\mathbf{C}^+_{47} = [\ 0 \quad 0 \quad 0 \quad -1 \quad -1 \quad 1 \quad 0].$$

Their sum

$$\mathbf{C}^+_{17} = [-1 \quad -1 \quad 1 \quad 0 \quad -1 \quad 1 \quad 0]$$

is the row matrix of the path between vertices (1) and (7).

* In the case of non-directed graphs, the same result may be obtained by using the modulo-2 sum, for which $1 \oplus 1 = 0$

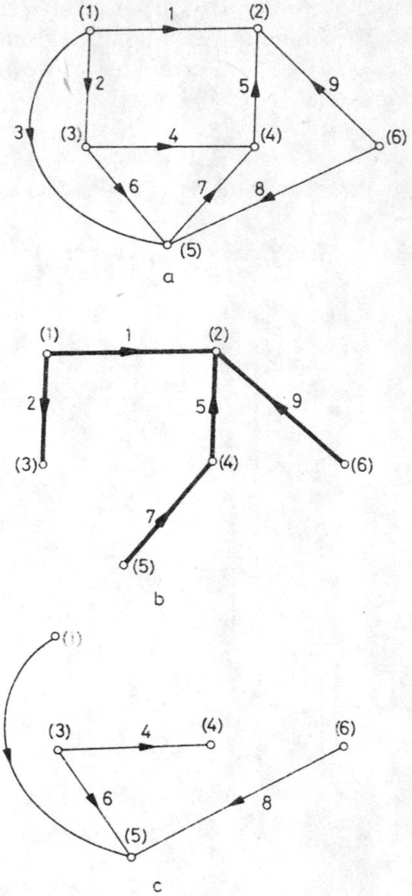

Fig. 1.11

The complementary subgraph (Fig. 1.11, c) of a tree (Fig. 1.11, b) of a connected graph (Fig. 1.11, a) is called a *cotree*. The cotree does not in general contain all vertices of the graph. The sum of the numbers of edges in a tree and cotree gives the number of edges in the graph. The edges of a cotree are called *chords*.* Let b denote the number of edges in a connected graph, n the number of vertices, and m the number of edges in a cotree, then

$$m+n-1=b, \tag{1.5}$$

since, as has been shown, the number of edges in a tree is $n-1$.

* Also called *links*.

A tree of a connected graph is always connected by definition, while the cotree may be connected (Fig. 1.11, c). or it may consist of several connected subgraphs (Fig. 1.12, a is a graph, of which Fig. 1.12, b is a tree, while Fig. 1.12, c the corresponding cotree).

A tree can be associated with each connected subgraph of a non-connected graph (Fig. 1.13, a). Selecting a tree from each component, these trees form a forest (Fig. 1.13, b).

Let a non-connected graph consist of c components. Let the number of vertices in the i-th component be n_i, the number of edges b_i, the number of chords m_i. Now, since components are connected, Eq. (1.5) may be written:

$$b_i = m_i + n_i - 1. \tag{1.6}$$

Writing these for each component and summing:

$$\sum_{i=1}^{c} b_i = \sum_{i=1}^{c} m_i + \sum_{i=1}^{c} (n_i - 1), \tag{1.7}$$

hence

$$b = m + n - c, \tag{1.8}$$

where $b = \sum_{i=1}^{c} b_i$ is the number of edges, while $m = \sum_{i=1}^{c} m_i$ is the number of chords, and $n = \sum_{i=1}^{c} n_i$ is the number of vertices in the graph. The number $n-c$ is called the

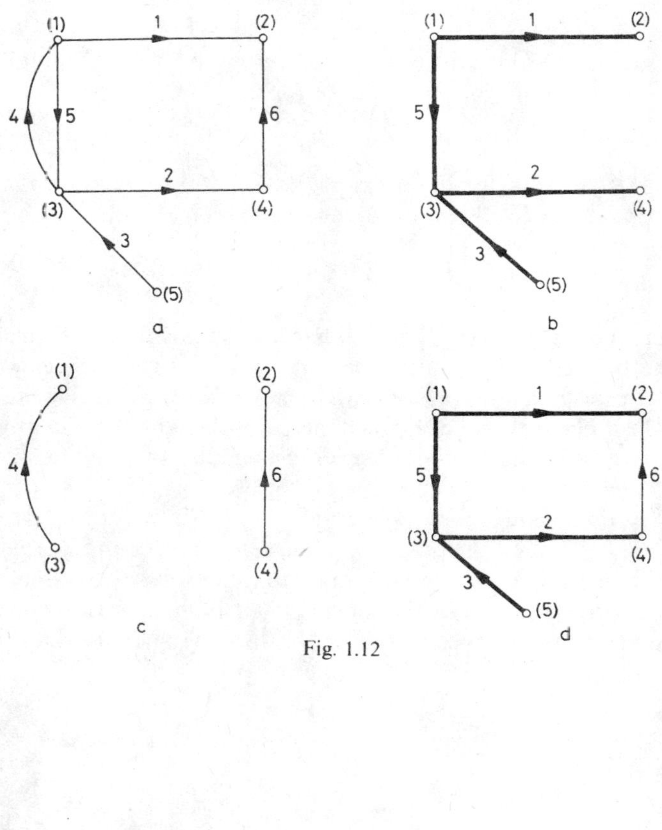

Fig. 1.12

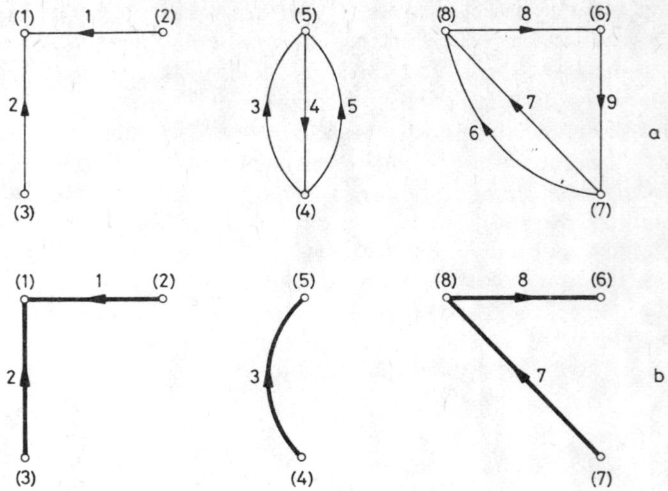

Fig. 1.13

rank (r) of the graph, while m its *nullity*. The rank of the graph equals the number of edges in all of its forests:

$$r=n-c. \tag{1.9}$$

The nullity equals the number of chords. The sum of the rank and nullity equals the number of edges in the graph according to (1.8):

$$b=r+m. \tag{1.10}$$

If there is an isolated vertex in the graph, its deletion decreases both n and c by one, i.e. $n-c$ does not vary. Thus an isolated vertex has no effect upon the rank or nullity of the graph.

Cutset

A *cutset* is a set of open edges of the graph, whose removal reduces the rank of the graph by exactly one, while by the reinsertion of any edge of the cutset the rank equals that of the original graph. Notice that open edges, i.e. edges without vertices, appear in cutsets. The removal of a cutset increases the number of components c in the graph by one, while the number of vertices n remains the same, so that the rank $r=n-c$ decreases by one.

Let us first examine a non-separable graph (Fig. 1.14, a). The edges of the cutset may be selected by cutting the graph by a curve v not crossing any vertex of the graph and dividing the graph into two connected subgraphs. The cutset is formed by the collection of open edges cut by the curve v (Fig. 1.14, b). On removing these edges the graph is split into two connected subgraphs. It is possible that an isolated vertex appears among the connected subgraphs of the graph split by a cutset. In this case the rank of the other subgraph equals that of the graph obtained by the removal of the cutset.

In case of a graph consisting of several components or blocks (Fig. 1.15, a), a cutset may only be formed by open edges belonging to one component or block (Fig. 1.15, b), otherwise it cannot satisfy the definition of a cutset. The removal of the open edges of a cutset from the graph divides this subgraph or block into two components, while other subgraphs remain intact (Fig. 1.15, c). Thus removing the cutset in Fig. 1.15, b from the graph drawn in Fig. 1.15, a, the four component graph shown in Fig. 1.15, c is obtained.

In a directed graph an orientation may be assigned to the cutset, the orientation pointing from one of the connected subgraphs created by the removal of the cutset towards the other. The directions of some edges of the cutset may coincide with that of the cutset, while others may be opposite to it.

A row matrix

$$\mathbf{Q}_i^+ = [x_1 \ x_2 \ \dots \ x_j \ \dots \ x_b] \tag{1.11}$$

can also be associated with a cutset, specifying the edges of the graph which are included in the cutset.

The cutset shown in Fig. 1.16 is represented by row matrix

$$\mathbf{Q}_{10}^+ = [0 \ 0 \ 1 \ 1 \ 0 \ 1 \ 1 \ 1 \ 0],$$

or with orientations indicated:

$$\mathbf{Q}_1^+ = [0 \ 0 \ -1 \ 1 \ 0 \ 1 \ 1 \ -1 \ 0].$$

Cutsets and their associated row matrices will be employed for writing the generalized node-equations of a network (Kirchhoff's current law).

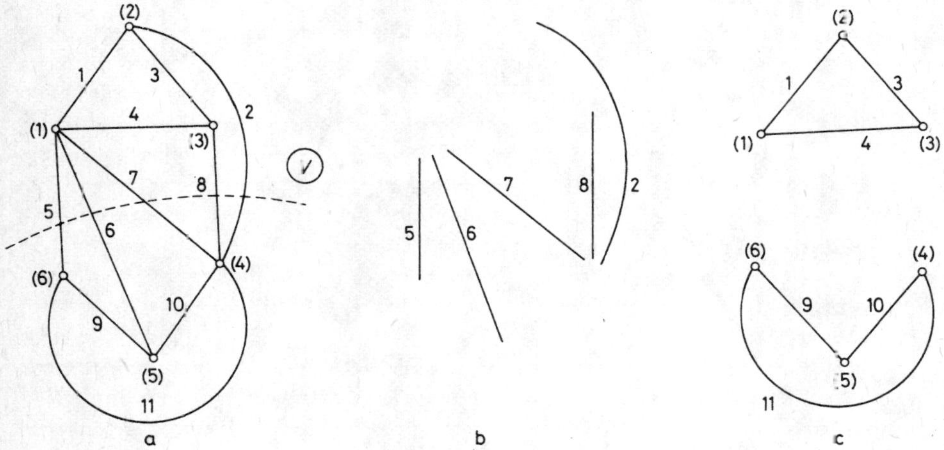

Fig. 1.14

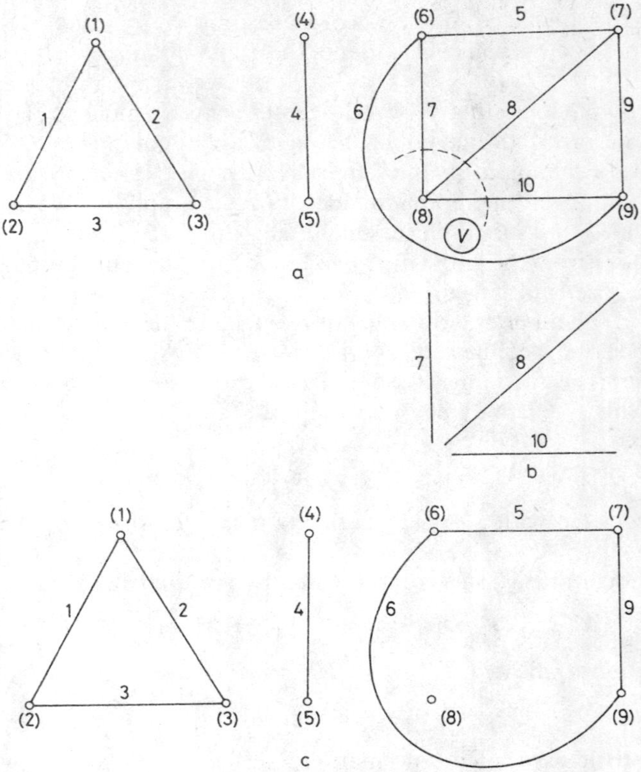

Fig. 1.15

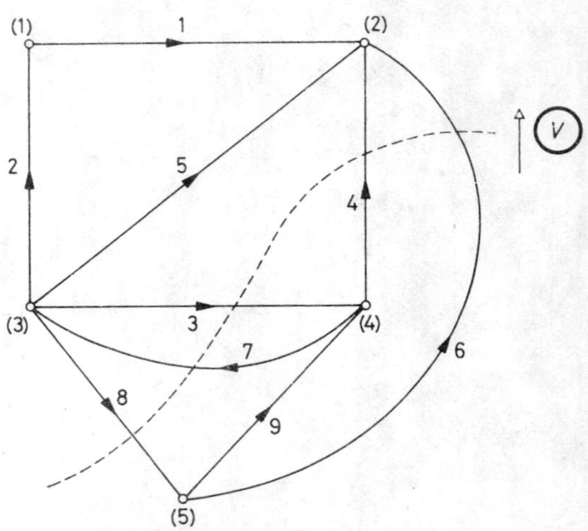

Fig. 1.16

Vertex

More than one edge may be incident with a vertex of the graph. An orientation may be assigned to a vertex, pointing by agreed convention away from the vertex. The vertex may be characterized by a row matrix

$$\mathbf{A}_i^+ = [x_1 \ x_2 \ \ldots \ x_j \ \ldots \ x_b]. \tag{1.12}$$

E.g. the row matrix of vertex (4) of the graph drawn in Fig. 1.16 is

$$\mathbf{A}_{40}^+ = [0 \ 0 \ 1 \ 1 \ 0 \ 0 \ 1 \ 0 \ 1],$$

or indicating orientation:

$$\mathbf{A}_4^+ = [0 \ 0 \ -1 \ 1 \ 0 \ 0 \ 1 \ 0 \ -1].$$

The row matrix of an end-vertex has all but one elements zero. The row matrix of an isolated vertex has all elements zero.

If the removal of the open edges of a cutset creates an isolated vertex, the row matrix of the cutset is identical to that of the isolated vertex, except that if the cutset is directed towards the isolated vertex, all the row matrix elements are multiplied by minus one.

The set of open edges incident with a cutpoint does not form one cutset, but several. Their removal from the graph reduces the rank of the graph by at least two. Accordingly, the row matrix of a cutpoint does not represent a cutset. Those edges incident with a cutpoint which belong to one block of the graph do form a cutset.

Fundamental set of cutsets

Several distinct cutsets of a given graph form a set of cutsets. A set of cutsets is linearly independent if and only if no row matrix of any cutset in the set can be expressed as a linear combination of the row matrices of the other cutsets. A *fundamental set of cutsets* of a graph is a linearly independent set of cutsets, which ceases to be linearly independent on the inclusion of any additional cutset of the graph. The row matrix of any cutset of the graph can be expressed as a linear combination of the row matrices of the cutsets in the fundamental set. In the following the fundamental set of cutsets will be associated with a particular tree or forest, and it will be called the set of cutsets generated by that tree or forest.

An arbitrary cutset of a connected graph contains at least one open edge of any tree in the graph. This may be shown as follows. Let us examine two vertices, one in each of two disjoint connected subgraphs of the graph split by the cutset. In any tree of the graph there is a path between the two vertices selected. After the removal of the cutset, however, no path exists between the two vertices. This is only possible if at least one open edge of the path in the tree is included in the cutset, and together with the cutset this edge (or these edges) must have been removed.

For any tree of a connected graph (Fig. 1.17, a), a cutset containing only one tree-branch of the tree can always be found. Let a tree be selected (Fig. 1.17, b). The tree is a connected graph, with each of its edges forming a cutset, since the removal of any open tree-branch splits the tree into two parts. The cutset sought contains this tree-branch, and all the chords forming a path between the two vertices of this tree-branch along with other suitable tree-edges (Fig. 1.17, c). It is easily verified that the collection of the chords thus obtained together with the tree-branch satisfies the definition of the cutset and contains only one edge of the tree selected.

The set of cutsets generated by a tree (Fig. 1.18, b) of a graph (Fig. 1.18, a) is obtained by associating a cutset with every tree-branch of the tree selected as

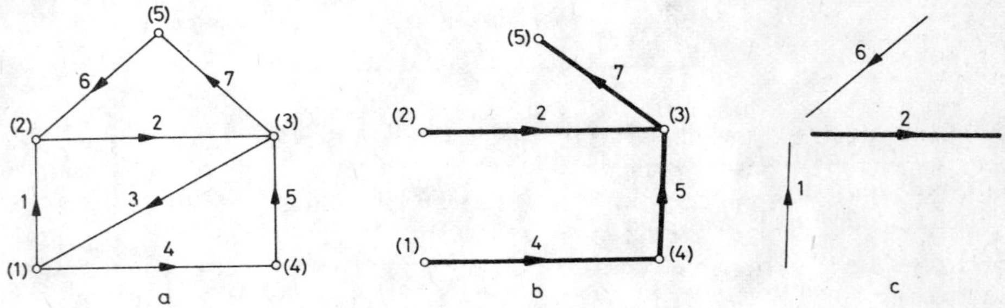

Fig. 1.17

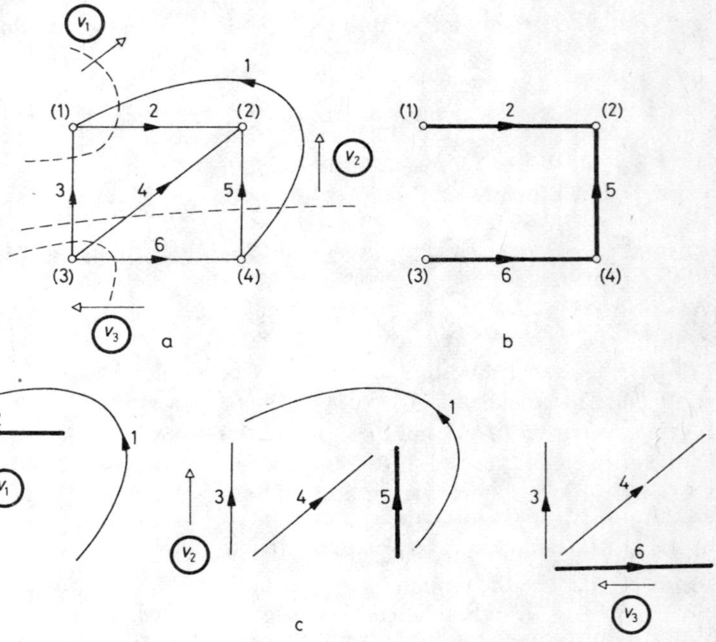

Fig. 1.18

previously explained. Each cutset of this set contains only one edge of the tree chosen, each edge of the tree is included in one cutset of the set only, and the set of cutsets obtained contains every edge of the tree (Fig. 1.18, c).

In this way $n-c$ cutsets may be selected in a graph consisting of c connected parts. These are evidently linearly independent, since in the row matrix of any cutset the 1 or -1 corresponding to the tree-branch can not be expressed as any linear

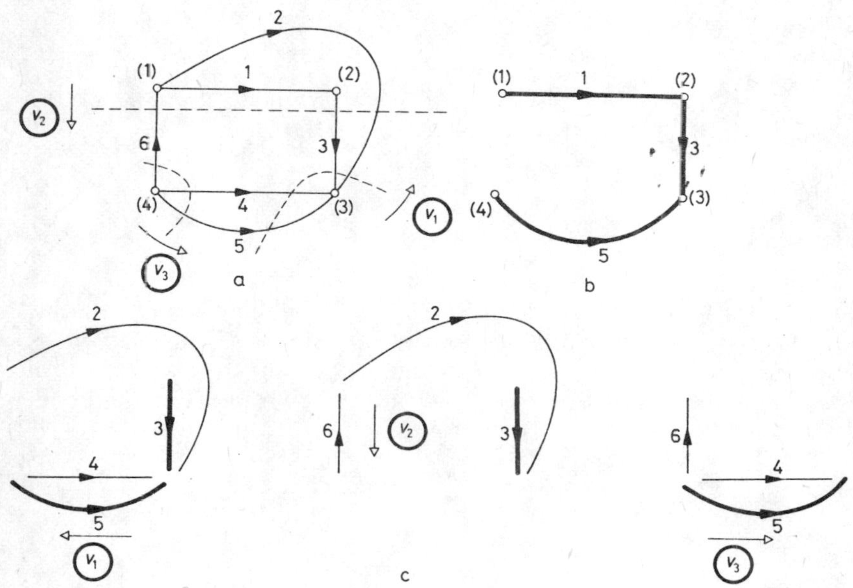

Fig. 1.19

combination of the zeros representing the same edge in the row matrices of the other cutsets. So the fundamental set of cutsets contains at least as many cutsets as the number of tree-branches or forest-branches, i.e. $(n-1)$ in a connected graph, or $(n-c)$ in a graph consisting of c components.

For example, let us examine the directed graph shown in Fig. 1.19, a, and choose one of its trees (Fig. 1.19, b). Let us select one cutset including two tree-branches of this tree and two cutsets each containing one of these two tree-branches (Fig. 1.19, c). Writing the row matrices of the cutsets, the row matrix of the first cutset is seen to be a linear combination of those of the last two.

It should be noted that fundamental sets of cutsets of the graph can also exist which cannot be determined by a tree or forest of the graph in the way described above.

Let us form the cutsets in a non-separable graph from the edges incident with each vertex (Fig. 1.20). Let their orientation be such that the edges pointing away from the vertex have positive signs, while those pointing inwards have negative signs. The matrices of the cutsets thus selected coincide with the row matrices of the vertices defined in Eq. (1.12). Thus the vertex row matrix $\mathbf{A}_i^+$ may be considered as a special cutset row matrix. If the corresponding elements of all such cutset row matrices are summed, zero is obtained in each case, since every edge is incident with two vertices, and thus in the row matrices the element corresponding to the edge is once $+1$, once -1 and zero in other cases. So:

$$\sum_{i=1}^{n} \mathbf{A}_i^+ = \mathbf{0}. \tag{1.13}$$

i.e.

$$\mathbf{A}_j^+ + \sum_{\substack{i=1 \\ i \neq j}}^{n} \mathbf{A}_i^+ = \mathbf{0}, \tag{1.14}$$

and thus

$$\mathbf{A}_j^+ = - \sum_{\substack{i=1 \\ i \neq j}}^{n} \mathbf{A}_i^+ . \tag{1.15}$$

Thus in case of a non-separable graph, among the n cutsets specially selected as above any one is not linearly independent of the others, and so the number of linearly independent cutsets can be $n-1$ at most. So if the cutset surrounding one of the vertices is omitted, each edge incident with this vertex appears in only one cutset.

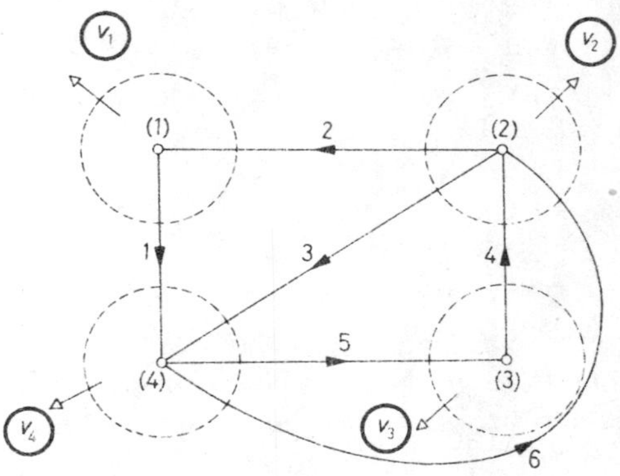

Fig. 1.20

After omitting the row matrix of the cutset formed by the edges incident with vertex (i), there is at least one among the remaining row matrices linearly independent of the others. One is the row matrix of the cutset formed by the edges incident with vertex (j), if vertices (i) and (j) are joined, say by edge k. The k-th element in this matrix is ± 1, while it is 0 in the other row matrices. Since ± 1 cannot be expressed as any linear combination of zeros, the row matrix of vertex (j) is linearly independent of the others. It can be shown by similar reasoning that the row matrices of all vertices adjacent to vertex (j) are linearly independent. It will be shown later that the number of linearly independent cutsets of a connected graph is exactly $n-1$.

In the example drawn in Fig. 1.20, on removing cutset v_1 the row matrices of the cutsets of the fundamental set are:

$$\mathbf{A}_2^+ = [\ 0\ 1\ \ 1\ -1\ \ 0\ -1],$$
$$\mathbf{A}_3^+ = [\ 0\ 0\ \ 0\ \ 1\ -1\ \ 0],$$
$$\mathbf{A}_4^+ = [-1\ 0\ -1\ \ 0\ \ 1\ \ 1].$$

$\mathbf{A}_4^+$ cannot be expressed as a linear combination of $\mathbf{A}_2^+$ and $\mathbf{A}_3^+$, since the first element of $\mathbf{A}_4^+$ is -1 which cannot equal any linear combination of the first elements of $\mathbf{A}_2^+$ and $\mathbf{A}_3^+$.

In the case of graphs consisting of several non-separable components, the number of linearly independent cutsets in component p with n_p vertices is n_p-1. If the graph consists of c components there are exactly

$$\sum_{p=1}^{c}(n_p-1)=n-c \tag{1.16}$$

linearly independent cutsets in the graph, where n is the number of vertices in the graph. The number of cutsets in a fundamental set equals the rank of the graph, i.e. the number of edges in a forest of the graph.

A fundamental set of cutsets in a non-separable graph may thus be obtained by forming the cutsets around $n-1$ vertices.

Fundamental set of loops

Several loops of a graph form a set of loops. A set of loops is linearly independent if and only if no row matrix of any loop in the set can be expressed as a linear combination of the row matrices of further loops. A *fundamental set of loops* of the graph is a linearly independent set of loops, such that the row matrix of any loop of the graph is expressible as a linear combination of the row matrices of the loops in the set.

A fundamental set of loops in a connected graph may be selected with the aid of any tree of the graph and is called the set of loops generated by the tree. If the appropriate chord is inserted between two vertices of the tree, the chord together with certain edges of the tree forms a loop (e.g. chord 6 in Fig. 1.12, d), since the chord creates a path between the two vertices (vertices (2) and (4) in Fig. 1.12, d), and a path between the same two vertices must exist in the tree as well. The union of these two distinct paths form a loop. Accordingly, the number of distinct loops belonging to a single tree is equal to the number of chords. As will be seen, no more linearly independent loops can be chosen in a graph. Thus these loops constitute a fundamental set of loops. If the graph is not connected, and consists of c components, linearly independent loops in each component may be selected as

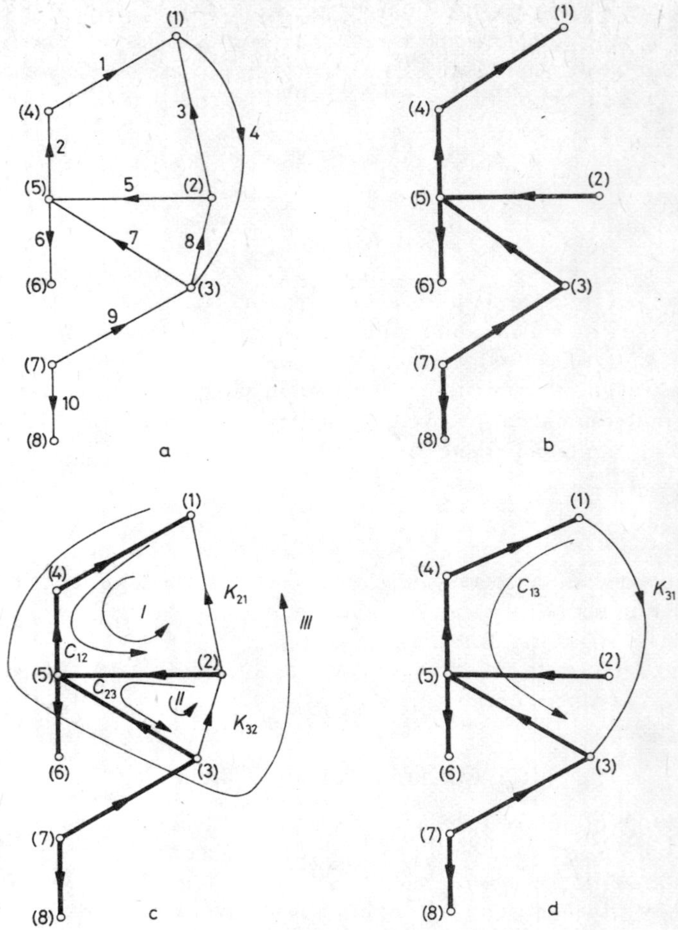

Fig. 1.21

above. Thus the number of fundamental loops equals the number of edges in a cotree, which is the nullity:

$$m=b-n+c. \tag{1.17}$$

The number of independent loops in a connected graph is $m=b-n+1$.

As an example let us examine the connected directed graph in Fig. 1.21, a. One of its trees is shown in Fig. 1.21, b. The path in the tree pointing from (i) to (k) is denoted by C_{ik}, its row matrix by $\mathbf{C}_{ik}^{+}$, the chord between vertices (i) and (k) by K_{ik}, and the corresponding row matrix by $\mathbf{K}_{ik}^{+}$. This is permissible, since the chord forms a path between the two appropriate vertices. Let us add chords K_{21} then K_{32} to the tree

(Fig. 1.21, c). In the first case K_{21} constitutes a loop with path C_{21}. Let this loop be denoted by I, and its row matrix by $\mathbf{B}_I^+$. Thus

$$\mathbf{B}_I^+ = \mathbf{C}_{12}^+ + \mathbf{K}_{21}^+,$$

where summation is carried out with orientations observed. Similarly, path C_{23} and chord K_{32} form a loop. Its row matrix is

$$\mathbf{B}_2^+ = \mathbf{C}_{23}^+ + \mathbf{K}_{32}^+.$$

On adding both chords K_{21} and K_{32} to the tree, a third loop is also created with matrix

$$\mathbf{B}_3^+ = \mathbf{K}_{32}^+ + \mathbf{K}_{21}^+ + \mathbf{C}_{12}^+ + \mathbf{C}_{23}^+.$$

(Note that interchanging indices in a path row matrix denotes a reversal of direction, thus effecting a change of signs of the elements of the matrix.) This loop, however, is seen to be linearly dependent upon the two former ones, its matrix being the sum of their matrices:

$$\mathbf{B}_3^+ = \mathbf{B}_1^+ + \mathbf{B}_2^+. \tag{1.18}$$

Thus the row matrices of the loops not in the fundamental set may be expressed as a linear combination of the row matrices of the loops in the fundamental set. This is true for any number of vertices and chords. The row matrices of loops I and II chosen as above are in fact linearly independent, since they contain distinct chords.

To sum up: A fundamental set of loops in the graph can be determined with the aid of a tree or forest of the graph. On insertion of a chord to the tree, a loop is created. By inserting the chords in the tree one at a time, a fundamental set of loops is formed by the loops consisting of each chord and the appropriate tree-branches. Thus the number of linearly independent loops equals the number of chords, i.e. nullity. This theorem which was demonstrated above will be proved later.

It is noted that fundamental sets of loops in a graph may exist which cannot be associated with a tree or forest of the graph in the above way. For normal applications, sets of loops and cutsets generated by a tree (forest) will be employed, since they can be selected by the simple procedure outlined above.

Isomorphic and dual graphs

Two graphs G and G' are isomorphic if there exists a one-to-one correspondence between the edges of G and G' as well as the vertices of G and G', with the correspondence preserving incidence. For the isomorphism of directed graphs orientations should also correspond. The edges and vertices of the two seemingly different graphs shown in Fig. 1.22 are seen to correspond to each other in a one-to-one way, i.e. the two graphs are isomorphic.

Graphs G_1 and G_2 are 2-isomorphic, if they can be made isomorphic by the following two operations (possibly by their successive application, or by one of the operations):

1. Cutting a connected graph or component consisting of several blocks at a cutpoint, with the blocks thus becoming components of a non-connected graph.

2. Splitting connected graphs or components into two complementary subgraphs connected in the original graph at two vertices (vertices (*i*) and (*j*) in Fig. 1.23, a), then

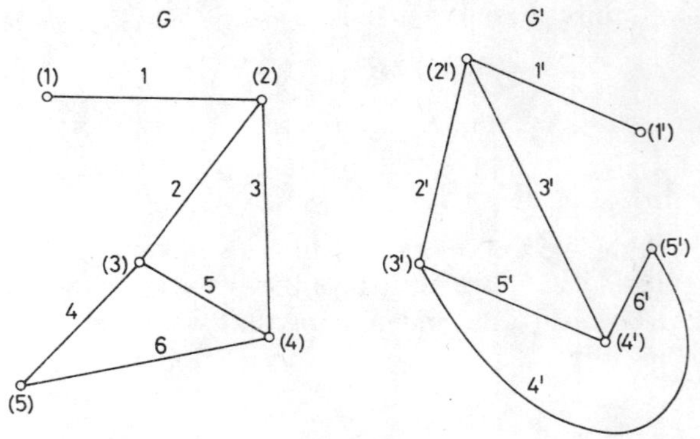

Fig. 1.22

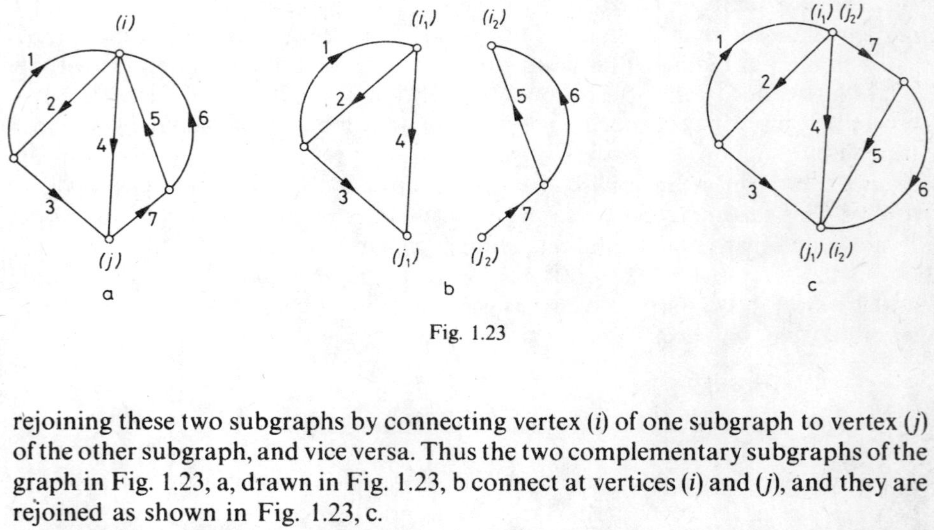

Fig. 1.23

rejoining these two subgraphs by connecting vertex (*i*) of one subgraph to vertex (*j*) of the other subgraph, and vice versa. Thus the two complementary subgraphs of the graph in Fig. 1.23, a, drawn in Fig. 1.23, b connect at vertices (*i*) and (*j*), and they are rejoined as shown in Fig. 1.23, c.

In the foregoing figures graphs have been drawn so that the common points of edges have always also been vertices. Those graphs for which this is possible when they are drawn on a two-dimensional surface are called *planar graphs*.

If two edges with one common vertex of degree 2 (series edges) are replaced by one edge, a graph *homeomorphic* with the original is obtained. Nonplanar graphs arise from two basic types. One of them contains 5 vertices, with any two of them connected by an edge (Fig. 1.24). The other basic type has 6 vertices, with 3 of them belonging to one group, and the further 3 to an other. Any vertex in one group is connected by one branch to all the vertices in the other (Fig. 1.25). A graph is a planar graph if and only if it has no subgraph homeomorphic with one of these two types. This theorem will not be proved. These two basic types of nonplanar graphs are called *Kuratowski graphs.*

A dual planar graph G'' may be associated with any non-separable planar graph G' as follows. Let G' be drawn with its vertices as the only common points of its edges (Fig. 1.26, a). Let us select the loops in G' not having a vertex in their inner regions (loops *I, II, III, IV* in Fig. 1.26, a)*, and the one not having a vertex in its outer region (loop *V* in Fig. 1.26, a). Let us draw points inside the former loops (*I, II, III, IV*), ((1)', (2)', (3)', (4)' in Fig. 1.26, b) and one outside the latter (*V*) ((5)' in Fig. 1.26, b). Let us consider these points the vertices of G'' and connect them by edges, with each edge now drawn intersecting one edge of G', and each original edge cut by a new edge (Fig. 1.26, b). The graph G'' thus obtained is a *dual graph* of G'. Any graph isomorphic with the dual graph is a dual of the original graph.

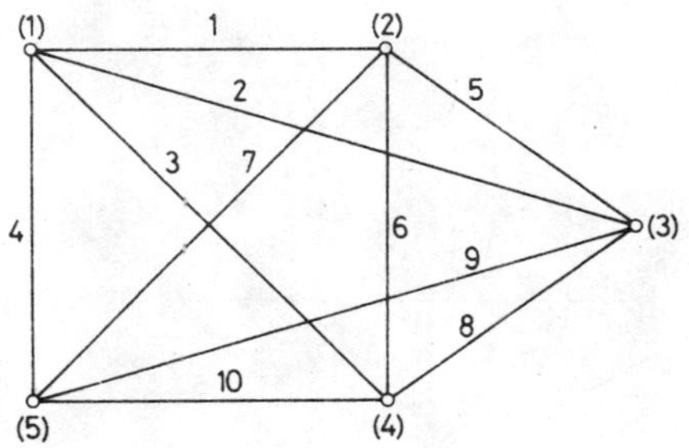

Fig. 1.24

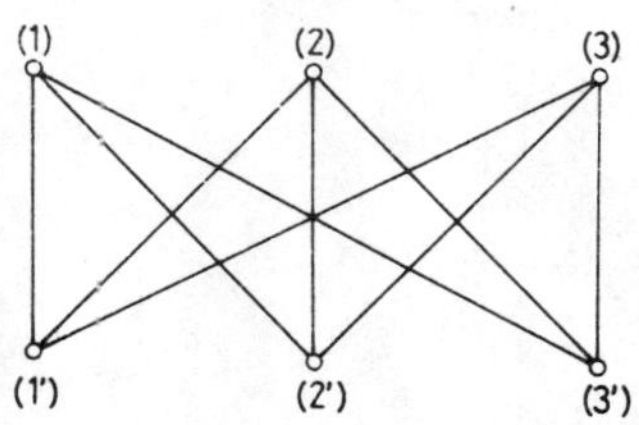

Fig. 1.25

G'' has the same number of edges as G'. The duals of the edges in G' forming the loops selected for the construction are the edges incident with the vertices in G'' associated with these loops. A vertex of the original graph corresponds to a loop in the dual graph.

* Such loops are often called *meshes.*

The dual of the dual graph is isomorphic or 2-isomorphic with the original graph.

In case of directed graphs the construction of the dual graph is carried out in the same way as for non-directed graphs. However, determination of the directions of the edges in the dual graph requires the assignment of orientations to the loops selected for the construction (Fig. 1.26, a). The orientations of the loops without inner vertices (*I*, *II*, *III*, *IV* in Fig. 1.26, a) are chosen to coincide (e.g. all clockwise), and opposite to the outer loop orientation (in Fig. 1.26 a loop *V* is anti-clockwise). In the dual graph a vertex is associated with each loop, such that edges directed away from the vertex in the dual graph correspond to edges of the original graph having the same orientation as the loop, while edges directed towards the vertex correspond to those having the opposite orientation. The edge in the dual graph corresponding to an edge of the original graph is the one which cuts that edge in the construction.

In the course of the construction of the dual graph the number of vertices is one more than the number of fundamental loops. This is illustrated by an example. Let

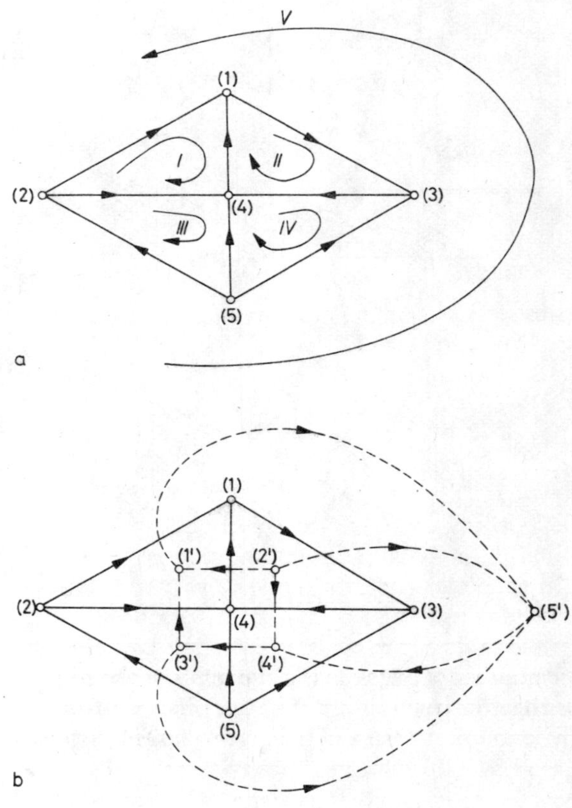

Fig. 1.26

us draw a tree of the graph in Fig. 1.27, a (Fig. 1.27, b). By drawing an edge of the cotree, a loop is created (Fig. 1.27, c). Each chord forms a new loop with the tree-branches. In any case there are as many independent loops as there are chords. At the construction of the dual graph a vertex has been associated with a further loop, the one formed by the outer edges of the graph. This cannot be considered as a formal proof, since for some planar graphs it is not possible to find a tree for which the loops correspond to the meshes.

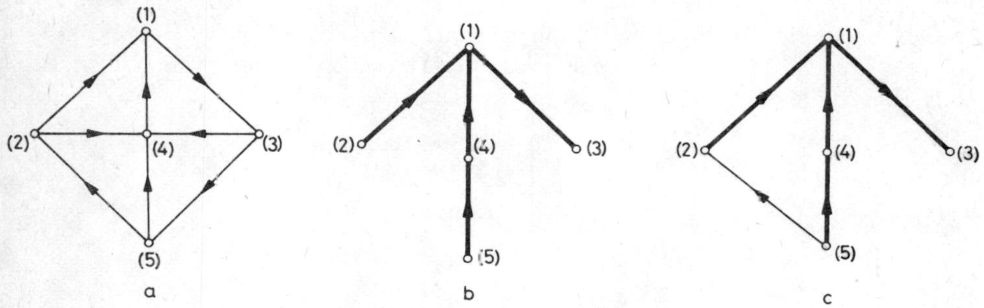

Fig. 1.27

Thus the dual G'' of a graph G' with m_1 fundamental loops has $n_2 = m_1 + 1$ vertices, i.e. the number of vertices in G'' is one more than the nullity of G':

$$n_2 = m_1 + 1. \tag{1.19}$$

Since G'' is connected, its rank is

$$r_2 = n_2 - 1 = m_1, \tag{1.20}$$

i.e. the rank of the dual graph equals the nullity of the original graph.

The number of edges in the original (G') and the dual (G'') graphs is equal:

$$b_1 = b_2. \tag{1.21}$$

Thus according to (1.10):

$$m_1 + r_1 = m_2 + r_2. \tag{1.22}$$

Hence, using (1.20):

$$r_1 = m_2, \tag{1.23}$$

i.e. the nullity of the dual graph equals the rank of the original graph.

The matrices characterizing the graph

Incidence matrix

In the foregoing the most important concepts for describing the interconnection of network elements have been examined. On the basis of these, matrices characterizing the structure of networks in the case of both directed and non-directed graphs can be defined.

One of the characterizations of the connection of a network is the *incidence matrix*, denoted by $\mathbf{A}_t$. The rows of the incidence matrix are the row matrices of the vertices in the order of their numbers. Let the j-th element in the row matrix of the i-th vertex be denoted by a_{ij} instead of x_j. Thus

$$\mathbf{A}_t = \begin{bmatrix} \mathbf{A}_1^+ \\ \mathbf{A}_2^+ \\ \vdots \\ \mathbf{A}_n^+ \end{bmatrix} = \begin{bmatrix} a_{11} & a_{12} & \dots & a_{1b} \\ a_{21} & a_{22} & \dots & a_{2b} \\ \vdots & \vdots & \dots & \vdots \\ a_{n1} & a_{n2} & \dots & a_{nb} \end{bmatrix}. \tag{1.24}$$

Each column corresponds to a branch, i.e. the number of columns in incidence matrix $\mathbf{A}_t$ is b, while the number of rows is n. The element a_{ij} of matrix $\mathbf{A}_t$ specifies the incidence of vertex (i) and edge j. Namely $a_{ij} = 1$, if (i) and j are incident, and $a_{ij} = 0$ otherwise. In a matrix indicating orientations, $a_{ij} = +1$ if edge j is incident with vertex (i) and the direction of j points away from (i) while $a_{ij} = -1$ if j points towards (i). As before, $a_{ij} = 0$ if j is not incident with (i).

In the network calculations which follow the forms indicating orientations are most often used both in the case of the incidence matrix and other network matrices.

The incidence matrix of the graph shown in Fig. 1.28 is

$$\mathbf{A}_t = \begin{matrix} \\ (1) \\ (2) \\ (3) \\ (4) \\ (5) \end{matrix} \begin{matrix} \begin{matrix} 1 & 2 & 3 & 4 & 5 & 6 & 7 & 8 & 9 \end{matrix} \\ \begin{bmatrix} 1 & -1 & 0 & 1 & 0 & 0 & 0 & 0 & 0 \\ 0 & 1 & 1 & 0 & -1 & -1 & 0 & 0 & 0 \\ -1 & 0 & -1 & 0 & 0 & 0 & -1 & 1 & 0 \\ 0 & 0 & 0 & -1 & 1 & 0 & 0 & 0 & -1 \\ 0 & 0 & 0 & 0 & 0 & 1 & 1 & -1 & 1 \end{bmatrix} \end{matrix}.$$

In each column of the matrix A_t there are two non-zero elements.

The graph can be easily derived from a given matrix $\mathbf{A}_t$. Vertices are selected to correspond to each row, and edge j is drawn between the vertices indicated by 1 and -1 in the column of j, directed from the vertex corresponding to 1 towards the one corresponding to -1.

The rank of the incidence matrix is $n-c$. (The rank of a matrix equals the maximal number of its linearly independent rows.) It was shown in the discussion of the fundamental set of cutsets that exactly $n-c$ of the vertex row matrices are linearly independent, i.e. the number of linearly independent rows in matrix $\boldsymbol{A}_t$ is $n-c$. It follows from this that the rank of $\boldsymbol{A}_t$ is given by

$$r(\boldsymbol{A}_t)=n-c, \tag{1.25}$$

and for a connected graph:

$$r(\boldsymbol{A}_t)=n-1, \tag{1.26}$$

equal in both cases to the rank of the graph. Therefore, there are c rows of the incidence matrix which contain information additional to the other $n-c$ rows, so c rows of the matrix A_t may be deleted. These c rows are selected such that any one

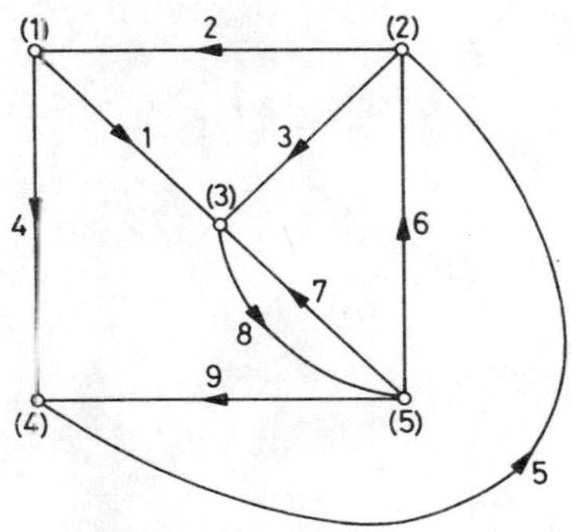

Fig. 1.28

among the rows corresponding to each of the c components is removed. The matrix thus obtained is the *basis incidence matrix* $\boldsymbol{A}$*. For example one of the basis incidence matrices of the graph in Fig. 1.28 is

$$\boldsymbol{A}=\begin{array}{c} \\ (1) \\ (2) \\ (4) \\ (5) \end{array}\begin{array}{c} \begin{array}{ccccccccc} 1 & 2 & 3 & 4 & 5 & 6 & 7 & 8 & 9 \end{array} \\ \left[\begin{array}{ccccccccc} 1 & -1 & 0 & 1 & 0 & 0 & 0 & 0 & 0 \\ 0 & 1 & 1 & 0 & -1 & -1 & 0 & 0 & 0 \\ 0 & 0 & 0 & -1 & 1 & 0 & 0 & 0 & -1 \\ 0 & 0 & 0 & 0 & 0 & 1 & 1 & -1 & 1 \end{array}\right] \end{array}.$$

* Also called the reduced node-branch incidence matrix, while $\boldsymbol{A}_t$ is also called the augmented incidence matrix.

The basis incidence matrix of a connected graph may be completed by a row to obtain the incidence matrix. The completion is effected by adding a new row with elements -1 or $+1$ in those columns containing a single non-zero element of $+1$ or -1 and zero in the other columns. In the case of graphs consisting of several components, the completion of the basis incidence matrix should be carried out separately for each component. In such cases it is expedient to choose the vertices in one component to have successive order numbers. Since the incidence matrix can be reconstructed from the basis incidence matrix, the graph can also be drawn, in the way described previously.

Loop matrix

The loop matrix is another way of representing the graph. To write it requires a knowledge of all loops in the graph. The rows of the loop matrix correspond to loops, its columns to edges of the graph. Accordingly, the loop matrix is formed by the row matrices (1.3) representing the loops. A loop matrix is denoted by $\boldsymbol{B}_t$. Writing b_{ij} for the j-th element of the i-th row:

$$\boldsymbol{B}_t = \begin{bmatrix} \mathbf{B}_1^+ \\ \mathbf{B}_2^+ \\ \vdots \\ \mathbf{B}_h^+ \end{bmatrix} = \begin{bmatrix} b_{11} & b_{12} & \dots & b_{1b} \\ b_{21} & b_{22} & \dots & b_{2b} \\ . & . & \dots & . \\ b_{h2} & b_{h2} & \dots & b_{hb} \end{bmatrix} \tag{1.27}$$

where h denotes the number of loops in the graph. We do not concern ourselves with the question of how to determine all loops of the graph, since this will not be needed in what follows.

As an example let us write the loop matrix of the graph shown in Fig. 1.29:

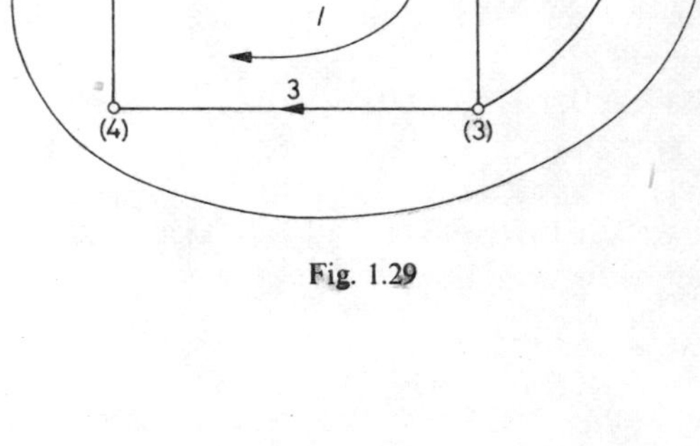

Fig. 1.29

$$B_t = \begin{array}{c} \\ I \\ II \\ III \end{array} \begin{array}{c} \begin{matrix} 1 & 2 & 3 & 4 & 5 \end{matrix} \\ \begin{bmatrix} 1 & 1 & 1 & -1 & 0 \\ -1 & -1 & 0 & 0 & 1 \\ 0 & 0 & -1 & 1 & -1 \end{bmatrix} \end{array}.$$

Let the order of the edges be the same for the incidence and loop matrices of the graph. In this case multiplying one of the matrices by the transpose of the other, gives a zero matrix:

$$\boldsymbol{A}_t \boldsymbol{B}_t^+ = \boldsymbol{0} \quad \text{and} \quad \boldsymbol{B}_t \boldsymbol{A}_t^+ = \boldsymbol{0}. \tag{1.28}$$

To prove this, substitute (1.24) and (1.27) into (1.28):

$$\boldsymbol{A}_t \boldsymbol{B}_t^+ = \begin{bmatrix} \mathbf{A}_1^+ \\ \mathbf{A}_2^+ \\ \vdots \\ \mathbf{A}_n^+ \end{bmatrix} [\mathbf{B}_2 \ldots \mathbf{B}_h] = \begin{bmatrix} \mathbf{A}_1^+\mathbf{B}_1 & \mathbf{A}_1^+\mathbf{B}_2 & \ldots & \mathbf{A}_1^+\mathbf{B}_h \\ \mathbf{A}_2^+\mathbf{B}_1 & \mathbf{A}_2^+\mathbf{B}_2 & \ldots & \mathbf{A}_2^+\mathbf{B}_h \\ \ldots & \ldots & \ldots & \ldots \\ \mathbf{A}_n^+\mathbf{B}_1 & \mathbf{A}_n^+\mathbf{B}_2 & \ldots & \mathbf{A}_n^+\mathbf{B}_h \end{bmatrix} = \boldsymbol{0}. \tag{1.29}$$

This evidently holds, if

$$\mathbf{A}_i^+ \mathbf{B}_k = 0, \quad (i = 1, 2, \ldots, n; \; k = 1, 2, \ldots, h), \tag{1.30}$$

or in scalar form

$$\sum_{j=1}^{b} a_{ij} b_{kj} = 0. \tag{1.31}$$

Equation (1.31) can be shown to be valid for matrices of directed graphs as follows: $a_{ij} \neq 0$ means that vertex (i) is incident with edge j, while $b_{kj} \neq 0$ indicates that loop k contains edge j. Thus the following can be stated about the elements of the sum in (1.31):

$a_{ij} b_{kj} = \pm 1$, if edge j is incident with vertex (i) and loop k crosses edge j. The product is positive if the orientation of loop k along edge j points away from vertex (i) and negative if it points towards the vertex;

$a_{ij} b_{kj} = 0$, if edge j is not incident with vertex (i), or loop k does not cross edge j.

If loop k crosses vertex (i), it enters the vertex along one edge and leaves it along an other. Thus among the products $a_{ij} b_{kj}$ $(j = 1, 2, \ldots, b)$ one is $+1$, one is -1 and the others are zero, i.e.

$$c_{ik} = \sum_{j=1}^{b} a_{ij} b_{jk} = 1 - 1 + 0 + \ldots + 0 = 0. \tag{1.32}$$

If loop k is not incident with vertex (i), each term in c_{ik} equals zero. Thus $\mathbf{A}_i^+ \mathbf{B}_k = 0$, i.e. every element of $\boldsymbol{A}_t \boldsymbol{B}_t^+$ is zero:

$$\boldsymbol{A}_t \boldsymbol{B}_t^+ = \boldsymbol{0}. \tag{1.33}$$

Or transposing:

$$\boldsymbol{B}_t \boldsymbol{A}_t^+ = \boldsymbol{0}^+ . \tag{1.34}$$

It can be shown on the basis of this orthogonal relation, that the rank of $\boldsymbol{B}_t$ equals the nullity of the graph:

$$r(\boldsymbol{B}_t) = m. \tag{1.35}$$

This means that the number of linearly independent loops in a graph is exactly m, as has already been stated.

Let us select a tree of the graph and the fundamental set of m loops generated by it, so that m rows of the matrix $\boldsymbol{B}_t$ correspond to this set of loops. These rows are linearly independent, since each is associated with a loop containing a different chord. Thus the rank of $\boldsymbol{B}_t$ is at least m, i.e.:

$$r(\boldsymbol{B}_t) \geq m. \tag{1.36}$$

Let the vertices be numbered in a way to allow the following partitioning of $\boldsymbol{A}_t$:

$$\boldsymbol{A}_t = \begin{bmatrix} \boldsymbol{A}_{11} & \boldsymbol{A}_{12} \\ \boldsymbol{A}_{21} & \boldsymbol{A}_{22} \end{bmatrix}, \tag{1.37}$$

where $\boldsymbol{A}_{11}$ is a non-singular square matrix with order $n-c$. Accordingly the number of rows in $\boldsymbol{A}_{21}$ and $\boldsymbol{A}_{22}$ is c, while the number of columns in $\boldsymbol{A}_{12}$ and $\boldsymbol{A}_{22}$ is $m = b - n + c$, and further the number of rows in $\boldsymbol{A}_{12}$ and that of columns in $\boldsymbol{A}_{21}$ is $n-c$. Let the loop matrix $\boldsymbol{B}_t$ be also partitioned:

$$\boldsymbol{B}_t = [\boldsymbol{B}_1 \quad \boldsymbol{B}_2], \tag{1.38}$$

where the number of columns in $\boldsymbol{B}_1$ is $n-c$ and thus m in $\boldsymbol{B}_2$.

Substituting (1.37) and (1.38) into (1.33):

$$\boldsymbol{A}_t \boldsymbol{B}_t^+ = \begin{bmatrix} \boldsymbol{A}_{11} & \boldsymbol{A}_{12} \\ \boldsymbol{A}_{21} & \boldsymbol{A}_{22} \end{bmatrix} \begin{bmatrix} \boldsymbol{B}_1^+ \\ \boldsymbol{B}_2^+ \end{bmatrix} = \boldsymbol{0}, \tag{1.39}$$

i.e.

$$\boldsymbol{A}_{11} \boldsymbol{B}_1^+ + \boldsymbol{A}_{12} \boldsymbol{B}_2^+ = \boldsymbol{0} \tag{1.40}$$

holds, and hence

$$\boldsymbol{B}_1^+ = -\boldsymbol{A}_{11}^{-1} \boldsymbol{A}_{12} \boldsymbol{B}_2^+ . \tag{1.41}$$

Therefore $n-c$ rows of matrix $\boldsymbol{B}_t^+$ may be expressed as linear combinations of m other rows of the matrix. This yields for the rank of the matrices:

$$r(\boldsymbol{B}_t) = r(\boldsymbol{B}_t^+) = r(\boldsymbol{B}_2^+) \leq m. \tag{1.42}$$

Comparing (1.36) with (1.42):

$$r(\boldsymbol{B}_t)=m, \tag{1.43}$$

i.e. the rank of $\boldsymbol{B}_t$ exactly equals the nullity.

Accordingly let us form the *basis loop matrix* $\boldsymbol{B}$ from loop matrix $\boldsymbol{B}_t$ by deleting the rows representing the loops not in the fundamental set of loops. The rows deleted are linearly dependent upon the others. Thus the rank of $\boldsymbol{B}_t$ equals the rank of $\boldsymbol{B}$, both equalling the nullity m.

Graphs described by the same basis loop matrix are isomorphic or 2-isomorphic.

A tree of the graph shown in Fig. 1.29 has been drawn in Fig. 1.30. It can be

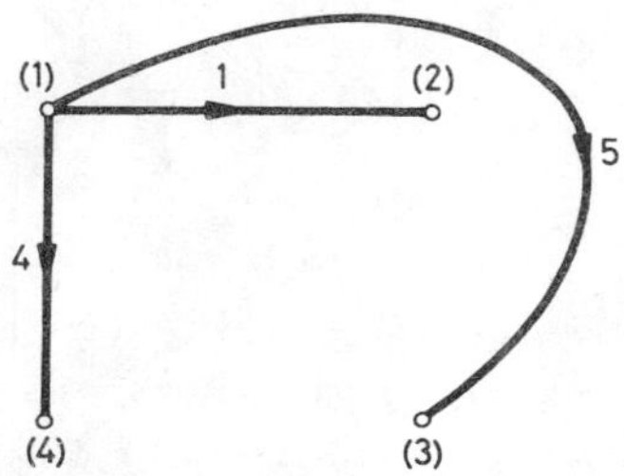

Fig. 1.30

established that there are two independent loops in the graph. The basis loop matrix associated with the tree shown is

$$\boldsymbol{B}=\begin{matrix} & 1 & 2 & 3 & 4 & 5 \\ II & -1 & -1 & 0 & 0 & 1 \\ III & 0 & 0 & -1 & 1 & -1 \end{matrix}.$$

Similarly to the proof of (1.28), it can be shown that the orthogonal relations

$$\boldsymbol{AB}^{+}=\boldsymbol{0} \quad \text{and} \quad \boldsymbol{BA}^{+}=\boldsymbol{0} \tag{1.44}$$

hold for basis incidence and loop matrices if the edges are ordered in a corresponding sequence.

Selecting a tree (forest) of a graph to be examined, let the edges of the graph be numbered so as to assign order numbers 1, 2, . . ., m to the chords. Let us choose the fundamental set of loops of the graph generated by the tree, with loop 1 containing edge 1, loop 2 edge 2, . . . loop m edge m with the orientations of the loops corresponding to that of the associated chords. Now the basis loop matrix may be partitioned as follows:

$$\boldsymbol{B}=[\boldsymbol{1}_m \quad \boldsymbol{F}], \tag{1.45}$$

where $\boldsymbol{1}_m$ is a unit matrix of order m. (1.45) is the *normal form of the basis loop matrix.*

Cutset matrix

The third matrix characterizing the graph is the *cutset matrix*. The cutset matrix provides information about the inclusion of edges in cutsets. The rows of the cutset matrix correspond to the cutsets of the graph and its columns to edges of the graph. Thus the cutset matrix is formed by the row matrices (1.11) representing the cutsets, with the elements of the matrices arranged in a given order of the edges. The j-th element in the row matrix of the i-th cutset is denoted by q_{ij}. Thus the cutset matrix, denoted by $\boldsymbol{Q}_t$ is:

$$\boldsymbol{Q}_t = \begin{bmatrix} \mathbf{Q}_1^+ \\ \mathbf{Q}_2^+ \\ \vdots \\ \mathbf{Q}_p^+ \end{bmatrix} = \begin{bmatrix} q_{11} & q_{12} & \cdots & q_{1b} \\ q_{21} & q_{22} & \cdots & q_{2b} \\ \cdots & \cdots & \cdots & \cdots \\ q_{p1} & q_{p2} & \cdots & q_{pb} \end{bmatrix}, \tag{1.46}$$

where p denotes the number of cutsets. It has been shown that the rows of the incidence matrix are row matrices describing vertices and these equal the row matrices of certain cutsets. The cutset matrix of a non-separable graph, or one consisting of such components only includes all rows of the incidence matrix.

We intend to prove that the number of linearly independent cutsets is $n-c$ exactly. To this end the orthogonal relations

$$\boldsymbol{Q}_t \boldsymbol{B}_t^+ = \boldsymbol{0} \quad \text{and} \quad \boldsymbol{B}_t \boldsymbol{Q}_t^+ = \boldsymbol{0} \tag{1.47}$$

will be employed, where the order of edges in $\boldsymbol{Q}_t$ and $\boldsymbol{B}_t$ corresponds.

$$\boldsymbol{Q}_t \boldsymbol{B}_t^+ = \begin{bmatrix} \mathbf{Q}_1^+ \\ \mathbf{Q}_2^+ \\ \vdots \\ \mathbf{Q}_p^+ \end{bmatrix} [\mathbf{B}_1 \; \mathbf{B}_2 \; .. \; \mathbf{B}_h] = \begin{bmatrix} \mathbf{Q}_1^+\mathbf{B}_1 & \mathbf{Q}_1^+\mathbf{B}_2 & .. & \mathbf{Q}_1^+\mathbf{B}_h \\ \mathbf{Q}_2^+\mathbf{B}_1 & \mathbf{Q}_2^+\mathbf{B}_2 & .. & \mathbf{Q}_2^+\mathbf{B}_h \\ .. & .. & .. & .. \\ \mathbf{Q}_p^+\mathbf{B}_1 & \mathbf{Q}_p^+\mathbf{B}_2 & .. & \mathbf{Q}_p^+\mathbf{B}_h \end{bmatrix} = \boldsymbol{0} \tag{1.48}$$

holds if each element of the matrix equals zero, i.e.:

$$\mathbf{Q}_i^+ \mathbf{B}_j = \sum_{k=1}^{b} q_{ik} b_{jk} = 0. \tag{1.49}$$

The elements of the sum may be $+1$, -1 or 0, according to the following conditions:

$q_{ik} b_{jk} = \pm 1$, if edge k is included both in cutset i and loop j, i.e. cutset i cuts loop j. The value of the product is $+1$ if the orientations of cutset i and loop j correspond on edge k, and is -1 otherwise;

$q_{ik} b_{jk} = 0$, if edge k is not included in cutset i or loop j.

A cutset either does not cut a loop (Fig. 1.31, loop *I*) or the number of cuts is even (Fig. 1.31, loop *II*). At the cut the orientations of the cutset and the loop correspond (Fig. 1.31, edge 5) as many times as they are opposite (Fig. 1.31, edge 1). Therefore, as many terms in the sum (1.49) are +1 as −1, while the remaining terms are zero. This means that the orthogonal relation (1.47) $\boldsymbol{Q}_t\boldsymbol{B}_t^+ = \boldsymbol{0}$ holds, provided that the order of the edges is the same for both matrices. Writing its transpose:

$$(\boldsymbol{Q}_t\boldsymbol{B}_t^+)^+ = \boldsymbol{B}_t\boldsymbol{Q}_t^+ = \boldsymbol{0}. \tag{1.50}$$

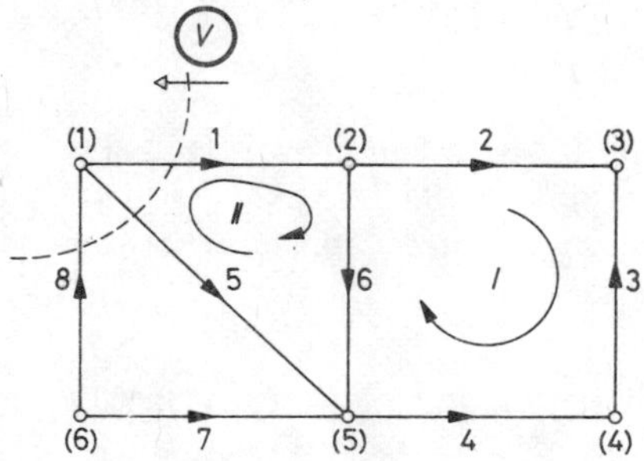

Fig. 1.31

It can be shown on the basis of this orthogonality relation that the rank of $\boldsymbol{Q}_t$ is $n-c$, i.e. the number of linearly independent cutsets in the graph is $n-c$ exactly. Let the loops be ordered in a way to allow the following partitioning of loop matrix $\boldsymbol{B}_t$:

$$\boldsymbol{B}_t = \begin{bmatrix} \boldsymbol{B}_{11} & \boldsymbol{B}_{12} \\ \boldsymbol{B}_{21} & \boldsymbol{B}_{22} \end{bmatrix}, \tag{1.51}$$

where $\boldsymbol{B}_{11}$ is a square, non-singular matrix of order m. Such a matrix certainly exists, since the rank of $\boldsymbol{B}_t$ is m. Accordingly, the number of columns in $\boldsymbol{B}_{12}$ is $n-c$, and that of its rows is m. Matrix $\boldsymbol{Q}_t$ is also partitioned:

$$\boldsymbol{Q}_t = [\boldsymbol{Q}_1 \ \ \boldsymbol{Q}_2], \tag{1.52}$$

where the number of columns in $\boldsymbol{Q}_1$ is m, and in $\boldsymbol{Q}_2$ is $n-c$. Substituting (1.51) and (1.52) into (1.47):

$$\boldsymbol{B}_t\boldsymbol{Q}_t^+ = \begin{bmatrix} \boldsymbol{B}_{11} & \boldsymbol{B}_{12} \\ \boldsymbol{B}_{21} & \boldsymbol{B}_{22} \end{bmatrix} \begin{bmatrix} \boldsymbol{Q}_1^+ \\ \boldsymbol{Q}_2^+ \end{bmatrix} = \boldsymbol{0}. \tag{1.53}$$

I.e.

$$\boldsymbol{B}_{11}\boldsymbol{Q}_1^+ + \boldsymbol{B}_{12}\boldsymbol{Q}_2^+ = \boldsymbol{0}, \tag{1.54}$$

hence

$$\boldsymbol{Q}_1^+ = -\boldsymbol{B}_{11}^{-1}\boldsymbol{B}_{12}\boldsymbol{Q}_2^+ . \tag{1.55}$$

Thus m rows of matrix $\boldsymbol{Q}_t^+$ may be expressed as linear combinations of $n-c$ different rows. Hence the rank of $\boldsymbol{Q}_t$ is

$$r(\boldsymbol{Q}_t)=r(\boldsymbol{Q}_t^+)\leq n-c. \tag{1.56}$$

It has been shown that the graph has at least as many linearly independent cutsets as the number of tree-branches in the graph, i.e. the number of linearly independent rows in $\boldsymbol{Q}_t$ is at least $n-c$:

$$r(\boldsymbol{Q}_t)\geq n-c. \tag{1.57}$$

Comparing (1.56) with (1.57):

$$r(\boldsymbol{Q}_t)=n-c, \tag{1.58}$$

i.e. the number of linearly independent cutsets is $n-c$ exactly.

With the aid of the foregoing results the *basis cutset matrix* $\boldsymbol{Q}$ may be derived, with its $n-c$ rows given by the row matrices representing the cutsets of a fundamental set of cutsets of the graph.*

The orthogonality relations

$$\boldsymbol{B}\boldsymbol{Q}^+=\boldsymbol{0} \qquad \text{and} \qquad \boldsymbol{Q}\boldsymbol{B}^+=\boldsymbol{0} \tag{1.59}$$

involving the basis matrices can similarly be obtained.

Graphs described by the same basis cut matrix are isomorphic or 2-isomorphic.

Let us construct a fundamental set of cutsets with the aid of a selected tree (forest) of the graph. Let the edges be numbered so as to associate the order numbers $1, 2, \ldots, m$ with chords. In the fundamental set cutset 1 contains edge $m+1$, cutset 2 edge $m+2$, etc. The orientation of the cutset and that of the corresponding tree-branch correspond. Now the basis cutset matrix may be partitioned as follows:

$$\boldsymbol{Q}=[\boldsymbol{Q}_e \quad \boldsymbol{1}_{n-c}], \tag{1.60}$$

where $\boldsymbol{1}_{n-c}$ is a unit matrix of order $n-c$.**

This is the *normal form of the basis cutset matrix.*

Interrelations between the characterizing matrices and the graph

In order to write the basis matrices of the fundamental sets of loops and cutsets generated by a particular tree of a given graph, the edges are numbered so as to give order numbers $1, 2, \ldots, m$ to the chords, and the remaining order numbers to tree-branches. Now, provided that suitable orientations are chosen for the loops and branches the normal forms (1.45) and (1.60) of the matrices have been seen to be

* From now on, unless otherwise mentioned, the incidence, loop and cutset matrices denote basis matrices.

** The order of the matrices $\boldsymbol{1}$ and $\boldsymbol{0}$ as well as their numbers of rows and columns will not in general be indicated from now on, these being evident from the context.

obtained. Substituting these into the orthogonality relation (1.59):

$$QB^+ = [Q_e \ 1_{n-c}] \begin{bmatrix} 1_m \\ F^+ \end{bmatrix} = 0. \tag{1.61}$$

Hence

$$Q_e = -F^+, \tag{1.62}$$

i.e.

$$F = -Q_e^+. \tag{1.63}$$

These formulas show the matrices B and Q to be easily reconstructible one from the other, provided that the numbering and orientations of edges, cutsets and loops are chosen as above.

Matrices B and Q of 2-isomorphic graphs may be identical for an appropriate choice of order numbers. Blocks and components are not distinguishable on the basis of B or Q.

Several distinct matrices B and Q may be written for a given graph, but all are obtainable from one another. Basis loop matrices or cutset matrices may differ either owing to different numbering or orientations of the edges, loops or cutsets or because they describe different fundamental sets of loops or cutsets. Opposite orientation results in multiplication by -1 of the corresponding column or row. A matrix of a fundamental set of loops or cutsets yields those of further fundamental sets of loops or cutsets, by substitution of one of its rows by a suitable linear combination of other rows containing 1, -1 or 0 as elements. If the graph is separable, a linear combination of rows describing loops or cutsets in the same block or component of the graph may be employed as new rows. The structure of matrices B and Q ensures their convertibility into normal form by the operations mentioned above (change of columns, rows, their multiplication by -1, substitutions of rows by a linear combination of others).

The normal forms of matrices B and Q satisfy the following two criteria [35, 42]:

(a) Matrix F does not include matrix

$$N = \begin{bmatrix} \pm 1 & 0 & \pm 1 & \pm 1 \\ \pm 1 & \pm 1 & 0 & \pm 1 \\ \pm 1 & \pm 1 & \pm 1 & 0 \end{bmatrix} \tag{1.64}$$

or its transpose as a submatrix after the above operations.

(b) Matrix B must not include as a submatrix the basis cutset matrix of any Kuratowski graph or matrix Q the loop matrix of any Kuratowski graph.

We shall not concern ourselves with the verification of the criteria. Condition (b) may be shown by taking into account the fact that Kuratowski graphs have no dual.

It can be proved that the two criteria (a) and (b) form a necessary and sufficient condition for B and Q to represent a graph.

The graph has been seen to be reconstructible from a given incidence matrix A. Since the incidence matrix of a graph without a cutpoint is a special cutset matrix, any cutset matrix of such a graph may be converted into an incidence matrix, i.e. into a matrix having at most one element +1 and one −1 in each column, with the remaining elements zero. Thus non-separable graphs may be drawn from a given cutset matrix as well. If the loop matrix is given, this is first brought to normal form, from which the normal form of the cutset matrix is determined. Knowing the cutset matrix the graph can be drawn as described above.

Methods enabling the graph to be derived directly from $\boldsymbol{B}$ or $\boldsymbol{Q}$ are also known. These, however, will not be discussed here [22].

Matrices characterizing graphs with end-edges or self-loops

Applications of graphs with end-edges and with self-loops will be discussed in chapters 4 and 7. Directed graphs with end-edges will be seen to be characterized by matrices $\frac{1}{2}(\boldsymbol{A}_t+\boldsymbol{A}_{t0})$ and $\frac{1}{2}(\boldsymbol{A}_t-\boldsymbol{A}_{t0})$, derived from incidence matrices $\boldsymbol{A}_t$ including orientations and $\boldsymbol{A}_{t0}$ disregarding orientations. In both matrices columns correspond to edges, and rows to vertices. The j-th element in the i-th row of $\frac{1}{2}(\boldsymbol{A}_t+\boldsymbol{A}_{t0})$ is 1, if edge j is incident with vertex (i), with its direction pointing away from the vertex, otherwise this element is zero. The j-th element of the i-th row in $\frac{1}{2}(\boldsymbol{A}_t-\boldsymbol{A}_{t0})$ is 1, if edge j is incident with vertex (i) with its direction pointing towards the vertex, otherwise the element is zero.

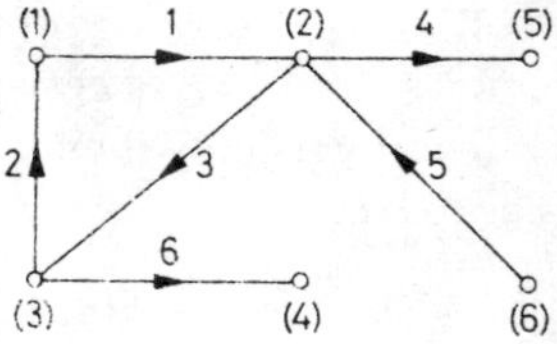

Fig. 1.32

For the graph shown in Fig. 1.32:

$$\boldsymbol{A}_t=\begin{bmatrix} 1 & -1 & 0 & 0 & 0 & 0\\ -1 & 0 & 1 & 1 & -1 & 0\\ 0 & 1 & -1 & 0 & 0 & 1\\ 0 & 0 & 0 & 0 & 0 & -1\\ 0 & 0 & 0 & -1 & 0 & 0\\ 0 & 0 & 0 & 0 & 1 & 0 \end{bmatrix},$$

$$A_{t0} = \begin{bmatrix} 1 & 1 & 0 & 0 & 0 & 0 \\ 1 & 0 & 1 & 1 & 1 & 0 \\ 0 & 1 & 1 & 0 & 0 & 1 \\ 0 & 0 & 0 & 0 & 0 & 1 \\ 0 & 0 & 0 & 1 & 0 & 0 \\ 0 & 0 & 0 & 0 & 1 & 0 \end{bmatrix},$$

$$\frac{1}{2}(A_t + A_{t0}) = \begin{bmatrix} 1 & 0 & 0 & 0 & 0 & 0 \\ 0 & 0 & 1 & 1 & 0 & 0 \\ 0 & 1 & 0 & 0 & 0 & 1 \\ 0 & 0 & 0 & 0 & 0 & 0 \\ 0 & 0 & 0 & 0 & 0 & 0 \\ 0 & 0 & 0 & 0 & 1 & 0 \end{bmatrix},$$

$$\frac{1}{2}(A_t - A_{t0}) = \begin{bmatrix} 0 & 1 & 0 & 0 & 0 & 0 \\ 1 & 0 & 0 & 0 & 1 & 0 \\ 0 & 0 & 1 & 0 & 0 & 0 \\ 0 & 0 & 0 & 0 & 0 & 1 \\ 0 & 0 & 0 & 1 & 0 & 0 \\ 0 & 0 & 0 & 0 & 0 & 0 \end{bmatrix}.$$

If the graph includes directed self-loops as well it is characterized as follows.

The element corresponding to the self-loop in matrix $\frac{1}{2}(A_t + A_{t0})$ as well as in matrix $\frac{1}{2}(A_t - A_{t0})$ is zero. Let us construct a further matrix L with rows corresponding to vertices and columns to edges. The j-th element of the i-th row, $l_{ij} = 1$ if self-loop j is incident with vertex (i), otherwise $l_{ij} = 0$. Graphs including self-loops will be taken into account in our calculations by matrices

$$M_t = \frac{1}{2}(A_t + A_{t0}) + L, \tag{1.65}$$

$$N_t = \frac{1}{2}(A_t - A_{t0}) + L. \tag{1.66}$$

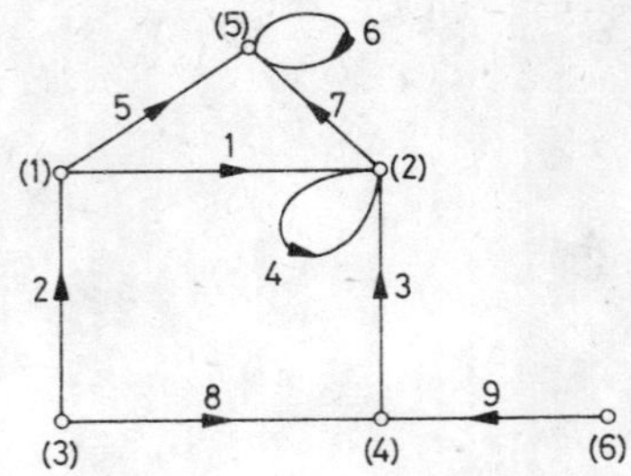

Fig. 1.33

For the graph drawn in Fig. 1.33:

$$\boldsymbol{L}=\begin{bmatrix}0&0&0&0&0&0&0&0&0\\0&0&0&1&0&0&0&0&0\\0&0&0&0&0&0&0&0&0\\0&0&0&0&0&0&0&0&0\\0&0&0&0&0&1&0&0&0\\0&0&0&0&0&0&0&0&0\end{bmatrix},$$

and

$$\boldsymbol{M}_t=\begin{bmatrix}1&0&0&0&1&0&0&0&0\\0&0&0&1&0&0&1&0&0\\0&1&0&0&0&0&0&1&0\\0&0&1&0&0&0&0&0&0\\0&0&0&0&0&1&0&0&0\\0&0&0&0&0&0&0&0&1\end{bmatrix},$$

$$\boldsymbol{N}_t=\begin{bmatrix}0&1&0&0&0&0&0&0&0\\1&0&1&1&0&0&0&0&0\\0&0&0&0&0&0&0&0&0\\0&0&0&0&0&0&0&1&1\\0&0&0&0&1&1&1&0&0\\0&0&0&0&0&0&0&0&0\end{bmatrix}.$$

Self-loops 4 and 6 are attached to vertices (2) and (5), respectively. Accordingly the 4th element of the 2nd row and the 6th element of the 5th row equal one in both the matrices $\boldsymbol{M}_t$ and $\boldsymbol{N}_t$.

Examples

A few examples will be presented to illustrate the foregoing discussion.

1. Let us first construct the dual of the graph drawn in Fig. 1.34, a.

The directed loops necessary for the construction have been indicated in Fig. 1.34, b, and the dual has been drawn in Fig. 1.34, c.

It should be noted that "series" edges, e.g. 1 and 2 (Fig. 1.35, a) correspond to "parallel" edges of the dual graph (1′ and 2′, Fig. 1.35, b). Similarly "parallel" edges 4 and 5 in the original graph (Fig. 1.35, c) correspond to "series" edges in the dual graph (Fig. 1.35, d).

Edges 2, 3 and 6 (similarly edges 3, 7 and 8, etc.) in the original graph form a "three-pointed star" (Fig. 1.36, a). The edges corresponding to these in the dual graph form a "triangle"* (Fig. 1.36, b).

Several such "triangles" may be found in the original graph. For example, consider the one formed by edges 4, 7, 8 (Fig. 1.36, c), forming loop *IV* selected for the construction of the dual. These correspond to the "three-pointed star" at vertex (4′) in the dual graph corresponding to loop *IV*.

2. The non-separable graph characterized by the basis loop matrix

$$\boldsymbol{B}_1 = \begin{array}{c} \\ I \\ II \\ III \\ IV \\ V \\ VI \end{array} \begin{array}{c} \begin{array}{cccccccccc} 1 & 2 & 3 & 4 & 5 & 6 & 7 & 8 & 9 & 10 \end{array} \\ \begin{bmatrix} 1 & 1 & 1 & 0 & 0 & 0 & 0 & 0 & 0 & 0 \\ -1 & 0 & 0 & 1 & 0 & 1 & 0 & 0 & 0 & 0 \\ 0 & 0 & 0 & 0 & -1 & 1 & -1 & 0 & 0 & 0 \\ 0 & 1 & 0 & 0 & -1 & 1 & 0 & 1 & 0 & 0 \\ -1 & 0 & 0 & 0 & -1 & 1 & 0 & 0 & 1 & 0 \\ 0 & -1 & 0 & 0 & 1 & -1 & 0 & 0 & 0 & -1 \end{bmatrix} \end{array} .$$

will now be determined.

To this end let us transform $\boldsymbol{B}_1$ into a matrix allowing the partitioning of (1.45). This is attained, for example, by writing the columns corresponding to edges 3, 4, 7, 8, 9, 10 as the first six columns, then multiplying the rows corresponding to loops *III* and *VI* by -1. The matrix so obtained is

* "Three-pointed star" and "triangle" are often called "wye" and "delta", respectively.

$$
\boldsymbol{B}=\begin{bmatrix}
1 & 0 & 0 & 0 & 0 & 0 & 1 & 1 & 0 & 0\\
0 & 1 & 0 & 0 & 0 & 0 & -1 & 0 & 0 & 1\\
0 & 0 & 1 & 0 & 0 & 0 & 0 & 0 & 1 & -1\\
0 & 0 & 0 & 1 & 0 & 0 & 0 & 1 & -1 & 1\\
0 & 0 & 0 & 0 & 1 & 0 & -1 & 0 & -1 & 1\\
0 & 0 & 0 & 0 & 0 & 1 & 0 & 1 & -1 & 1
\end{bmatrix}=[\boldsymbol{1}_6 \;\; \boldsymbol{F}].
$$

Thus according to (1.60) and (1.62) a basis cutset matrix of the graph is

$$
\boldsymbol{Q}=[-\boldsymbol{F}^+ \;\; \boldsymbol{1}_4]=
\begin{array}{c}
\\ v_1 \\ v_2 \\ v_3 \\ v_4
\end{array}
\begin{array}{c}
\begin{array}{cccccccccc}
3 & 4 & 7 & 8 & 9 & 10 & 1 & 2 & 5 & 6
\end{array}\\
\begin{bmatrix}
-1 & 1 & 0 & 0 & 1 & 0 & 1 & 0 & 0 & 0\\
-1 & 0 & 0 & -1 & 0 & -1 & 0 & 1 & 0 & 0\\
0 & 0 & -1 & 1 & 1 & 1 & 0 & 0 & 1 & 0\\
0 & -1 & 1 & -1 & -1 & -1 & 0 & 0 & 0 & 1
\end{bmatrix}
\end{array}.
$$

On transposal and multiplication the relation $\boldsymbol{B}\boldsymbol{Q}^+=\boldsymbol{0}$ is seen to hold for the matrices $\boldsymbol{B}$ and $\boldsymbol{Q}$ so written. If the columns of $\boldsymbol{Q}$ are now ordered in the original sequence of the branches, and certain rows are replaced by linear combinations with other rows, a matrix $\boldsymbol{Q}_1$, also a basis cutset matrix of the graph, can be obtained, with $\boldsymbol{B}_1\boldsymbol{Q}_1^+=\boldsymbol{0}$. The following is such a matrix:

$$
\boldsymbol{Q}_1=\begin{bmatrix}
1 & 0 & -1 & 1 & 0 & 0 & 0 & 0 & 1 & 0\\
0 & -1 & 1 & 0 & 0 & 0 & 0 & 1 & 0 & 1\\
0 & 0 & 0 & 0 & 1 & 0 & -1 & 1 & 1 & 1\\
0 & 0 & 0 & -1 & 0 & 1 & 1 & -1 & -1 & -1
\end{bmatrix}.
$$

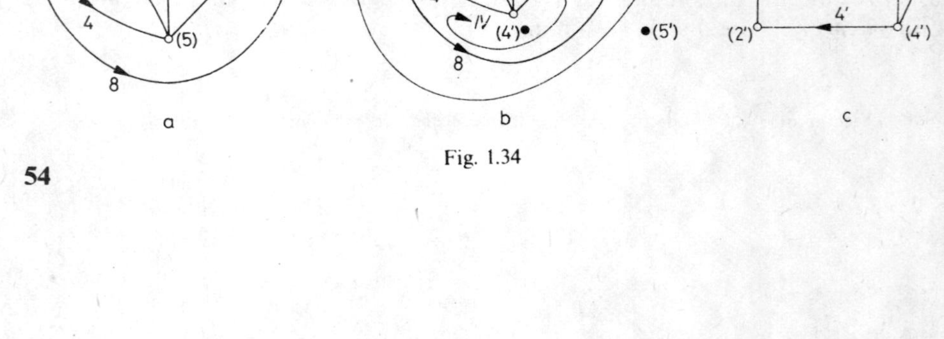

Fig. 1.34

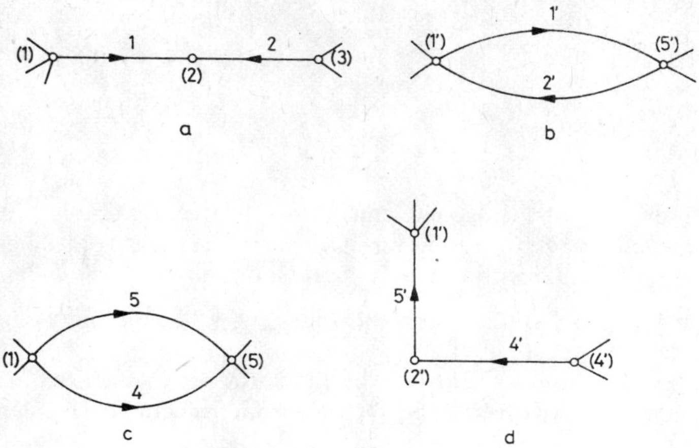

Fig. 1.35

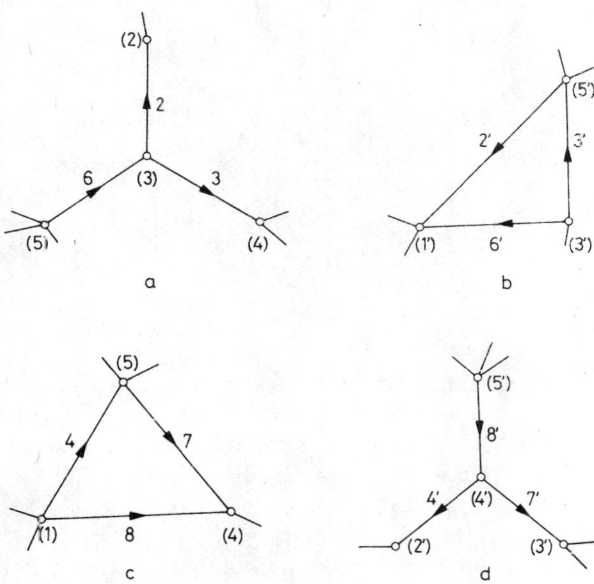

Fig. 1.36

This was derived from $\boldsymbol{Q}$ by arranging the columns according to the order numbers of the branches and multiplying the row of v_2 by -1.

Further cutset matrices may similarly be obtained from $\boldsymbol{Q}$ or $\boldsymbol{Q}_1$. For example, by replacing the fourth row of $\boldsymbol{Q}_1$ by the sum of the third and fourth rows, and multiplying the third row by -1, a special cutset matrix is derived:

$$A = \begin{bmatrix} 1 & 0 & -1 & 1 & 0 & 0 & 0 & 0 & 1 & 0 \\ 0 & -1 & 1 & 0 & 0 & 0 & 0 & 1 & 0 & 1 \\ 0 & 0 & 0 & 0 & -1 & 0 & 1 & -1 & -1 & -1 \\ 0 & 0 & 0 & -1 & 1 & 1 & 0 & 0 & 0 & 0 \end{bmatrix}.$$

In each column here at most one element is $+1$, at most one is -1, and the remaining elements are zero, i.e. it is a basis incidence matrix. Accordingly a graph may be drawn (Fig. 1.37) with one of its basis loop matrices being $\boldsymbol{B}_1$.

3. Let us select all loops of the graph shown in Fig. 1.38 containing edge 1 of the graph.

Along with edge 1 loops are formed by the paths between its two vertices, i.e. vertices (1) and (5). Such paths may be obtained from the complete incidence matrix of the graph:

$$A_{t0} = \begin{bmatrix} 1 & 1 & 0 & 0 & 1 & 0 & 0 & 0 & 0 \\ 0 & 1 & 1 & 1 & 0 & 0 & 0 & 0 & 0 \\ 0 & 0 & 0 & 1 & 1 & 1 & 1 & 1 & 0 \\ 0 & 0 & 0 & 0 & 0 & 0 & 1 & 0 & 1 \\ 1 & 0 & 1 & 0 & 0 & 1 & 0 & 1 & 1 \end{bmatrix}.$$

Let us select an edge incident with (1), such as edge 2. This is indicated by the second element in the first row of the incidence matrix being 1. Progressing along the edge vertex (2) is reached. This has been indicated by the line between the two unity elements in the second column. In vertex (2) edges represented by 1 in the second row are incident with edge 2, such as edge 3, the other vertex of which is just (5). Thus edges 2 and 3 form a path between (1) and (5).

The paths between (1) and (5) are collected from the incidence matrix by selecting a column, in the row corresponding to one of the vertices of edge 1 (vertex (1)), corresponding to an edge other than 1, incident with this vertex. Starting from this element 1, the other element 1 in this column is sought. In its row an element 1 is sought again, in the column of the latter another element 1, etc. If a row already crossed since starting is reached, the edges selected form a loop not containing edge 1. Otherwise vertex (5) is reached, and the edges so far indicated form a loop along with edge 1. When an appropriate systematic procedure is employed, all loops sought are obtained.

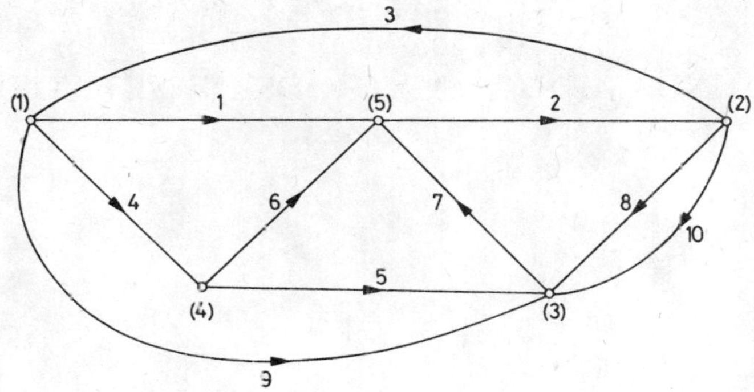

Fig. 1.37

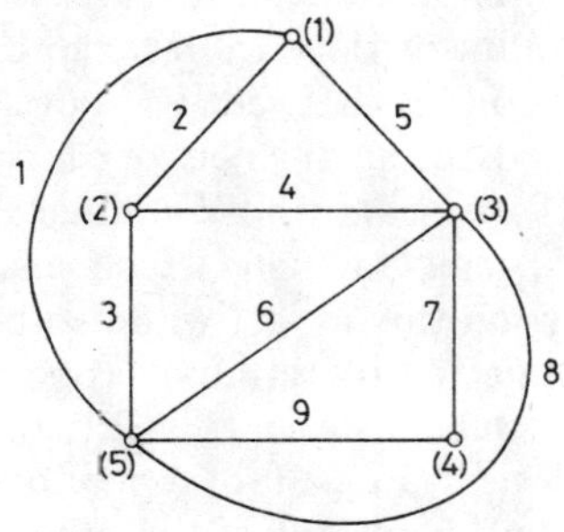

Fig. 1.38

In the former incidence matrix the paths thus selected have been indicated, and these form the loops represented by row matrices

$$\mathbf{B}_{10}^{+}=[1\quad 1\quad 1\quad 0\quad 0\quad 0\quad 0\quad 0\quad 0],$$
$$\mathbf{B}_{20}^{+}=[1\quad 1\quad 0\quad 1\quad 0\quad 1\quad 0\quad 0\quad 0],$$
$$\mathbf{B}_{30}^{+}=[1\quad 1\quad 0\quad 1\quad 0\quad 0\quad 1\quad 0\quad 1],$$
$$\mathbf{B}_{40}^{+}=[1\quad 1\quad 0\quad 1\quad 0\quad 0\quad 0\quad 1\quad 0],$$
$$\mathbf{B}_{50}^{+}=[1\quad 0\quad 1\quad 1\quad 1\quad 0\quad 0\quad 0\quad 0],$$
$$\mathbf{B}_{60}^{+}=[1\quad 0\quad 0\quad 0\quad 1\quad 1\quad 0\quad 0\quad 0],$$
$$\mathbf{B}_{70}^{+}=[1\quad 0\quad 0\quad 0\quad 1\quad 0\quad 1\quad 0\quad 1],$$
$$\mathbf{B}_{80}^{+}=[1\quad 0\quad 0\quad 0\quad 1\quad 0\quad 0\quad 1\quad 0]$$

together with edge 1.

CHAPTER 2

FUNDAMENTAL METHODS OF NETWORK ANALYSIS

Electrical networks [3, 16, 24, 25, 26, 30, 36, 37, 40, 42, 43, 44, 45] consist of an interconnection of network elements. A knowledge of the network element voltages, currents and their interdependence is sufficient for the investigation of the network from an electrical point of view. In order to calculate the currents in the branches of the network, the voltages between its nodes, and their interdependence, some model of the network is used, e.g. the network is considered to be constructed from interconnected network elements whose voltage and current or voltages and currents are related in a specific manner which characterizes each particular network element. From now on this model will also be called a network.

Network elements are connected by terminals (poles). A network element with two terminals is called *two-terminal element* (Fig. 2.1, a), while that with n terminals is referred to as *n-terminal element* (Fig. 2.1, b). A pair of terminals is a *port*, if all the current into one terminal flows back out of the other (Fig. 2.2, a). One part of a

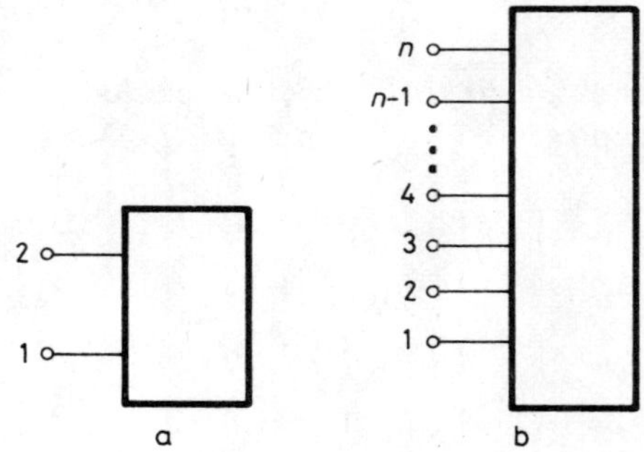

Fig. 2.1

network may connect to other parts by several ports, such an *n-port* being shown in Fig. 2.2, b. *Two-ports* as shown in Fig. 2.2, c are commonly employed.

Special network elements called generators (sources) supply electrical energy to the network, while energy is consumed or stored by passive elements. Parts of the network consisting of passive elements only are also passive.

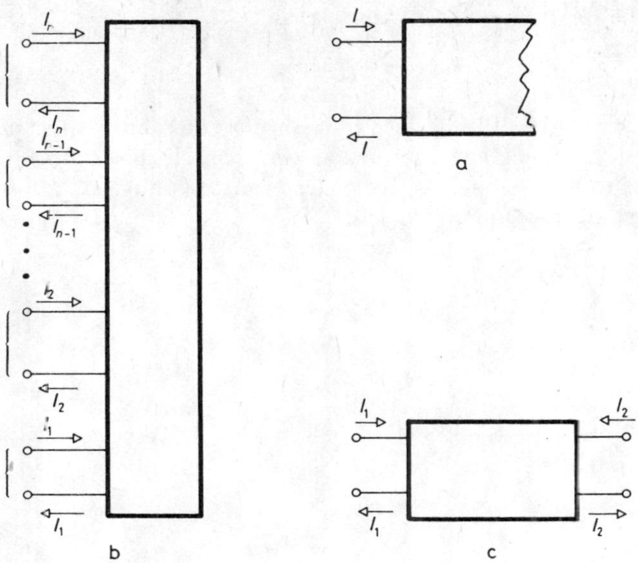

Fig. 2.2

A network element is said to be *linear* if the relationships between its currents and voltages are linear. The set of linear equations describing the relationship between the currents and voltages of a passive linear element and a network consisting of such elements only is homogeneous, while for linear networks containing active linear elements as well as passive elements it is inhomogeneous. If the relationship between currents and voltages of a network element is not linear, the element and the network containing it is called nonlinear. If the coefficients in the equations representing the relationship between currents and voltages are independent of time, the network element as well as the network are *time-invariant*. In our calculations networks are approximated by linear, time-invariant models. We shall discuss those problems for which this approximation is valid.

In this chapter are presented methods suitable for the determination of branch-currents and voltages in linear, time-invariant networks that can be modelled by a connection of the following types of network elements. The two-terminal elements considered are:

(a) *voltage-source* (Fig. 2.3), whose voltage u_g, the *source-voltage*, is independent of its current as well as of other currents and voltages of the network;

(b) *current-source* (Fig. 2.4), whose current i_g, the *source-current*, is independent of its voltage as well as of other currents and voltages of the network;

(c) *resistor* (Fig. 2.5), whose current i_R and voltage u_R are proportional:

$$u_R = Ri_R \qquad \text{or} \qquad i_R = Gu_R, \tag{2.1}$$

where R is the *resistance* of the two-pole, and

$$G = 1/R \tag{2.2}$$

is its *conductance*. A resistor of zero resistance is commonly called a short-circuit, and one of zero conductance is termed an open-circuit. The short-circuit may also be considered to be a voltage-source of zero source-voltage, and an open-circuit a current-source of zero source-current:

Fig. 2.3

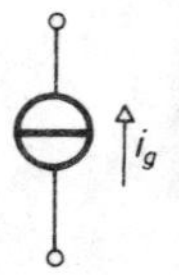

Fig. 2.4

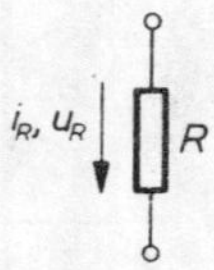

Fig. 2.5

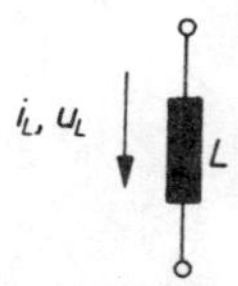

Fig. 2.6

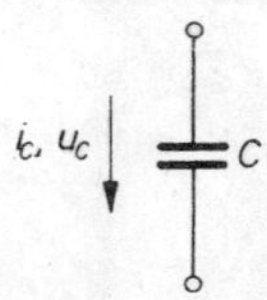

Fig. 2.7

(d) *inductor* (Fig. 2.6), whose voltage u_L is proportional to the time derivative of its current i_L:

$$u_L = L\frac{\mathrm{d}i_L}{\mathrm{d}t}, \tag{2.3}$$

where L is the *self-inductance*;

(e) *capacitor* (Fig. 2.7), whose current i_C is proportional to the time-derivative of its voltage u_C:

$$i_C = C\frac{\mathrm{d}u_C}{\mathrm{d}t}, \tag{2.4}$$

where C is the *capacitance*.

Positive-valued resistors, inductors and capacitors are passive two-poles.

In addition to the above two-poles, the network may also contain inductively-coupled two-terminal elements (Fig. 2.8, a), whose voltages $u_1, u_2, \ldots, u_j, \ldots, u_n$ and currents $i_1, i_2, \ldots, i_k, \ldots, i_n$ are related by

$$u_j = \sum_{k=1}^{n} L_{jk}\frac{\mathrm{d}i_k}{\mathrm{d}t}, \qquad j = 1, 2, \ldots, n, \tag{2.5}$$

where L_{jj} is the self-inductance of the j-th element, while L_{jk} is the *mutual inductance* between the j-th and k-th elements ($j, k = 1, 2, \ldots, n; j \neq k$). For the specific case $n = 2$ (Fig. 2.8, b) relation (2.5) has the form:

$$u_1 = L_{11}\frac{di_1}{dt} + L_{12}\frac{di_2}{dt},$$

$$u_2 = L_{21}\frac{di_1}{dt} + L_{22}\frac{di_2}{dt}. \qquad (2.6)$$

The coupled two-terminal elements are *reciprocal* if $L_{jk} = L_{kj}$ ($j \neq k$). Calculation of networks including non-reciprocal two-terminal elements will be discussed in Chapter 5. In the present chapter as well as in chapters 3 and 4 it is assumed that any such inductively coupled two-terminal elements appearing in the networks under discussion are reciprocal.

By interconnecting elements of the types listed above, two-terminal elements, n-terminal elements, n-ports and further networks may be derived.

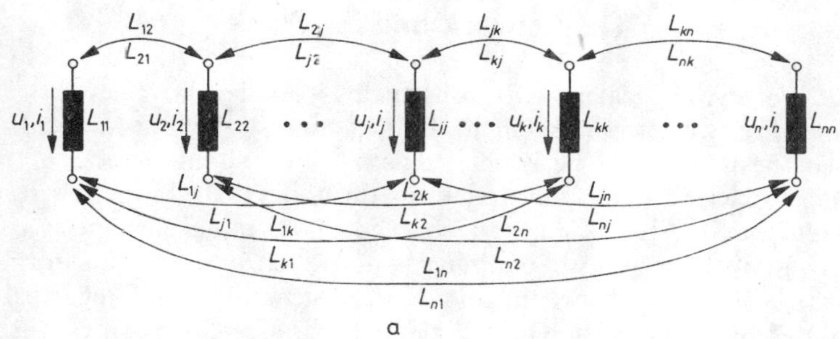

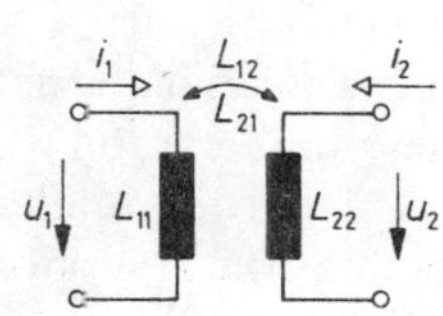

Fig. 2.8

The branches of a real network will be modelled by suitable combinations of the elements described above. Thus, in many cases, the connection in series of an inductor and a resistor may be considered a suitable model of a coil, and a parallel connection of a capacitor and a resistor that of a real capacitor. Linear non-ideal generators are modelled by Thevenin and Norton equivalents. The former is a connection in series of a voltage-source and a passive two-terminal element (Fig.

2.9), while the latter is a parallel connection of a current-source and a passive two-terminal element (Fig. 2.10). From now on, such non-ideal generators will be called generators, while the term source (voltage-source, current-source) will be reserved for ideal generators.

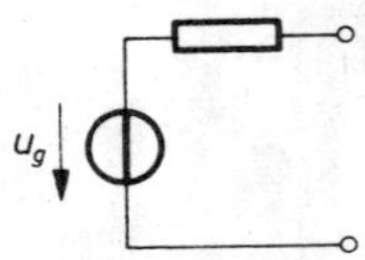

Fig. 2.9

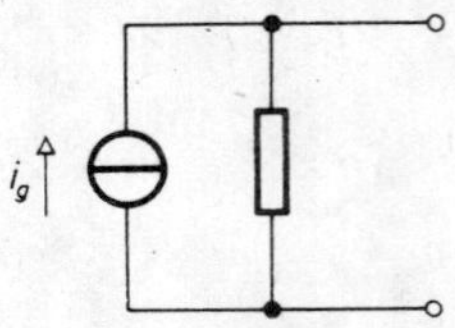

Fig. 2.10

In following chapters in addition to the two-terminal elements and inductively coupled two-terminal elements so far described calculations for networks containing other types of two-terminal elements, coupled two-terminal elements and further network elements will also be touched upon.

The network and its graph

In a model of an electrical network constructed from the above network elements every branch is a two-terminal element. In the course of the following calculations the model of the network will be associated with a graph so that an edge of the graph corresponds to each branch of the model, and the nodes of the branch correspond to the vertices of the edge. Each element of coupled two-terminal elements is associated with one edge of the graph, i.e. coupling is not taken into account in the graph. Unless otherwise stated Thevenin and Norton generators are regarded as one branch. Later on problems involving elements not yet mentioned will also be discussed where the graph will be associated with the network in a different manner.

Let us define the column matrices consisting of branch-currents and branch-voltages of the network. For this the branches of the network and the corresponding edges of the graph are assigned order numbers, and the current and voltage of each branch is given an index identical to the order number of the corresponding edge. The orientation of each edge is chosen as the reference direction of the current and voltage of the associated branch. If the network consists of b branches, the branch-currents are $I_1, I_2, \ldots, I_b$. It should be noted that the branch-current of a branch corresponding to a Norton generator is the sum of the currents of the current-source and the parallel passive two-terminal element (Fig. 2.11, a). The current column matrix

$$\mathbf{I} = \begin{bmatrix} I_1 \\ I_2 \\ \vdots \\ I_b \end{bmatrix} \tag{2.7}$$

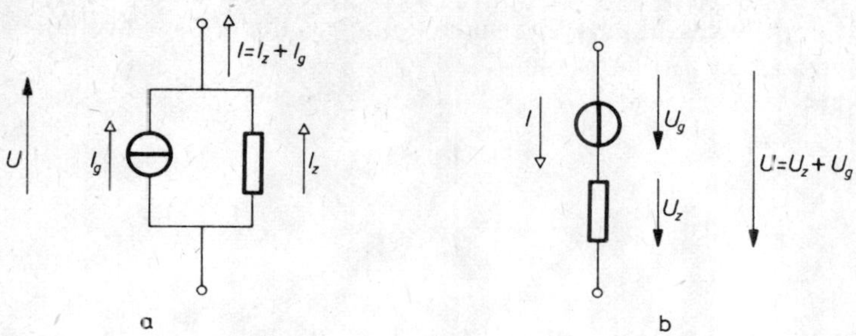

Fig. 2.11

is formed from these branch-currents. The column matrix of the currents of passive two-terminal elements

$$\mathbf{I}_z = \begin{bmatrix} I_{z1} \\ I_{z2} \\ \vdots \\ I_{zb} \end{bmatrix} \tag{2.8}$$

may also be defined. Here the element of the matrix corresponding to a branch representing a Norton generator is the current of the parallel passive two-terminal element.

The currents of the sources are given in a column matrix

$$\mathbf{I}_g = \begin{bmatrix} I_{g1} \\ I_{g2} \\ \vdots \\ I_{gb} \end{bmatrix}. \tag{2.9}$$

The elements of this matrix corresponding to branches without a current-source naturally equal zero. In accordance with their definitions the sum of matrices $\mathbf{I}_g$ and $\mathbf{I}_z$ is matrix $\mathbf{I}$:

$$\mathbf{I}_g + \mathbf{I}_z = \mathbf{I}. \tag{2.10}$$

The voltage column matrix

$$\mathbf{U} = \begin{bmatrix} U_1 \\ U_2 \\ \vdots \\ U_b \end{bmatrix} \tag{2.11}$$

is formed from branch voltages. A branch voltage is the voltage drop between the two nodes incident with the branch.

The column matrix

$$\mathbf{U}_z = \begin{bmatrix} U_{z1} \\ U_{z2} \\ \vdots \\ U_{zb} \end{bmatrix} \tag{2.12}$$

is constructed from the voltage drops of branch-impedances. Here $U_{z1}, U_{z2}, \ldots, U_{zb}$ are the voltages across the passive two-terminal elements in branches $1, 2, \ldots, b$ respectively, with reference directions coinciding with edge orientations. The elements of matrix (2.12) corresponding to branches representing Thevenin generators are the voltages across the series passive two-terminal elements.

The voltages of the voltage-sources constitute a column matrix

$$\mathbf{U}_g = \begin{bmatrix} U_{g1} \\ U_{g2} \\ \vdots \\ U_{gb} \end{bmatrix}. \tag{2.13}$$

$U_{g1}, U_{g2}, \ldots, U_{gb}$ are the voltages of voltage-sources in branches $1, 2, \ldots, b$ respectively, with reference directions identical with edge orientations. The elements of matrix $\mathbf{U}_g$ associated with branches without voltage-sources are equal to zero.

In accordance with the above definitions the relation

$$\mathbf{U}_g + \mathbf{U}_z = \mathbf{U} \tag{2.14}$$

holds between matrices $\mathbf{U}$, $\mathbf{U}_z$ and $\mathbf{U}_g$ (Fig. 2.11, b).

The above equations retain their validity in the case when Norton and Thevenin generators are each considered to consist of two branches, one of them being a source. Certain analysis methods are not applicable to networks which contain branches which are ideal sources and therefore have either zero resistance or zero conductance. Consequently in those cases where the network contains no branches representing sources, open-circuits or short-circuits, it is expedient to consider non-ideal generators as forming one branch.

Ohm's law

Kirchhoff's laws relate to instantaneous values of currents and voltages. It is easily verified that Kirchhoff's laws remain valid for complex peak or effective values representing steady-state responses in networks with sinusoidal or other periodic excitation, and for Laplace transforms in the case of excitation with general functions of time. In the present chapter as well as in Chapters 3, 4 and 5 complex peak values or Laplace transforms are used while in Chapter 6 a method is presented for the case of currents with general time variation where calculations are carried out with the aid of time-functions.

To obtain a generalized form of Ohm's law for signals of various forms of time-dependence, and to write Kirchhoff's laws, the fundamental relations required for the conditions mentioned above are now reviewed:

(a) in *direct current networks* the relation between the voltage U and current I of a resistance R is given by

$$U = RI. \tag{2.15}$$

In direct current networks inductors are regarded as short-circuits in compliance with (2.3), while capacitors are considered open-circuits according to (2.4).

(b) The *sinusoidal* real functions of voltage

$$u(t) = U_0 \cos(\omega t + \psi_u) \tag{2.16}$$

and of current

$$i(t) = I_0 \cos(\omega t + \psi_i) \tag{2.17}$$

with angular frequency ω are replaced in our calculations by their complex peak values

$$U = U_0 e^{j\psi_u} \quad \text{and} \quad I = I_0 e^{j\psi_i}. \tag{2.18}$$

The quotient of the complex peak values of the voltage and current of the steady-state response of a passive two-terminal element is the complex impedance between the two terminals:

$$Z = |Z| e^{j\varphi} = \frac{U}{I} = \frac{U_0}{I_0} e^{j(\psi_u - \psi_i)}. \tag{2.19}$$

For sinusoidal signals (voltages, currents) with angular frequency ω the impedances of a resistance R, inductance L and capacitance C are given by

$$Z_R = R; \qquad Z_L = j\omega L; \qquad Z_C = \frac{1}{j\omega C}, \tag{2.20}$$

respectively. The reciprocal of the impedance is the admittance Y:

$$Y = \frac{1}{Z} = |Y| e^{-j\varphi}. \tag{2.21}$$

If inductively coupled two-terminal elements are also present in the network mutual impedances or admittances appear in the equations describing the relationships between currents and voltages. Thus the voltage of the i-th branch and the current of the k-th branch are

$$U_i = \sum_j Z_{ij} I_j; \qquad I_k = \sum_l Y_{kl} U_l, \tag{2.22}$$

where Z_{ij} is the mutual impedance between the i-th and j-th branch, Y_{kl} is the mutual admittance between the k-th and l-th branch, Z_{ii} is the self-impedance of the i-th branch, Y_{kk} is the self-admittance of the k-th branch, I_j is the current of the j-th branch and U_l is the voltage of the l-th branch.

(c) Signals with *periodic time variation* may be expanded into Fourier series. A function $f(t)$ is periodic with respect to T if

$$f(t) = f(t+nT), \qquad (n=1, 2, \ldots), \tag{2.23}$$

where T is the period of the function. The Fourier series of a periodic, bounded and Riemann integrable function is

$$f(t) = \sum_{k=0}^{\infty} (A_k \cos k\omega t + B_k \sin k\omega t). \tag{2.24}$$

Fourier coefficients A_k and B_k are given by the well-known formulas:

$$A_0 = \frac{1}{T}\int_0^T f(t)\,dt,$$

$$A_k = \frac{2}{T}\int_0^T f(t)\cos k\omega t\,dt,$$

$$B_k = \frac{2}{T}\int_0^T f(t)\sin k\omega t\,dt, \qquad k = 1, 2, \ldots, \qquad \omega = \frac{2\pi}{T}. \tag{2.25}$$

For network calculations the terms of the above Fourier expansion are conveniently expressed as the real part of a complex quantity:

$$A_k \cos k\omega t + B_k \sin k\omega t = \mathrm{Re}\,(A_k - jB_k)e^{jk\omega t} = \mathrm{Re}\; C_k e^{j(k\omega t + \psi_k)}, \tag{2.26}$$

where

$$C_k = \sqrt{A_k^2 + B_k^2} \tag{2.27}$$

and

$$\psi_k = -\arctan\frac{B_k}{A_k}, \tag{2.28}$$

or according to (2.24)

$$f(t) = \mathrm{Re}\sum_{k=0}^{\infty} C_k\, e^{j(k\omega t + \psi_k)} = \sum_{k=0}^{\infty} C_k \cos(k\omega t + \psi_k). \tag{2.29}$$

Voltages and currents in networks with periodic excitation are written in this form:

$$u(t) = \sum_{k=0}^{\infty} U_k \cos(k\omega t + \psi_k) = \mathrm{Re}\sum_{k=0}^{\infty} U_k e^{j(k\omega t + \psi_k)}, \tag{2.30}$$

$$i(t) = \sum_{k=0}^{\infty} I_k \cos(k\omega t + \xi_k) = \mathrm{Re}\sum_{k=0}^{\infty} I_k e^{j(k\omega t + \xi_k)}. \tag{2.31}$$

Since only linear networks are being considered, analysis may be carried out independently for each term of the sums i.e. for direct current components, the fundamental and the higher harmonics. Each of these analyses may be carried out by methods suitable for direct currents and sinusoidal signals, respectively. By superposition, the time-function of the voltage or current sought is the sum of the time-functions of the separate responses so determined.

A method is next presented by means of which all harmonics are simultaneously calculated with the aid of hypermatrices. To this end the following notation is introduced:

$$\sum_{k=0}^{\infty} U_k e^{j(k\omega t+\psi_k)} = \mathbf{E}^+\mathbf{U},$$

$$\sum_{k=0}^{\infty} I_k e^{j(k\omega t+\xi_k)} = \mathbf{E}^+\mathbf{I}, \tag{2.32}$$

where

$$\mathbf{E}^+ = [1 \; e^{j\omega t} \; e^{j2\omega t} \; \dots \; e^{jk\omega t} \; \dots], \tag{2.33}$$

$$\mathbf{U} = \begin{bmatrix} U_0 \\ U_1 e^{j\psi_1} \\ U_2 e^{j\psi_2} \\ \vdots \\ U_k e^{j\psi_k} \\ \vdots \end{bmatrix}, \qquad \mathbf{I} = \begin{bmatrix} I_0 \\ I_1 e^{j\xi_1} \\ I_2 e^{j\xi_2} \\ \vdots \\ I_k e^{j\xi_k} \\ \vdots \end{bmatrix}. \tag{2.34}$$

U and **I** are the complex-valued column matrices of voltage (2.30) and of current (2.31), respectively.

The voltage column matrix $\mathbf{U}_i$ of the passive two-terminal element forming the i-th branch is given by

$$\mathbf{U}_i = \sum_j \mathbf{Z}_{ij}\mathbf{I}_j, \tag{2.35}$$

where Z_{ij} $(i \neq j)$ is a diagonal matrix whose elements are the mutual impedances between branches i and j for angular frequencies $0, \omega, 2\omega, \dots$, i.e. for direct current, fundamental and higher harmonics, while the diagonal matrix Z_{ii} is formed by the self-impedances of the i-th branch in the same manner. A similar formula may be written for the current of the i-th branch:

$$\mathbf{I}_i = \sum_j Y_{ij}\mathbf{U}_j \tag{2.36}$$

with diagonal matrices Y_{ij} formed by the mutual admittances between the branches or branch-admittances for direct current, fundamental and higher harmonics.

The method introduced above retains its simplicity if only a few terms in the Fourier series of the applied signals are taken into account.

(d) In the case of *general time variation* of voltages and currents the events occurring may include switching on or off, or switching over connections of the network, all of which are commonly termed transient processes.

For switch-on transients, the signal is instantaneously applied to the network which typically contains no stored energy prior to this. Given the time-function of the applied signal, the response in any branch of the network can be determined with the aid of the Laplace transformation. Between the voltage $u(t)$ and current $i(t)$ of a passive two-pole, the relation

$$U(s)=Z(s)I(s) \tag{2.37}$$

holds for switch-on transients, where $U(s)=\mathscr{L}u(t)$ and $I(s)=\mathscr{L}i(t)$ are the Laplace transforms of voltage and current respectively, while $Z(s)$ is the operational impedance of the two-pole. The operational impedance of a resistance R, inductance L and capacitance C are given by

$$Z_R(s)=R; \qquad Z_L(s)=sL; \qquad Z_C(s)=\frac{1}{sC}. \tag{2.38}$$

The transform of excitation $u(t)$ is calculated from

$$U(s)=\int_{-\infty}^{\infty} 1(t)\,u(t)\,e^{-st}\,\mathrm{d}t\,, \tag{2.39}$$

or is obtained from tables of Laplace transforms. Here*

$$1(t)=\begin{cases}0 & \text{if} \quad t<0\\ 1 & \text{if} \quad t>0\,,\end{cases} \tag{2.40}$$

is the unit step function. The time-function of a response is determined from its Laplace transform, a function of s, with the aid of the inverse Laplace transformation. In many problems of network analysis this inverse Laplace transformation is carried out by means of the expansion theorem. When the Laplace transform is a proper rational function with simple poles,** i.e. the Laplace transform of the response is

$$I(s)=\frac{M(s)}{N(s)}=\frac{a_m s^m+a_{m-1}s^{m-1}+\ldots+a_2s^2+a_1s+a_0}{b_n s^n+b_{n-1}s^{n-1}+\ldots+b_2s^2+b_1s+b_0}$$

$$m<n \tag{2.41}$$

* It might be noted that $H(t)$ or $V_{-1}(t)$ are commonly used symbols for $1(t)$.

** The poles are the roots of the determinator polynomial.

the time function of the response according to the expansion theorem is given by

$$i(t)=\mathscr{L}^{-1}\, I(s) = \sum_{k=1}^{n} 1(t)\frac{M(s_k)}{N'(s_k)}\, e^{s_k t}, \tag{2.42}$$

where s_k $(k=1, 2, \ldots, n)$ are the poles of $I(s)$, $N'(s)$ is the derivative of the denominator with respect to s. Since the coefficients b_i $(i=0, 1, \ldots, n)$ are real, the poles are of the form $s_k = -\dfrac{1}{\tau_k} \pm j\omega_k$, where $\tau_k > 0$ are time constants of the network*, and $\omega_k \geq 0$ are angular frequencies of oscillations occurring in the network. If $N(s)$ has multiple roots the determination of the time-function may be carried out by other methods given in the literature [19].

For the analysis of switching over and switching off transients the current $i(-0)$ of inductors and the voltage $u(-0)$ of capicators at the instant immediately before switching $(t=-0)$ must be known. For the time $t>0$ after the switching transient inductors may be replaced by Norton generators (Fig. 2.12) and capacitors by Thevenin-generators (Fig. 2.13), whose internal impedances $\left(sL \text{ and } \dfrac{1}{sC}\right)$ are formed of inductances L and capacitances C without stored energy at the instant before switching $(t=-0)$. The source-current of the Norton generator is $1(t)\, i(-0)$, while the source-voltage of the Thevenin generator is $1(t)\, u(-0)$. In the subsequent analysis these generators are taken into account in exactly the same way as any other generators of the network. Events in the network thus transformed are calculated as if they were switch-on transients.

The interdependence between the voltage of the impedance in the i-th branch and the currents of the network is given by

$$U_{zi} = \sum_{j=1}^{b} Z_{ij} I_{zj}, \tag{2.43}$$

where Z_{ij} $(i \neq j)$ is the mutual impedance of the i-th and j-th branch, while Z_{ii} is the self-impedance of the i-th branch. The impedances Z_{ij} and Z_{ii} are determined in

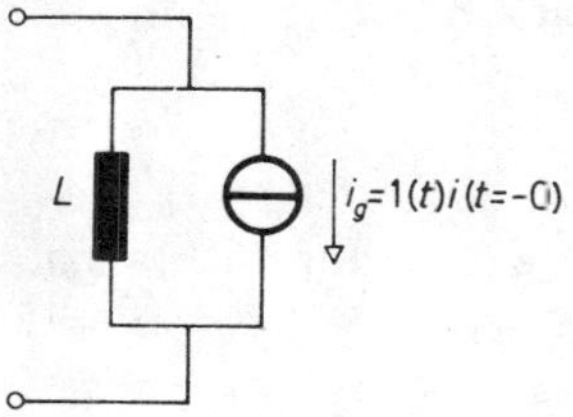

Fig. 2.12

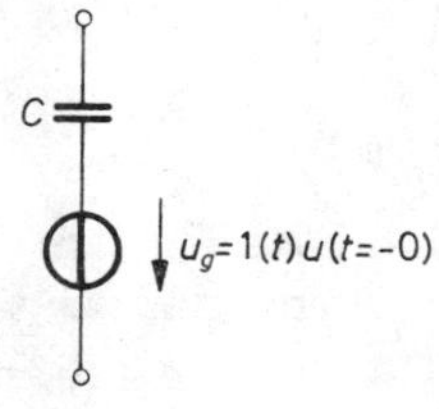

Fig. 2.13

* For passive networks, which must be stable, poles cannot occur in the right half of the s-plane, so the τ_k are non-negative. This restriction does not apply to active networks, which may be unstable.

different ways according to the time variation of the source-voltages and source-currents of generators in the network. Writing equation (2.43) for all branches, these are summarized in one matrix equation:

$$\mathbf{U}_z = \boldsymbol{Z}\mathbf{I}_z, \tag{2.44}$$

where

$$\boldsymbol{Z} = \begin{bmatrix} Z_{11} & Z_{12} & \dots & Z_{1b} \\ Z_{21} & Z_{22} & \dots & Z_{2b} \\ \dots & \dots & \dots & \dots \\ Z_{b1} & Z_{b2} & \dots & Z_{bb} \end{bmatrix} \tag{2.45}$$

is the *impedance matrix*. Equation (2.44) may be regarded as a matrix form of Ohm's law.

The inverse of the impedance matrix is the admittance matrix $\boldsymbol{Y}$:

$$\boldsymbol{Y} = \boldsymbol{Z}^{-1}. \tag{2.46}$$

If there are no mutual impedances in the network $\boldsymbol{Y}$ and $\boldsymbol{Z}$ are diagonal matrices and can be written directly from a knowledge of the elements of the network. If known mutual impedances are present in the network $\boldsymbol{Y}$ may be determined as the inverse of $\boldsymbol{Z}$.

The equations are of a similar form for signals with various different time forms of variation.

For direct current networks the elements on the diagonal of matrix $\boldsymbol{Z}$ (2.45) are the values of the resistance in the branch corresponding to the row (column) of the matrix, while all elements off the diagonal are zero.

$$\boldsymbol{Z} = \boldsymbol{R} = \boldsymbol{G}^{-1}. \tag{2.47}$$

For the steady state response to excitation with sinusoidal functions of time the elements of $\boldsymbol{Z}$ are the complex impedances of branches and the complex mutual impedances between branches.

For networks with other forms of periodic excitation, the column matrix

$$\mathbf{U}_{zi} = \sum_{j=1}^{b} \mathbf{Z}_{ij}\mathbf{I}_{zj} \tag{2.48}$$

is used instead of (2.43). Equation (2.48) may be written for all branches of the network containing passive elements, and thus hypermatrices appear in the equation corresponding to (2.44):

$$\mathbf{U} = \begin{bmatrix} \mathbf{U}_{z1} \\ \mathbf{U}_{z2} \\ \vdots \\ \mathbf{U}_{zb} \end{bmatrix}; \quad \mathbf{I}_z = \begin{bmatrix} \mathbf{I}_{z1} \\ \mathbf{I}_{z2} \\ \vdots \\ \mathbf{I}_{zb} \end{bmatrix}. \tag{2.49}$$

where $\mathbf{U}_{z1}, \mathbf{U}_{z2}, \ldots, \mathbf{U}_{zb}, \mathbf{I}_{z1}, \mathbf{I}_{z2}, \ldots, \mathbf{I}_{zb}$ are the column matrices constructed from the direct current component, fundamental and higher harmonics of voltages and currents in branches 1, 2, ..., b, while

$$\boldsymbol{Z}=\begin{bmatrix} \boldsymbol{Z}_{11} & \boldsymbol{Z}_{12} & \ldots & \boldsymbol{Z}_{1b} \\ \boldsymbol{Z}_{21} & \boldsymbol{Z}_{22} & \ldots & \boldsymbol{Z}_{2b} \\ \ldots & \ldots & \ldots & \ldots \\ \boldsymbol{Z}_{b1} & \boldsymbol{Z}_{b2} & \ldots & \boldsymbol{Z}_{bb} \end{bmatrix}. \tag{2.50}$$

$\boldsymbol{Z}_{ij}$ denotes the matrix defined in (2.35).

The column matrices $\mathbf{U}_{g1}, \mathbf{U}_{g2}, \ldots, \mathbf{U}_{gb}, \mathbf{I}_{g1}, \mathbf{I}_{g2}, \ldots, \mathbf{I}_{gb}$ characterizing the source-voltages and source-currents of sources in branches may similarly be written. The matrices of source-voltages and source-currents of the network sources are thus given by

$$\mathbf{U}_g=\begin{bmatrix} \mathbf{U}_{g1} \\ \mathbf{U}_{g2} \\ \vdots \\ \mathbf{U}_{gb} \end{bmatrix}; \quad \mathbf{I}_g=\begin{bmatrix} \mathbf{I}_{g1} \\ \mathbf{I}_{g2} \\ \vdots \\ \mathbf{I}_{gb} \end{bmatrix}. \tag{2.51}$$

For the analysis of networks with periodic excitation with the aid of hypermatrices, the matrices $\boldsymbol{A}$, $\boldsymbol{B}$ and $\boldsymbol{Q}$ of the network have to be modified so that the equations including these matrices describe all harmonics. Therefore 1 is replaced by a unit matrix $\boldsymbol{1}$, while a square zero matrix $\boldsymbol{0}$ replaces 0 in them, and the order of these matrices is determined by the number of harmonics present in the excitations.

For signals of general time-variation (including switching on, switching over and switching off transients), the analysis is carried out with the aid of Laplace transformation. For switch-on transients column matrices $\mathbf{U}_z, \mathbf{U}_g, \mathbf{I}_z, \mathbf{I}_g$ are formed by the Laplace transforms of voltages and currents, while impedance matrix $\boldsymbol{Z}$ is constructed of the operational impedances of branches:

$$\mathbf{U}_z(s) = \boldsymbol{Z}(s)\mathbf{I}_z(s). \tag{2.52}$$

For the calculation of switching over and switching off transients each inductor and capacitor of the network (i.e. the Thevenin and Norton generators replacing them) should be represented by one edge of the graph. Thus, switching over and switching off phenomena are converted to switch-on transients for the purpose of analysis.

Kirchhoff's laws

According to Kirchhoff's current law the algebraic sum of currents in branches incident with node (j) equals zero:

$$\sum i_k(t)=0, \qquad \sum I_k=0. \tag{2.53}$$

Here $i_k(t)$ denotes the time functions of the currents in branches incident with the node, with reference direction pointing away from the node, while I_k is the complex peak value for sinusoidal currents, the column matrix of complex amplitudes for periodic currents, and the Laplace transform for currents of general time variation.

It has been shown that a row matrix $\mathbf{A}_j^+$ can be assigned to each node (see Chapter 1, (1.12)). Thus, using (2.10), (2.53) for node (j) may be written in the form

$$\mathbf{A}_j^+(\mathbf{I}_g+\mathbf{I}_z)=0. \tag{2.54}$$

This equation holds for each node of the network. The system of equations thus obtained is summarized as

$$\boldsymbol{A}_t(\mathbf{I}_g+\mathbf{I}_z)=\mathbf{0}. \tag{2.55}$$

(2.55) is equivalent to n algebraic equations. Among these equations $n-c$ are linearly independent, i.e. c node equations are unnecessary. Therefore, equation

$$\boldsymbol{AI}=\boldsymbol{A}(\mathbf{I}_g+\mathbf{I}_z)=\mathbf{0} \tag{2.56}$$

is sufficient instead of (2.55), where $\boldsymbol{A}$ is the basis incidence matrix. (2.56) consists of Kirchhoff's linearly independent node equations of the network.

Kirchhoff's current law is a special case of equation

$$\sum_i I_i=0, \tag{2.57}$$

where the sum refers to a cutset of the graph, with I_i positive if its direction coincides with the orientation of the cutset, and negative otherwise. Thus Kirchhoff's current law may be written not only for nodes, but also for cutsets of the network. A fundamental set of cutsets is formed by $n-c$ linearly independent cutsets. Kirchhoff's equations written for independent nodes are equivalent to Kirchhoff's equations for a fundamental set of cutsets. The latter may be summarized in the matrix equation

$$\boldsymbol{QI}=\boldsymbol{Q}(\mathbf{I}_g+\mathbf{I}_z)=\mathbf{0}, \tag{2.58}$$

where $\boldsymbol{Q}$ is the basis cutset matrix. The incidence matrix is a special case of the cutset matrix, and so (2.58) is more general than (2.56).

Kirchhoff's loop equation

$$\sum u_j(t)=0, \qquad \sum U_j=0 \tag{2.59}$$

is valid for any loop of a network, where the sum refers to each branch of the loop with the orientation of the loop taken into account. $u_j(t)$ is the time-function of the voltage in branch j of the loop, U_j is the complex peak value for sinusoidal voltages, the column matrix of complex amplitudes for periodic voltages, or the Laplace transform of voltages of general time-variation.

The j-th loop of the network is characterized by row matrix $\mathbf{B}_j^+$ (see Chapter 1, (1.3)). With the aid of column matrices $\mathbf{U}_g$ and $\mathbf{U}_z$, (2.59) takes the form

$$\mathbf{B}_j^+(\mathbf{U}_g+\mathbf{U}_z)=0. \tag{2.60}$$

Similar equations may be written for each loop of the network. For the m loops of a fundamental set of loops these are summarized in the matrix equation

$$\boldsymbol{B}\mathbf{U}=\boldsymbol{B}(\mathbf{U}_g+\mathbf{U}_z)=\mathbf{0}, \tag{2.61}$$

where $\boldsymbol{B}$ is the basis loop matrix of the fundamental set of loops. (2.61) includes all of Kirchhoff's loop equations for an independent loop-set of the network.

If the fundamental set of loops and cutsets in the network are chosen so as to derive the normal forms of $\boldsymbol{Q}$ and $\boldsymbol{B}$, (2.58) and (2.61) may be written in the forms

$$[-\boldsymbol{F}^+ \quad \boldsymbol{1}]\,\mathbf{I}=\mathbf{0}, \qquad [-\boldsymbol{F}^+ \quad \boldsymbol{1}]\,(\mathbf{I}_g+\mathbf{I}_z)=\mathbf{0} \tag{2.62}$$

and

$$[\boldsymbol{1} \quad \boldsymbol{F}]\,\mathbf{U}=\mathbf{0}, \qquad [\boldsymbol{1} \quad \boldsymbol{F}]\,(\mathbf{U}_g+\mathbf{U}_z)=\mathbf{0}, \tag{2.63}$$

respectively.

The existence and uniqueness of the solution to network analysis problems

To obtain the solution to a network analysis problem the voltage and current of each branch forming the network has to be determined, given a knowledge of the structure of the linear, time-invariant network, the interdependence between currents and voltages of the coupled and uncoupled passive two-terminal elements modelling the branches, and the characteristics of Thevenin and Norton equivalent-circuits of generators or of sources.

The number of branch-currents and branch-voltages to be determined is $2b$. The above characteristics of the two-terminal elements yield b equations for the currents and voltages of the two-terminal elements forming the network. The number of linearly independent cutset equations is $n-c$, and of loop equations is m, i.e. a further $n-c+m=b$ equations are obtained from Kirchhoff's laws. Thus, on the whole, $2b$ equations may be written for the determination of $2b$ unknowns, and so the problem is solvable unless the equations are inconsistent or redundant.

Redundancy or inconsistency may appear in Kirchhoff's equations if two-terminal elements with fixed voltages form a loop or if two-poles with fixed currents form a cutset. Of the network elements being discussed, voltage-sources, short-circuits and, at the initial instant of transient processes, capacitors have their voltage specified. Loops consisting only of such elements are commonly called capacitive loops, especially in the analysis of transient processes. Elements with prescribed current are current-sources, open-circuits, and at the initial instant of transient processes, inductors. Cutsets composed only of such elements are inductive cutsets.

If capacitive loops or inductive cutsets are present in the network, Kirchhoff's laws with the voltages or currents specified are either satisfied or not. In the former instance, since one or more of Kirchhoff's equations are automatically satisfied, the set of equations written for the network is redundant. A set of equations with a unique solution is obtained if any one branch of each capacitive loop is replaced by an open-circuit, any one branch of each inductive cutset is replaced by a short-circuit, and the equations are written for the network so derived. On adding an arbitrary current to edge currents of capacitive loops and an arbitrary voltage to edge voltages of inductive cutsets a solution still satisfying Kirchhoff's laws is derived. This indicates that the network model being employed does not satisfactorily approximate the real network.

If Kirchhoff's law for a capacitive loop or inductive cutset of the network is not satisfied, a fundamental requirement of network theory is violated, and the network model cannot be used for the solution of any real problem. It is possible that a minor change of a source-voltage or source-current could reduce this problem to the former (i.e. to a redundant problem), or alternatively it may be that the generators of the network cannot be modelled properly by ideal sources.

To determine whether the network contains a capacitive loop or inductive cutset the following procedure may be used. A tree of the network graph should be chosen in which each voltage-source, short-circuit, and, in the case of transient processes, each capacitor, corresponds to a tree-branch, while each current-source, open-circuit, and, in the case of transient processes, each inductor, corresponds to a chord. If such a classification is possible, neither capacitive loops nor inductive cutsets are present in the network, otherwise the network contains at least one such loop or cutset.

In the course of our calculations it will be assumed that neither capacitive loops nor inductive cutsets appear in the network, unless otherwise stated.

The principle of duality

As has been shown, if the graph of a network is non-separable and planar, its dual may be constructed. Each cutset of the graph corresponds to a loop in the dual graph, each loop to a cutset, and each edge to an edge. The quantities of the network and its graph will be marked by a prime ($'$), while those of the dual graph and a network whose graph is the dual by a double prime ($''$). This new network is the dual of the original. Let us examine a network without mutual impedances.

The cutset equation for a cutset of the original network is

$$\sum_{k=1}^{r} I'_k = 0, \tag{2.64}$$

where r is the number of branches in the cutset. In the dual network a loop consisting of r branches is associated with this cutset. For this:

$$\sum_{k=1}^{r} U''_k = 0. \quad (2.65)$$

Comparing the two equations it can be seen that branch-currents of the original network may be associated with branch-voltages of the dual network.

Let us now write the loop equation for a loop of the original network. If the loop consists of s branches, then

$$\sum_{j=1}^{s} U'_j = 0. \quad (2.66)$$

In the dual network a cutset consisting of s branches corresponds to this loop. The cutset equation for this cutset is

$$\sum_{j=1}^{s} I''_j = 0. \quad (2.67)$$

Thus, the dual of branch-voltage is branch-current and that of branch-current is branch-voltage.

The relation between voltage U' and current I' of an impedance Z' in the network is

$$Z' = \frac{U'}{I'}. \quad (2.68)$$

In the corresponding equation for the dual network U' is replaced by I'' and I' by U''. Thus the dual of U'/I' is I''/U'', and the latter is an admittance:

$$Y'' = \frac{I''}{U''}. \quad (2.69)$$

Therefore the dual of impedance Z' is admittance Y''. Similar considerations show that the dual of admittance Y' is impedance Z''. The quotient of two impedances of the original network equals the quotient of the two corresponding admittances in the dual network.

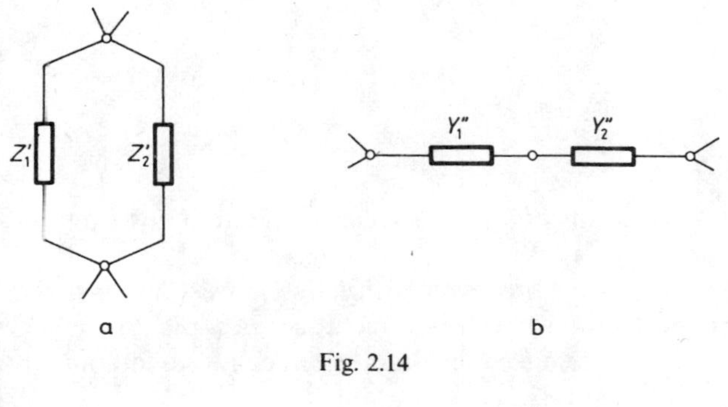

Fig. 2.14

Since the dual of voltage is current, the dual of a voltage-source with source-voltage U_g is a current-source with source-current I_g and vice versa.

It follows from the duality of branch-current and branch-voltage that the elements in the dual network corresponding to two parallel connected impedances in the original network (Fig. 2.14, a) are two admittances connected in series (Fig. 2.14, b). Two impedances connected in series in the original network (Fig. 2.15, a) may be associated with two admittances connected parallel in the dual network (Fig. 2.15, b).

Similarly, the duality of branch-current and branch-voltage indicates that the dual of a *Y*-connection (Fig. 2.16, a) is a delta-connection (Fig. 2.16, b), and conversely: the dual of a delta connection (Fig. 2.17, a) is a *Y*-connection (Fig. 2.17, b). In the dual network an edge corresponding to an impedance of the original

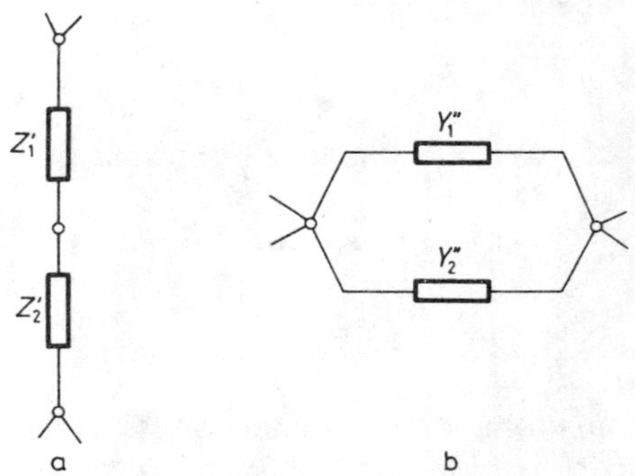

Fig. 2.15

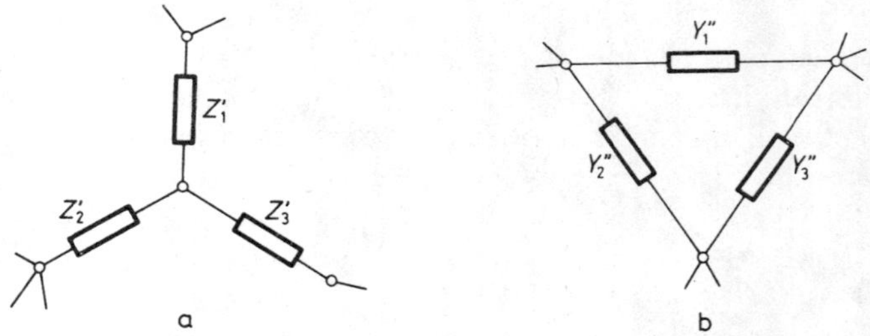

Fig. 2.16

network becomes an admittance, and that corresponding to an admittance becomes an impedance.

It follows from the preceeding explanation that a resistance is replaced in the dual network by a conductance, and a conductance by a resistance. The dual of self-inductance is capacitance, since the complex impedance of an inductor is a positive imaginary quantity, which corresponds in the dual network to a positive imaginary admittance, which can be realized by a capacitor. Similarly the dual of the negative imaginary impedance of a capacitor is an inductor with negative imaginary admittance.

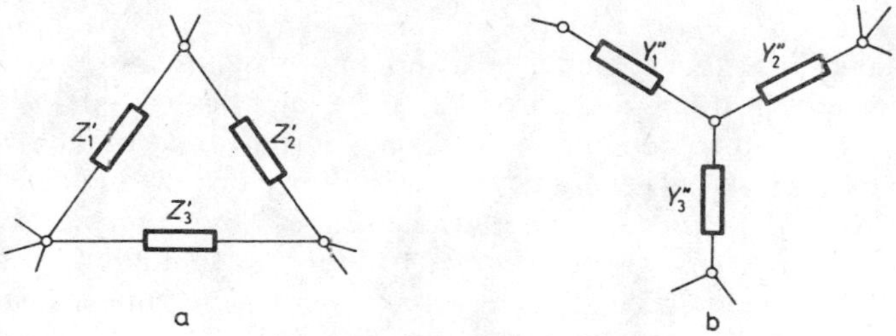

Fig. 2.17

Calculation of branch-currents and branch-voltages

Let us consider the following problem: the graph of the network as well as the location and characteristics of impedances, voltage-sources, current-sources are given (i.e. matrices $\boldsymbol{Q}$, $\boldsymbol{B}$, $\boldsymbol{Z}$, $\mathbf{U}_g$ and $\mathbf{I}_g$ are known), and the branch-currents and branch-voltages are to be determined. For the solution of this problem it is at first assumed that besides the previously introduced conditions regarding Kirchhoff's laws, no edges containing zero admittance or zero impedance are present in the network.

The following set of equations may be written in accordance with (2.58), (2.51) and (2.44):

$$\boldsymbol{Q}\mathbf{I}_z = -\boldsymbol{Q}\mathbf{I}_g \tag{2.70}$$

$$\boldsymbol{B}\mathbf{U}_z = \boldsymbol{BZ}\mathbf{I}_z = -\boldsymbol{B}\mathbf{U}_g, \tag{2.71}$$

or in one equation, using hypermatrices:

$$\begin{bmatrix} \boldsymbol{Q} \\ \boldsymbol{BZ} \end{bmatrix} \mathbf{I}_z = -\begin{bmatrix} \boldsymbol{Q}\mathbf{I}_g \\ \boldsymbol{B}\mathbf{U}_g \end{bmatrix}. \tag{2.72}$$

Matrices $\boldsymbol{Q}$ and $\boldsymbol{BZ}$ have b columns, while the number of rows of $\boldsymbol{Q}$ is $n-c$, and of $\boldsymbol{BZ}$ is m, so that the coefficient matrix of $\mathbf{I}_z$ contains $n-c+m=b$ rows, i.e. it is a square matrix. Provided that its inverse exists, $\mathbf{I}_z$ may be obtained from (2.72):

$$\mathbf{I}_z = -\begin{bmatrix} \boldsymbol{Q} \\ \boldsymbol{BZ} \end{bmatrix}^{-1} \begin{bmatrix} \boldsymbol{Q}\mathbf{I}_g \\ \boldsymbol{B}\mathbf{U}_g \end{bmatrix}. \tag{2.73}$$

Having calculated $\mathbf{I}_z$, branch currents are readily determined from (2.10).

The voltages of the passive elements of the network, using the notation $\boldsymbol{Z}^{-1} = \boldsymbol{Y}$ are

$$\mathbf{U}_z = \boldsymbol{Z}\mathbf{I}_z = -\begin{bmatrix} \boldsymbol{QY} \\ \boldsymbol{B} \end{bmatrix}^{-1} \begin{bmatrix} \boldsymbol{Q}\mathbf{I}_g \\ \boldsymbol{B}\mathbf{U}_g \end{bmatrix}. \tag{2.74}$$

If the network contains branches with zero impedance or zero admittance (voltage-sources, short-circuits, current-sources, or open-circuits), the following method may be used to determine the branch-currents and branch-voltages. Let us consider each non-ideal generator to consist of two distinct branches containing an ideal source and an impedance. Let us then classify the branches of the network into three groups. The first group is formed by branches containing current-sources as well as any branches which are open-circuits but which require their voltages to be determined. The branches with non-zero and finite impedances belong to the second group, and voltage-sources and short-circuits to the third. Let the branches be numbered in the order of the groups, i.e. order numbers 1, 2, ..., b_1 refer to branches in the first group, $b_1+1, b_1+2, \ldots, b_1+b_2$ to the second, and b_1+b_2+1, $b_1+b_2+2, \ldots, b_1+b_2+b_3=b$ to the third.

Let us choose a tree of the network graph, in which branches belonging to the first group are chords, and those in the third group are tree-branches. This can be attained in every realistic case, as it has already been shown. The system of loops and cutsets generated by this tree is used for the analysis. The orientations of loops coincide with those of the corresponding chords, and the directions of cutsets with those of the tree-branches which generate them. Let the loops be numbered in the order of the chords, and the cutsets in that of the tree-branches. The loop-matrix and cutset-matrix so derived are denoted by $\boldsymbol{B}$ and $\boldsymbol{Q}$, respectively.

Let **U** and **I** denote the column matrices of branch-voltages and branch-currents, and let us partition them in accordance with the three groups of edges:

$$\mathbf{U} = \begin{bmatrix} \mathbf{U}_1 \\ \mathbf{U}_2 \\ \mathbf{U}_3 \end{bmatrix} = \begin{bmatrix} \mathbf{U}_1 \\ \mathbf{U}_2 \\ \mathbf{U}_{g3} \end{bmatrix}; \tag{2.75}$$

$$\mathbf{I} = \begin{bmatrix} \mathbf{I}_1 \\ \mathbf{I}_2 \\ \mathbf{I}_3 \end{bmatrix} = \begin{bmatrix} \mathbf{I}_{g1} \\ \mathbf{I}_2 \\ \mathbf{I}_3 \end{bmatrix}, \tag{2.76}$$

where $\mathbf{U}_{g3}$ and $\mathbf{I}_{g1}$ are the column matrices formed, respectively, by the source-voltages of branches in the third group, and source-currents of branches in the first group.

To write the loop equations (2.61) of the network, let us partition matrix $\boldsymbol{B}$ in accordance with the classification of the branches. Taking (2.75) into account:

$$\begin{bmatrix} \boldsymbol{1} & \boldsymbol{B}_{11} & \boldsymbol{B}_{12} \\ \boldsymbol{0} & \boldsymbol{B}_{21} & \boldsymbol{B}_{22} \end{bmatrix} \begin{bmatrix} \mathbf{U}_1 \\ \mathbf{U}_2 \\ \mathbf{0} \end{bmatrix} = - \begin{bmatrix} \boldsymbol{1} & \boldsymbol{B}_{11} & \boldsymbol{B}_{12} \\ \boldsymbol{0} & \boldsymbol{B}_{21} & \boldsymbol{B}_{22} \end{bmatrix} \begin{bmatrix} \mathbf{0} \\ \mathbf{0} \\ \mathbf{U}_{g3} \end{bmatrix}, \tag{2.77}$$

where $\boldsymbol{1}$ is a unit matrix of order b_1, while the number of columns in $\boldsymbol{B}_{11}$ and $\boldsymbol{B}_{21}$ is b_2. $\boldsymbol{B}$ takes this form because of the above numbering of the loops. From this equation:

$$\mathbf{U}_1 + \boldsymbol{B}_{11}\mathbf{U}_2 = -\boldsymbol{B}_{12}\mathbf{U}_{g3}, \tag{2.78}$$

$$\boldsymbol{B}_{21}\mathbf{U}_2 = -\boldsymbol{B}_{22}\mathbf{U}_{g3}. \tag{2.79}$$

Let us also partition cutset matrix $\boldsymbol{Q}$ in accordance with our branch classification. Thus cutset equations (2.58) may be written in the following form with the aid of (2.76):

$$\begin{bmatrix} \boldsymbol{Q}_{11} & \boldsymbol{Q}_{12} & \boldsymbol{0} \\ \boldsymbol{Q}_{21} & \boldsymbol{Q}_{22} & \boldsymbol{1} \end{bmatrix} \begin{bmatrix} \mathbf{0} \\ \mathbf{I}_2 \\ \mathbf{I}_3 \end{bmatrix} = - \begin{bmatrix} \boldsymbol{Q}_{11} & \boldsymbol{Q}_{12} & \boldsymbol{0} \\ \boldsymbol{Q}_{21} & \boldsymbol{Q}_{22} & \boldsymbol{1} \end{bmatrix} \begin{bmatrix} \mathbf{I}_{g1} \\ \mathbf{0} \\ \mathbf{0} \end{bmatrix}, \tag{2.80}$$

where $\boldsymbol{1}$ is a unit matrix of order b_3, and the number of columns in $\boldsymbol{Q}_{11}$ and $\boldsymbol{Q}_{21}$ is b_1. Thus:

$$\boldsymbol{Q}_{12}\mathbf{I}_2 \qquad = -\boldsymbol{Q}_{11}\mathbf{I}_{g1}, \tag{2.81}$$

$$\boldsymbol{Q}_{22}\mathbf{I}_2 + \mathbf{I}_3 = -\boldsymbol{Q}_{21}\mathbf{I}_{g1}. \tag{2.82}$$

$\mathbf{U}_2$ and $\mathbf{I}_2$ are the column matrices formed by the voltages and currents of passive elements. Equation

$$\mathbf{U}_2 = \boldsymbol{Z}_2\mathbf{I}_2 \tag{2.83}$$

may be written for them, where $\boldsymbol{Z}_2 = \boldsymbol{Y}_2^{-1}$ is the impedance matrix of that part of the network consisting of the branches of finite, non-zero impedances. The main diagonal of $\boldsymbol{Z}_2$ consists of branch impedances, while off the main diagonal the mutual impedances are present. Thus, (2.79) and (2.81) may be summarized as

$$\begin{bmatrix} \boldsymbol{B}_{21}\boldsymbol{Z}_2 \\ \boldsymbol{Q}_{12} \end{bmatrix} \mathbf{I}_2 = - \begin{bmatrix} \boldsymbol{B}_{22}\mathbf{U}_{g3} \\ \boldsymbol{Q}_{11}\mathbf{I}_{g1} \end{bmatrix}, \tag{2.84}$$

and this yields the currents of the passive elements. The coefficient matrix of $\mathbf{I}_2$ in (2.84) is square, and if its inverse exists, then

$$\mathbf{I}_2 = - \begin{bmatrix} \boldsymbol{B}_{21}\boldsymbol{Z}_2 \\ \boldsymbol{Q}_{12} \end{bmatrix}^{-1} \begin{bmatrix} \boldsymbol{B}_{22}\mathbf{U}_{g3} \\ \boldsymbol{Q}_{11}\mathbf{I}_{g1} \end{bmatrix}. \tag{2.85}$$

Similarly, $\mathbf{U}_2$ may also be obtained:

$$\mathbf{U}_2 = - \begin{bmatrix} \boldsymbol{B}_{21} \\ \boldsymbol{Q}_{12}\boldsymbol{Y}_2 \end{bmatrix}^{-1} \begin{bmatrix} \boldsymbol{B}_{22}\mathbf{U}_{g3} \\ \boldsymbol{Q}_{11}\mathbf{I}_{g1} \end{bmatrix}. \tag{2.86}$$

Hence, knowing $\mathbf{U}_2$ and $\mathbf{I}_2$, if follows from (2.78) and (2.82) that

$$\mathbf{U}_1 = -\boldsymbol{B}_{11}\mathbf{U}_2 - \boldsymbol{B}_{12}\mathbf{U}_{g3} = \boldsymbol{B}_{11}\begin{bmatrix}\boldsymbol{B}_{21}\\ \boldsymbol{Q}_{12}\boldsymbol{Y}_2\end{bmatrix}^{-1}\begin{bmatrix}\boldsymbol{B}_{22}\mathbf{U}_{g3}\\ \boldsymbol{Q}_{11}\mathbf{I}_{g1}\end{bmatrix} - \boldsymbol{B}_{12}\mathbf{U}_{g3} \quad (2.87)$$

$$\mathbf{I}_3 = -\boldsymbol{Q}_{22}\mathbf{I}_2 - \boldsymbol{Q}_{21}\mathbf{I}_{g1} = \boldsymbol{Q}_{22}\begin{bmatrix}\boldsymbol{B}_{21}\boldsymbol{Z}_2\\ \boldsymbol{Q}_{12}\end{bmatrix}^{-1}\begin{bmatrix}\boldsymbol{B}_{22}\mathbf{U}_{g3}\\ \boldsymbol{Q}_{11}\mathbf{I}_{g1}\end{bmatrix} - \boldsymbol{Q}_{21}\mathbf{I}_{g1}, \quad (2.88)$$

i.e. the voltages of current-sources and currents of voltage-sources have been determined.

The method of loop-currents

The number of unknown variables in the equations so far written for the determination of branch-currents and branch-voltages is equal to the number of branches in the network. In the methods to be presented now, alternative unknown variables will be introduced, whose number is less than this. First, the method of loop-currents is discussed.

Let us form a fundamental set of loops in the network. Imagine that a current, the so-called *loop-current* flows around each loop, with is reference direction coinciding with the orientation of the loop. Each branch-current is given by the sum of the loop-currents flowing through the branch, with the orientations of branch- and loop-currents observed (Fig. 2.18).

The number of loop-currents is m, and equals the number of independent loops. Loop-currents automatically satisfy Kirchhoff's current law at every node, since a loop-current flowing into a node also flows out of it (Fig. 2.18). By the introduction of loop-currents $n-c$ equations are automatically satisfied, and only further m equations have to be satisfied for the determination of the $b=n-c+m$ edge-currents. Thus, the problem is capable of being solved with the aid of the loop-currents.

Let the loop-currents be denoted by $J_1, J_2, \ldots, J_m$, and their column matrix by $\mathbf{J}$:

$$\mathbf{J} = \begin{bmatrix} J_1 \\ J_2 \\ \vdots \\ J_m \end{bmatrix}. \quad (2.89)$$

The actual current of branch j is given by the superposition of the loop-currents linked with the branch. This may be written in a matrix equation:

$$I_j = \mathbf{B}_j^+ \mathbf{J}, \quad (2.90)$$

where row matrix $\mathbf{B}_j^+$ is the j-th row of the transpose of the basis loop-matrix $\boldsymbol{B}$. The current of each branch may be similarly expressed.

Summarizing the equations thus obtained:

$$\mathbf{I} = \mathbf{I}_z + \mathbf{I}_g = \begin{bmatrix} \mathbf{B}_1^+ \\ \mathbf{B}_2^+ \\ \vdots \\ \mathbf{B}_b^+ \end{bmatrix} \mathbf{J} = \boldsymbol{B}^+ \mathbf{J} . \tag{2.91}$$

The fact that loop-currents satisfy the node-current equations is also evident from (2.91). Let us substitute (2.91) into (2.58):

$$\boldsymbol{Q}(\mathbf{I}_z + \mathbf{I}_g) = \boldsymbol{QB}^+ \mathbf{J} = \mathbf{0}. \tag{2.92}$$

Since $\boldsymbol{QB}^+ \equiv \boldsymbol{0}$ (see Chapter 1 (1.59)), the cutset equation (2.58) is satisfied by the introduction of loop-currents in all cases.

Initially the calculation of loop-currents in networks without open-circuits and current-sources will be considered. To this end let us write $\mathbf{I}_z$ in accordance with (2.91):

$$\mathbf{I}_z = \boldsymbol{B}^+ \mathbf{J} - \mathbf{I}_g. \tag{2.93}$$

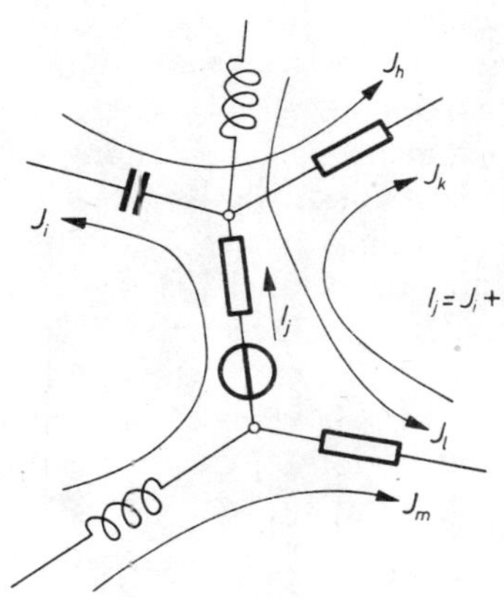

Fig. 2.18

Let us substitute $\mathbf{U}_z = \boldsymbol{Z}\mathbf{I}_z$ into loop equation (2.61), writing the above expression for $\mathbf{I}_z$:

$$\boldsymbol{B}\mathbf{U}_g + \boldsymbol{BZB}^+ \mathbf{J} - \boldsymbol{BZ}\mathbf{I}_g = \mathbf{0}. \tag{2.94}$$

On rearrangement, introducing

$$\boldsymbol{Z}_B = \boldsymbol{BZB}^+ \tag{2.95}$$

(2.94) yields:

$$\boldsymbol{Z}_B \mathbf{J} = \boldsymbol{B}(\boldsymbol{Z}\mathbf{I}_g - \mathbf{U}_g). \tag{2.96}$$

$\boldsymbol{Z}_B$, a square matrix of order m, is called the *loop-impedance matrix*. (2.96) consists of the loop equations for each loop of the set of loops chosen. Its left-hand side gives the voltages of the passive elements, while its right-hand side is formed by the source-voltages of Norton and Thevenin generators with negative sign in each loop.

(2.96) yields the following expression for the column matrix of loop-currents:

$$\mathbf{J} = \boldsymbol{Z}_B^{-1} \boldsymbol{B}(\boldsymbol{Z}\mathbf{I}_g - \mathbf{U}_g). \tag{2.97}$$

The currents of passive elements, according to (2.93) are:

$$\mathbf{I}_z = \boldsymbol{B}^+ \boldsymbol{Z}_B^{-1} \boldsymbol{B}(\boldsymbol{Z}\mathbf{I}_g - \mathbf{U}_g) - \mathbf{I}_g. \tag{2.98}$$

The loop impedance matrix $\boldsymbol{Z}_B$ may be written directly given the fundamental set of loops in the network. Z_{ij} $(i \neq j)$, the j-th element in the i-th row of $\boldsymbol{Z}_B$, is the sum of the common self-impedances and the mutual impedances of the i-th and j-th loops, with positive or negative signs, depending upon the relative orientation of the loops in the shared branches, while Z_{ii} is the sum of impedances in the i-th loop. The mutual impedance of two loops means the mutual impedance between a branch of one loop and a branch of the other.

If the network contains branches with zero admittance (current-sources, open-circuits), then to apply the method of loop-currents, consider any Norton generators to consist of two branches: one containing a current-source, and one formed by an impedance. In the network so derived let us connect an impedance Z_N, not yet defined, in parallel with branches of zero admittance. Regarding each of s Norton generators so created as one branch, let us write the equation for loop-currents on the basis of a set of loops generated by a tree in which the edges numbered 1, 2, . . ., s associated with the Norton generators are chords. Let us choose the first s loops in such a way that they contain chords 1, 2, . . ., s, and their orientations coincide with the directions of the source-currents. As a result (2.96) may be written in the following form:

$$\begin{bmatrix} \boldsymbol{Z}_{B1} & \boldsymbol{Z}_{B2} \\ \boldsymbol{Z}_{B3} & \boldsymbol{Z}_{B4} \end{bmatrix} \begin{bmatrix} \mathbf{J}_N \\ \mathbf{J}_E \end{bmatrix} = \begin{bmatrix} \boldsymbol{Z}_{11} & \boldsymbol{Z}_{12} \\ \boldsymbol{Z}_{21} & \boldsymbol{Z}_{22} \end{bmatrix} \begin{bmatrix} \mathbf{I}_{g1} \\ \mathbf{0} \end{bmatrix} - \begin{bmatrix} \boldsymbol{B}_{11} & \boldsymbol{B}_{12} \\ \boldsymbol{B}_{21} & \boldsymbol{B}_{22} \end{bmatrix} \mathbf{U}_g, \tag{2.99}$$

where $\boldsymbol{Z}_{B1}$, $\boldsymbol{Z}_{11}$ and $\boldsymbol{B}_{11}$ are square submatrices of order s of matrices $\boldsymbol{Z}_B$, $\boldsymbol{BZ}$ and $\boldsymbol{B}$, respectively, and $\mathbf{J}_N$ and $\mathbf{I}_{g1}$ consist of the first s elements of loop-currents $\mathbf{J}$ and source-currents $\mathbf{I}_g$. This equation may be resolved into the following equations:

$$\boldsymbol{Z}_{B1}\mathbf{J}_N + \boldsymbol{Z}_{B2}\mathbf{J}_E = \boldsymbol{Z}_{11}\mathbf{I}_{g1} - [\boldsymbol{B}_{11} \;\; \boldsymbol{B}_{12}]\mathbf{U}_g, \tag{2.100}$$

$$\boldsymbol{Z}_{B3}\mathbf{J}_N + \boldsymbol{Z}_{B4}\mathbf{J}_E = \boldsymbol{Z}_{21}\mathbf{I}_{g1} - [\boldsymbol{B}_{21} \;\; \boldsymbol{B}_{22}]\mathbf{U}_g. \tag{2.101}$$

If the value of the impedances connected, in parallel with the current-sources increases without limit, i.e. $Z_N \to \infty$, some elements of $\boldsymbol{Z}_{B1}$ and $\boldsymbol{Z}_{11}$ tend to infinity, and thus (2.100) cannot be used for the analysis. In this case, however $\mathbf{J}_N = \mathbf{I}_{g1}$, and so from (2.101):

$$\mathbf{J}_E = \boldsymbol{Z}_{B4}^{-1}\{(\boldsymbol{Z}_{21} - \boldsymbol{Z}_{B3})\mathbf{I}_{g1} - [\boldsymbol{B}_{21} \;\; \boldsymbol{B}_{22}]\mathbf{U}_g\}, \tag{2.102}$$

i.e. the unknown loop-currents have been expressed in terms of known quantities.

The method of cutset-voltages

Let us assign a voltage called a *cutset-voltage* to each of the linearly independent cutsets of the network, with the same reference direction as the orientation of the cutset. The voltage of a branch is the algebraic sum of cutset-voltages of cutsets containing the branch with the orientation of the branch and the cutsets taken into account.

The number of cutset-voltages is $n-c$, and the cutset-voltages automatically satisfy the loop equations. The cutset-voltages of a set of cutsets generated by a tree equal the voltages of tree-branches with positive or negative signs. The voltage of a chord is then given by the algebraic sum of voltages of those branches forming a loop with the chord. This implies that in the summation of the voltages around loop each cutset voltage either appears an even number of times or it does not appear at all, and in the former case its orientation is as many times identical to as it is opposite to that of the loop.

Let $\mathbf{V}_Q$ denote the column matrix formed by the cutset-voltages arranged in the order of the cutsets. The voltage of the i-th branch may thus be expressed:

$$U_i = \mathbf{Q}_i^+ \mathbf{V}_Q, \tag{2.103}$$

where $\mathbf{Q}_i^+$ is the transpose of the i-th column of matrix $\boldsymbol{Q}$ of the set of cutsets. The voltage of each branch can be similarly expressed, and these may be summarized in

$$\mathbf{U} = \begin{bmatrix} \mathbf{Q}_1^+ \\ \mathbf{Q}_2^+ \\ \vdots \\ \mathbf{Q}_b^+ \end{bmatrix} \mathbf{V}_Q = \boldsymbol{Q}^+ \mathbf{V}_Q. \tag{2.104}$$

Let us substitute this into the loop equation (2.61):

$$\boldsymbol{B}\mathbf{U} = \boldsymbol{B}\boldsymbol{Q}^+ \mathbf{V}_Q = \mathbf{0}, \tag{2.105}$$

which is an identity according to (1.59) of Chapter 1. (2.104) yields:

$$\mathbf{U}_z = \mathbf{U} - \mathbf{U}_g = \boldsymbol{Q}^+ \mathbf{V}_Q - \mathbf{U}_g. \tag{2.106}$$

To calculate the cutset-voltages of a network not having any zero impedance branches (voltage-sources, short-circuits), let us substitute this into (2.58):

$$\boldsymbol{Q}(\mathbf{I}_z+\mathbf{I}_g)=\boldsymbol{QY}\mathbf{U}_z+\boldsymbol{Q}\mathbf{I}_g=\boldsymbol{QYQ}^+\mathbf{V}_Q-\boldsymbol{QY}\mathbf{U}_g+\boldsymbol{Q}\mathbf{I}_g=\mathbf{0}. \tag{2.107}$$

On rearrangement:

$$\boldsymbol{QYQ}^+\mathbf{V}_Q=\boldsymbol{Q}(\boldsymbol{Y}\mathbf{U}_g-\mathbf{I}_g). \tag{2.108}$$

Let us introduce the *cutset-admittance matrix*:

$$\boldsymbol{Y}_Q=\boldsymbol{QYQ}^+. \tag{2.109}$$

So:

$$\mathbf{V}_Q=\boldsymbol{Y}_Q^{-1}\boldsymbol{Q}(\boldsymbol{Y}\mathbf{U}_g-\mathbf{I}_g). \tag{2.110}$$

Thus, the cutset-voltages have been determined.

Knowing $\mathbf{V}_Q$, all branch-voltages may be calculated. In accordance with (2.104):

$$\mathbf{U}=\boldsymbol{Q}^+\boldsymbol{Y}_Q^{-1}\boldsymbol{Q}(\boldsymbol{Y}\mathbf{U}_g-\mathbf{I}_g). \tag{2.111}$$

If the network contains no mutual impedances, cutset-admittance matrix $\boldsymbol{Y}_Q$ can be written directly given the fundamental set of cutsets in the network. Y_{ij} $(i \neq j)$, the j-th element in the i-th row of the matrix, is the sum of the admittances common to the i-th and j-th cutsets, with positive or negative sign depending upon the relative orientation of the cutsets in the shared branches, while Y_{ii} is the sum of admittances in the i-th cutset.

If there are branches in the network of zero impedance (voltage-sources, short-circuits), then to apply the method of cutset-voltages consider, for the sake of a simpler notation, Thevenin generators to consist of two branches: one containing a voltage-source and one formed by an impedance. In this network let us connect an admittance Y_T, not yet defined, in series with each branch of zero impedance. Each of the r Thevenin generators so obtained will be regarded as one branch. The equation for cutset-voltages is then written for a set of cutsets generated by a tree in which the branches numbered 1, 2, . . ., r, associated with the Thevenin generators, are tree-branches. The first r cutsets are chosen in such a way, that they contain tree-branches 1, 2, . . ., r and their orientations coincide with the directions of the source-voltages. As a result (2.108) may be written in the following form:

$$\begin{bmatrix}\boldsymbol{Y}_{Q1} & \boldsymbol{Y}_{Q2}\\ \boldsymbol{Y}_{Q3} & \boldsymbol{Y}_{Q4}\end{bmatrix}\begin{bmatrix}\mathbf{V}_T\\ \mathbf{V}_E\end{bmatrix}=\begin{bmatrix}\boldsymbol{Y}_{11} & \boldsymbol{Y}_{12}\\ \boldsymbol{Y}_{21} & \boldsymbol{Y}_{22}\end{bmatrix}\begin{bmatrix}\mathbf{U}_{g1}\\ \mathbf{0}\end{bmatrix}-\begin{bmatrix}\boldsymbol{Q}_{11} & \boldsymbol{Q}_{12}\\ \boldsymbol{Q}_{21} & \boldsymbol{Q}_{22}\end{bmatrix}\mathbf{I}_g, \tag{2.112}$$

where $\boldsymbol{Y}_{Q1}$, $\boldsymbol{Y}_{11}$ and $\boldsymbol{Q}_{11}$ are square submatrices of order r of matrices $\boldsymbol{Y}_Q$, $\boldsymbol{QY}$ and $\boldsymbol{Q}$ respectively, and $\mathbf{V}_T$ and $\mathbf{U}_{g1}$ consist of the first r elements of cutset-voltages $\mathbf{V}_Q$ and source-voltages $\mathbf{U}_g$. This may be resolved into

$$\boldsymbol{Y}_{Q1}\mathbf{V}_T+\boldsymbol{Y}_{Q2}\mathbf{V}_E=\boldsymbol{Y}_{11}\mathbf{U}_{g1}-[\boldsymbol{Q}_{11}\ \ \boldsymbol{Q}_{12}]\mathbf{I}_g, \tag{2.113}$$

$$\boldsymbol{Y}_{Q3}\mathbf{V}_T+\boldsymbol{Y}_{Q4}\mathbf{V}_E=\boldsymbol{Y}_{21}\mathbf{U}_{g1}-[\boldsymbol{Q}_{21}\ \ \boldsymbol{Q}_{22}]\mathbf{I}_g. \tag{2.114}$$

If $Y_T \rightarrow \infty$, some elements of $\boldsymbol{Y}_{Q1}$ and $\boldsymbol{Y}_{11}$ tend to infinity, while all other submatrices remain finite. In this case $\mathbf{V}_T = \mathbf{U}_{g1}$, and the unknown cutset-voltages may be determined from (2.114):

$$\mathbf{V}_E = \mathbf{Y}_{Q4}^{-1} \{(\boldsymbol{Y}_{21} - \boldsymbol{Y}_{Q3})\mathbf{U}_{g1} - [\boldsymbol{Q}_{21} \ \ \boldsymbol{Q}_{22}]\mathbf{I}_g\}, \tag{2.115}$$

which enables these unknown voltages to be expressed in terms of known quantities.

The method of node-voltages

The *node voltage* is the voltage of the node with reference to an arbitrary node whose potential is chosen as zero. If the network consists of several distinct connected parts, a node with zero potential may be chosen in each component. Let $\Phi_1, \Phi_2, \ldots, \Phi_{n-c}$ denote the voltages of the nodes not chosen at zero potential, and let these form a column matrix $\mathbf{\Phi}$:

$$\mathbf{\Phi} = \begin{bmatrix} \Phi_1 \\ \Phi_2 \\ \vdots \\ \Phi_{n-c} \end{bmatrix}. \tag{2.116}$$

Given the node voltages, the branch-voltages may be determined. The voltage of branch k is the difference of potential between nodes (i) and (j) incident with the branch, and

$$U_k = \Phi_i - \Phi_j \tag{2.117}$$

if the branch is directed from node (i) to node (j) and

$$U_k = \Phi_j - \Phi_i \tag{2.118}$$

if k is directed from node (j) to node (i). This is expressed by

$$U_k = \mathbf{A}_k^+ \mathbf{\Phi}, \tag{2.119}$$

where $\mathbf{A}_k^+$ is the transpose of the k-th column of incidence matrix $\boldsymbol{A}$. Expressing each element of column matrix $\mathbf{U}$ in the same way as (2.119), and summarizing these in one matrix equation:

$$\mathbf{U} = \begin{bmatrix} \mathbf{A}_1^+ \\ \mathbf{A}_2^+ \\ \vdots \\ \mathbf{A}_b^+ \end{bmatrix} \mathbf{\Phi} = \boldsymbol{A}^+ \mathbf{\Phi}. \tag{2.120}$$

In analysing a network with the aid of node-voltages Kirchhoff's loop equations are automatically satisfied. Indeed, the voltage between nodes (i) and (j) is $\Phi_i - \Phi_j$

calculated from the voltage of any path between (*i*) and (*j*). Thus the voltage of a loop containing nodes (*i*) and (*j*) is:

$$(\Phi_i - \Phi_j)_I - (\Phi_i - \Phi_j)_{II} = 0, \tag{2.121}$$

where indices *I* and *II* refer to two paths connecting nodes (*i*) and (*j*) (Fig. 2.19).

It has been shown that incidence matrix **A** may be considered to be a special cutset matrix. Taking this, as well as formulae (2.104) and (2.120), into account, the

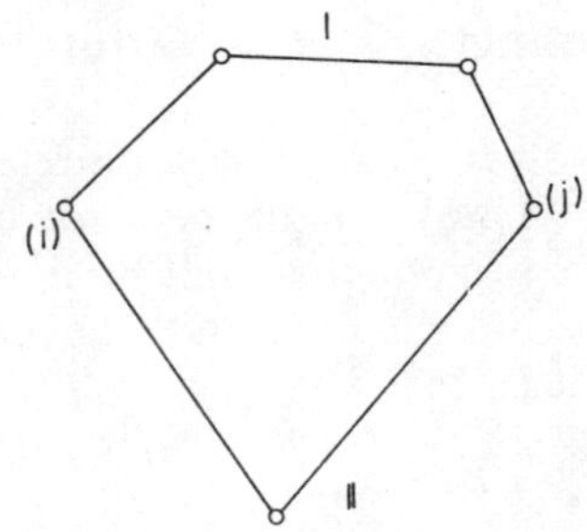

Fig. 2.19

calculation of node-voltages may be reduced to the determination of cutset-voltages. If the network contains no branches of zero impedance (voltage-sources, short-circuits), then according to (2.110):

$$\mathbf{\Phi} = \boldsymbol{Y}_A^{-1} \boldsymbol{A}(\boldsymbol{Y}\mathbf{U}_g - \mathbf{I}_g), \tag{2.122}$$

where

$$\boldsymbol{Y}_A = \boldsymbol{A}\boldsymbol{Y}\boldsymbol{A}^+ \tag{2.123}$$

is the *node-admittance matrix*. Here the *j*-th element in the *i*-th row ($i \neq j$) is the sum of admittances of branches connecting nodes (*i*) and (*j*) taken with negative sign, while the *i*-th element of the main diagonal is the sum of admittances of branches incident with node (*i*).

Analysis of networks containing current-sources and voltage-sources

For the methods described, the analysis of networks containing current-sources and open-circuits with the aid of loop-currents, as well as the analysis of networks containing voltage-sources and short-circuits with the aid of cutset-voltages may only be carried out after some special modifications, as described in previous sections. A method will now be presented which enables networks with zero admittance or impedance branches to be analysed by the inversion of a matrix whose order equals either the number of chords with finite admittance or of tree-branches with finite impedance [54].

A tree of the network graph is chosen for the analysis with each zero admittance branch corresponding to a chord and each zero impedance edge to a tree-edge. Other impedances and admittances may be freely associated with either chords or tree-branches. Branches are classified into four groups:

1. chords of zero admittance (current-sources, open-circuits), their number being denoted by b_1;
2. chords of finite admittance, in number b_2;
3. tree-branches of finite impedance, in number b_3;
4. tree-branches of zero impedance (voltage-sources, short-circuits), in number b_4.

Branches are numbered in accordance with the order of this classification. In the loop-set generated by the chosen tree loops are numbered in the order of the corresponding chords, cutsets in the order of the corresponding tree-branches, while orientations are chosen to coincide with those of the relevant chords and tree-branches.

The loop-equations of the network, written with matrices partitioned in accordance with the four groups of branches, are according to (2.63):

$$\begin{bmatrix} \mathit{1} & \mathit{0} & F_{11} & F_{12} \\ \mathit{0} & \mathit{1} & F_{21} & F_{22} \end{bmatrix} \begin{bmatrix} \mathbf{U}_1 \\ \mathbf{U}_2 \\ \mathbf{U}_3 \\ \mathbf{0} \end{bmatrix} = - \begin{bmatrix} \mathit{1} & \mathit{0} & F_{11} & F_{12} \\ \mathit{0} & \mathit{1} & F_{21} & F_{22} \end{bmatrix} \begin{bmatrix} \mathbf{0} \\ \mathbf{0} \\ \mathbf{0} \\ \mathbf{U}_{g4} \end{bmatrix}, \qquad (2.124)$$

where F_{11} has b_1 rows and b_3 columns, and $\mathbf{U}_{g4}$ contains the voltages of edges in the fourth group. (The dimensions of the unit and zero matrices have not been indicated, since they may be determined from the method of partitioning.) (2.124) in an expanded form is:

$$\mathbf{U}_1 + F_{11}\mathbf{U}_3 = -F_{12}\mathbf{U}_{g4}, \qquad (2.125)$$

$$\mathbf{U}_2 + F_{21}\mathbf{U}_3 = -F_{22}\mathbf{U}_{g4}. \qquad (2.126)$$

Let us write cutset-equations as well with the aid of partitioned matrices, and taking (2.62) into consideration:

$$\begin{bmatrix} -F_{11}^{+} & -F_{21}^{+} & \mathit{1} & \mathit{0} \\ -F_{12}^{+} & -F_{22}^{+} & \mathit{0} & \mathit{1} \end{bmatrix} \begin{bmatrix} \mathbf{0} \\ \mathbf{I}_2 \\ \mathbf{I}_3 \\ \mathbf{I}_4 \end{bmatrix} = - \begin{bmatrix} -F_{11}^{+} & -F_{21}^{+} & \mathit{1} & \mathit{0} \\ -F_{12}^{+} & -F_{22}^{+} & \mathit{0} & \mathit{1} \end{bmatrix} \begin{bmatrix} \mathbf{I}_{g1} \\ \mathbf{0} \\ \mathbf{0} \\ \mathbf{0} \end{bmatrix}, \qquad (2.127)$$

which yields:

$$-\boldsymbol{F}_{21}^{+}\mathbf{I}_2+\mathbf{I}_3=\boldsymbol{F}_{11}^{+}\mathbf{I}_{g1}\,, \tag{2.128}$$

$$-\boldsymbol{F}_{22}^{+}\mathbf{I}_2+\mathbf{I}_4=\boldsymbol{F}_{12}^{+}\mathbf{I}_{g1}\,. \tag{2.129}$$

Let us first restrict our analysis to the case where no mutual impedance exists between branches associated with tree-edges and chords. Then:

$$\mathbf{I}_2=\boldsymbol{Y}_2\mathbf{U}_2\,, \tag{2.130}$$

$$\mathbf{U}_3=\boldsymbol{Z}_3\mathbf{I}_3\,, \tag{2.131}$$

and thus, using (2.128):

$$\mathbf{U}_3=\boldsymbol{Z}_3\mathbf{I}_3=\boldsymbol{Z}_3\boldsymbol{F}_{11}^{+}\mathbf{I}_{g1}+\boldsymbol{Z}_3\boldsymbol{F}_{21}^{+}\boldsymbol{Y}_2\mathbf{U}_2\,. \tag{2.132}$$

Let us express $\mathbf{U}_2$ after substitution into (2.126):

$$\mathbf{U}_2=-[\boldsymbol{1}+\boldsymbol{F}_{21}\boldsymbol{Z}_3\boldsymbol{F}_{21}^{+}\boldsymbol{Y}_2]^{-1}[\boldsymbol{F}_{22}\mathbf{U}_{g4}+\boldsymbol{F}_{21}\boldsymbol{Z}_3\boldsymbol{F}_{11}^{+}\mathbf{I}_{g1}]\,. \tag{2.133}$$

Given $\mathbf{U}_2$, $\mathbf{U}_3$ may be determined from (2.132) and hence $\mathbf{U}_1$ from (2.125). Branch-currents may then be calculated from (2.130) and (2.131). The order of the matrix to be inverted in (2.133) is b_2.

Branch-currents and branch-voltages may also be calculated starting from (2.126). Using (2.130):

$$\mathbf{I}_2=-\boldsymbol{Y}_2\boldsymbol{F}_{21}\mathbf{U}_3-\boldsymbol{Y}_2\boldsymbol{F}_{22}\mathbf{U}_{g4}\,. \tag{2.134}$$

Substituting into (2.128) we obtain:

$$\mathbf{I}_3=[\boldsymbol{1}+\boldsymbol{F}_{21}^{+}\boldsymbol{Y}_2\boldsymbol{F}_{21}\boldsymbol{Z}_3]^{-1}[\boldsymbol{F}_{11}^{+}\mathbf{I}_{g1}-\boldsymbol{F}_{21}^{+}\boldsymbol{Y}_2\boldsymbol{F}_{22}\mathbf{U}_{g4}]\,. \tag{2.135}$$

Given $\mathbf{I}_3$, $\mathbf{I}_2$ may be determined from (2.134), and hence $\mathbf{I}_4$ from (2.129). Branch-voltages may then be derived from (2.130) and (2.131). The order of the matrix to be inverted in (2.135) is b_3.

If there are mutual impedances between chords and tree-branches, then:

$$\begin{bmatrix}\mathbf{U}_2\\ \mathbf{U}_3\end{bmatrix}=\begin{bmatrix}\boldsymbol{Z}_{22} & \boldsymbol{Z}_{23}\\ \boldsymbol{Z}_{32} & \boldsymbol{Z}_{33}\end{bmatrix}\begin{bmatrix}\mathbf{I}_2\\ \mathbf{I}_3\end{bmatrix}. \tag{2.136}$$

With the aid of (2.128), this yields, from (2.126):

$$\mathbf{I}_2=-[\boldsymbol{Z}_{22}+\boldsymbol{F}_{21}\boldsymbol{Z}_{32}+(\boldsymbol{Z}_{23}+\boldsymbol{F}_{21}\boldsymbol{Z}_{33})\boldsymbol{F}_{21}^{+}]^{-1}[\boldsymbol{F}_{22}\mathbf{U}_{g4}+(\boldsymbol{Z}_{23}+\boldsymbol{F}_{21}\boldsymbol{Z}_{33})\boldsymbol{F}_{11}^{+}\mathbf{I}_{g1}]. \tag{2.137}$$

Given $\mathbf{I}_2$, $\mathbf{I}_3$ may be calculated from (2.128) and $\mathbf{I}_4$ from (2.129). Hence, in accordance with (2.136), $\mathbf{U}_2$ and $\mathbf{U}_3$ and then, from (2.125), $\mathbf{U}_1$ may be determined.

Examples

The analysis methods presented are now illustrated by a few examples.

1. The switch in the network shown in Fig. 2.20 is initially open and is closed at the instant $t=0$. The capacitor is initially discharged. Because of direct current generator U_g, a time-varying current $i_3(t)$ flows into the capacitor. To calculate this current, let us choose branch 2 as the tree (Fig. 2.21). The matrices characterizing the network are:

$$\boldsymbol{Q}=[-1\ 1\ 1],$$

$$\boldsymbol{B}=\begin{bmatrix} 1 & 1 & 0 \\ 0 & -1 & 1 \end{bmatrix},$$

$$\boldsymbol{Z}(s)=\langle R_b\ R\ 1/sC\rangle,$$

$$\mathbf{I}_g=\mathbf{0},$$

$$\mathbf{U}_g(s)=\begin{bmatrix} -U_g/s \\ 0 \\ 0 \end{bmatrix},$$

$$\boldsymbol{Q}\mathbf{I}_g=\mathbf{0},$$

$$\boldsymbol{B}\mathbf{U}_g=\begin{bmatrix} -U_g/s \\ 0 \end{bmatrix}.$$

Substituting these into (2.73), we obtain

$$\begin{bmatrix} I_1(s) \\ I_2(s) \\ I_3(s) \end{bmatrix} = -\begin{bmatrix} -1 & 1 & 1 \\ R_b & R & 0 \\ 0 & -R & 1/sC \end{bmatrix}^{-1} \begin{bmatrix} 0 \\ -U_g/s \\ 0 \end{bmatrix} =$$

$$= \frac{1}{R_bR+(R+R_b)/sC} \begin{bmatrix} \cdot & \cdot & \cdot \\ \cdot & \cdot & \cdot \\ \cdot & -R & \cdot \end{bmatrix} \begin{bmatrix} 0 \\ -U_g/s \\ 0 \end{bmatrix}.$$

It is sufficient to calculate the second element of the third row of the inverse matrix, since only the third equation is needed to obtain $I_3(s)$, and the first and third elements are multiplied by zero.

Thus the Laplace transform of the required current is:

$$I_3(s) = \frac{U_g}{s}\,\frac{sCR}{R+R_b+sCRR_b} = \frac{U_g}{R_b}\,\frac{1}{s+\dfrac{R+R_b}{CRR_b}}.$$

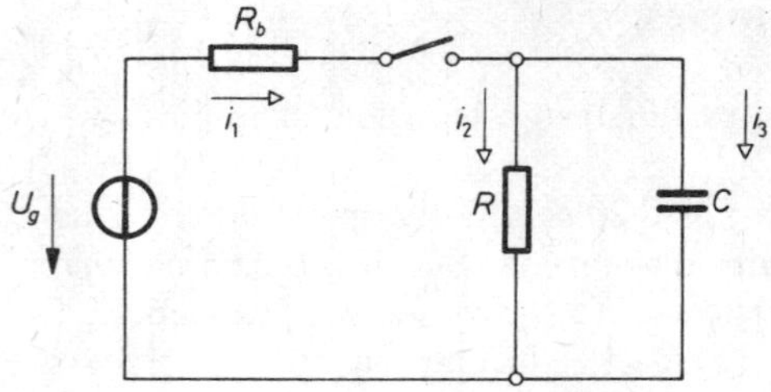

Fig. 2.20

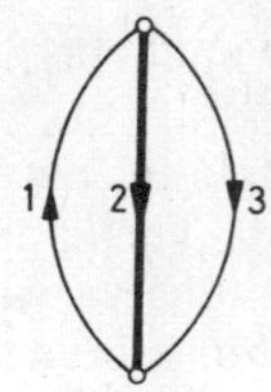

Fig. 2.21

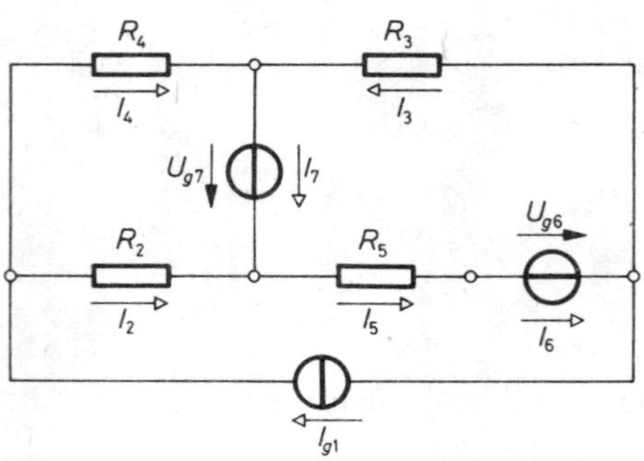

Fig. 2.22

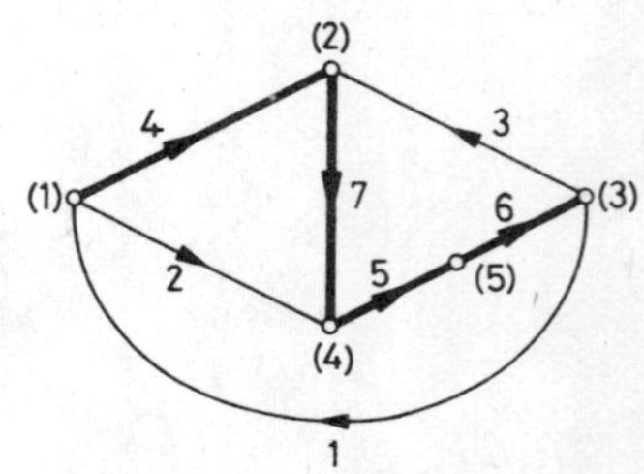

Fig. 2.23

As a function of time, the current is therefore given by:

$$i_3(t) = 1(t)\frac{U_g}{R_b} e^{-\frac{R+R_b}{CRR_b}t}.$$

2. The branch-currents of the network shown in Fig. 2.22 will be first calculated with the aid of (2.77) and (2.80). The graph of the network and the tree chosen for the analysis have been drawn in Fig. 2.23. (Tree-branches have been indicated by thick lines.) The first group of branches contains only branch 1, the second group contains branches 2, 3, 4, 5 while the third contains branches 6, 7. The matrices of sets of loops and cutsets generated by the tree chosen are

$$\boldsymbol{B} = \begin{bmatrix} \boldsymbol{1} & \boldsymbol{B}_{11} & \boldsymbol{B}_{12} \\ \boldsymbol{0} & \boldsymbol{B}_{21} & \boldsymbol{B}_{22} \end{bmatrix} = \left[\begin{array}{c|cccc|cc} 1 & 0 & 0 & 1 & 1 & 1 & 1 \\ \hline 0 & 1 & 0 & -1 & 0 & 0 & -1 \\ 0 & 0 & 1 & 0 & 1 & 1 & 1 \end{array}\right]$$

$$Q=\begin{bmatrix} Q_{11} & Q_{12} & 0 \\ Q_{21} & Q_{22} & 1 \end{bmatrix}=\left[\begin{array}{c|cccc|cc} -1 & 1 & 0 & 1 & 0 & 0 & 0 \\ -1 & 0 & -1 & 0 & 1 & 0 & 0 \\ \hline -1 & 0 & -1 & 0 & 0 & 1 & 0 \\ -1 & 1 & -1 & 0 & 0 & 0 & 1 \end{array}\right].$$

Partitioning has been indicated by dashed lines. The additional matrices required for the analysis are:

$$Z_2=\langle R_2\ \ R_3\ \ R_4\ \ R_5\rangle,$$

$$B_{21}Z_2=\begin{bmatrix} R_2 & 0 & -R_4 & 0 \\ 0 & R_3 & 0 & R_5 \end{bmatrix},$$

$$B_{22}\mathbf{U}_{g3}=\begin{bmatrix} -U_{g7} \\ U_{g6}+U_{g7} \end{bmatrix}, \qquad Q_{11}\mathbf{I}_{g1}=\begin{bmatrix} -I_{g1} \\ -I_{g1} \end{bmatrix}.$$

Substituting these into (2.85) we obtain:

$$\mathbf{I}_2=\begin{bmatrix} I_2 \\ I_3 \\ I_4 \\ I_5 \end{bmatrix}=-\begin{bmatrix} R_2 & 0 & -R_4 & 0 \\ 0 & R_3 & 0 & R_5 \\ 1 & 0 & 1 & 0 \\ 0 & -1 & 0 & 1 \end{bmatrix}^{-1}\begin{bmatrix} -U_{g7} \\ U_{g6}+U_{g7} \\ -I_{g1} \\ -I_{g1} \end{bmatrix}=$$

$$=\begin{bmatrix} \dfrac{1}{R_2+R_4}(U_{g7}+R_4I_{g1}) \\ -\dfrac{1}{R_3+R_5}(U_{g6}+U_{g7}+R_5I_{g1}) \\ -\dfrac{1}{R_2+R_4}(U_{g7}-R_2I_{g1}) \\ -\dfrac{1}{R_3+R_5}(U_{g6}+U_{g7}-R_3I_{g1}) \end{bmatrix}.$$

From (2.82):

$$\mathbf{I}_3=\begin{bmatrix} I_6 \\ I_7 \end{bmatrix}=-\begin{bmatrix} 0 & -1 & 0 & 0 \\ 1 & -1 & 0 & 0 \end{bmatrix}\mathbf{I}_2-\begin{bmatrix} -I_{g1} \\ -I_{g1} \end{bmatrix}$$

i.e.

$$\mathbf{I}_6 \frac{1}{R_3+R_5}(U_{g6}+U_{g7}-R_3 I_{g1})$$

$$\mathbf{I}_7=-\frac{1}{(R_2+R_4)(R_3+R_5)}[(R_2+R_4)U_{g6}+(R_2+R_3+R_4+R_5)U_{g7}+ \\ +(R_4R_5-R_2R_3)I_{g1}].$$

Thus the required currents have been determined. In the course of the calculation a matrix of order four had to be inverted.

The problem may also be solved on the basis of Eqs (2.124) and (2.127). Here

$$\boldsymbol{B}=\left[\begin{array}{c|cc|cc|cc} 1 & 0 & 0 & 1 & 1 & 1 & 1 \\ - & - & - & - & - & - & - \\ 0 & 1 & 0 & -1 & 0 & 0 & -1 \\ 0 & 0 & 1 & 0 & 1 & 1 & 1 \end{array}\right],$$

$$\boldsymbol{Y}_2=\langle 1/R_2 \quad 1/R_3\rangle, \qquad \boldsymbol{Z}_3=\langle \mathbf{R}_4 \quad \mathbf{R}_5\rangle .$$

From these, according to (2.135):

$$\mathbf{I}_3=\begin{bmatrix} I_4 \\ I_5 \end{bmatrix}=\begin{bmatrix} \frac{R_2+R_4}{R_2} & 0 \\ 0 & \frac{R_3+R_5}{R_3} \end{bmatrix}^{-1}\left\{\begin{bmatrix} I_{g1} \\ I_{g1} \end{bmatrix}-\begin{bmatrix} U_{g7}/R_2 \\ (U_{g6}+U_{g7})/R_3 \end{bmatrix}\right\}=$$

$$=\begin{bmatrix} \frac{1}{R_2+R_4}(R_2 I_{g1}-U_{g7}) \\ \frac{1}{R_3+R_5}(R_3 I_{g1}-U_{g6}-U_{g7}) \end{bmatrix}.$$

Here a matrix of order two had to be inverted. $\mathbf{I}_2$ is available from (2.134):

$$\mathbf{I}_2=\begin{bmatrix} I_2 \\ I_3 \end{bmatrix}=-\begin{bmatrix} 1/R_2 & 0 \\ 0 & 1/R_3 \end{bmatrix}\begin{bmatrix} -R_4 & 0 \\ 0 & R_5 \end{bmatrix}\mathbf{I}_3-\begin{bmatrix} 0 & -1/R_2 \\ 1/R_3 & 1/R_3 \end{bmatrix}\begin{bmatrix} U_{g6} \\ U_{g7} \end{bmatrix}=$$

$$=\begin{bmatrix} \frac{1}{R_2+R_4}(R_4 I_{g1}+U_{g7}) \\ -\frac{1}{R_3+R_5}(R_5 I_{g1}+U_{g6}+U_{g7}) \end{bmatrix},$$

in agreement with the results obtained by the previous procedure.

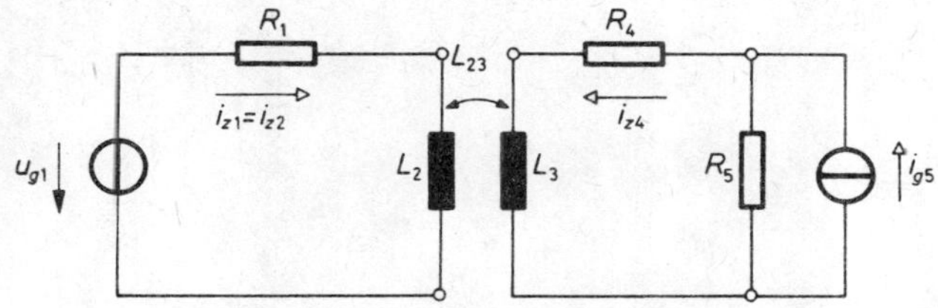

Fig. 2.24

3. The branch-currents of the network shown in Fig. 2.24 will be determined with the aid of loop-currents. The data of the network are:

$$u_{g1}(t)=[150\cos\omega t+50\cos(3\omega t+30°)]\ \text{V},$$

$$i_{g5}(t)=[10\cos(\omega t+50°)+5\cos(2\omega t-30°)]\ \text{A},$$

$$\omega=314\ s^{-1};$$

$$R_1=15\Omega,\quad R_4=5\Omega,\quad R_5=20\Omega,$$

$$L_2=0{\cdot}1\ \text{H},\quad L_3=0{\cdot}2\ \text{H},\quad L_{23}=L_{32}=0{\cdot}1\ \text{H}.$$

It can be directly determined from the structure of the network, that the fundamental set of loops is formed by two independent loops. Naturally, the same result follows from the investigation of a forest (Fig. 2.25, b) of the graph shown in Fig. 2.25, a. The orientation of the loops is chosen according to Fig. 2.25, a.

The matrices characterizing the network are:

$$\mathbf{U}_g=\begin{bmatrix}-\mathbf{U}_{g1}\\ \mathbf{0}\\ \mathbf{0}\\ \mathbf{0}\\ \mathbf{0}\end{bmatrix},\qquad \mathbf{I}_g=\begin{bmatrix}\mathbf{0}\\ \mathbf{0}\\ \mathbf{0}\\ \mathbf{0}\\ \mathbf{I}_{g5}\end{bmatrix},$$

where

$$\mathbf{U}_{g1}=\begin{bmatrix}150\\ 0\\ 50e^{j30°}\end{bmatrix}\text{V},\qquad \mathbf{I}_{g5}=\begin{bmatrix}10\,e^{j50°}\\ 5e^{-j30°}\\ 0\end{bmatrix}A,\qquad \mathbf{0}=\begin{bmatrix}0\\ 0\\ 0\end{bmatrix}.$$

$$Z=\begin{bmatrix} Z_{11} & Z_{12} & Z_{13} & Z_{14} & Z_{15} \\ Z_{21} & Z_{22} & Z_{23} & Z_{24} & Z_{25} \\ Z_{31} & Z_{32} & Z_{33} & Z_{34} & Z_{35} \\ Z_{41} & Z_{42} & Z_{43} & Z_{44} & Z_{45} \\ Z_{51} & Z_{52} & Z_{53} & Z_{54} & Z_{55} \end{bmatrix}.$$

Here

$$Z_{11}=\langle R_1\ R_1\ R_1\rangle=\langle 15\ 15\ 15\rangle\Omega,$$
$$Z_{22}=\langle j\omega L_2\ j2\omega L_2\ j3\omega L_2\rangle=\langle j31{\cdot}4\ j62{\cdot}8\ j94{\cdot}2\rangle\Omega,$$
$$Z_{23}=Z_{32}=\langle j\omega L_{23}\ j2\omega L_{23}\ j3\omega L_{23}\rangle=\langle j31{\cdot}4\ j62{\cdot}8\ j94{\cdot}2\rangle\Omega,$$
$$Z_{33}=\langle j\omega L_3\ j2\omega L_3\ j3\omega L_3\rangle=\langle j62{\cdot}8\ j125{\cdot}6\ j188{\cdot}4\rangle\Omega,$$
$$Z_{44}=\langle R_4\ R_4\ R_4\rangle=\langle 5\ 5\ 5\rangle\Omega,$$
$$Z_{55}=\langle R_5\ R_5\ R_5\rangle=\langle 20\ 20\ 20\rangle\Omega.$$

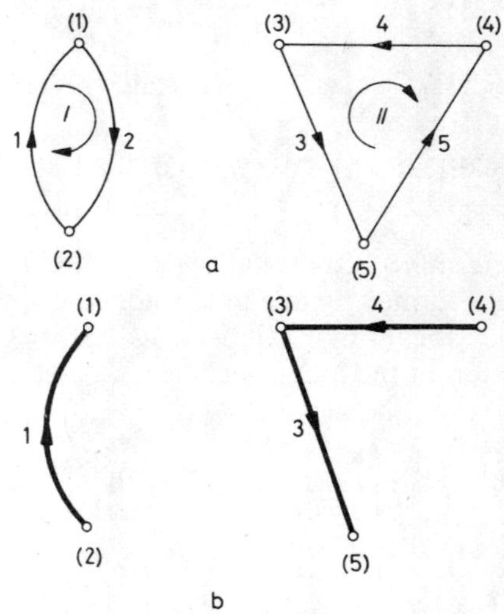

Fig. 2.25

For all cases not listed above,

$$Z_{ij}=0.$$

The loop matrix of the network is

$$B=\begin{bmatrix} 1 & 1 & 0 & 0 & 0 \\ 0 & 0 & -1 & -1 & -1 \end{bmatrix},$$

where $\boldsymbol{1}$ and $\boldsymbol{0}$ are square matrices of order three. The loop impedance matrix is:

$$\boldsymbol{Z}_B = \boldsymbol{BZB}^+ = \begin{bmatrix} \boldsymbol{Z}_{11}+\boldsymbol{Z}_{22} & -\boldsymbol{Z}_{23} \\ -\boldsymbol{Z}_{23} & \boldsymbol{Z}_{33}+\boldsymbol{Z}_{44}+\boldsymbol{Z}_{55} \end{bmatrix} =$$

$$= \begin{bmatrix} 15+j31{\cdot}4 & 0 & 0 & -j31{\cdot}4 & 0 & 0 \\ 0 & 15+j62{\cdot}8 & 0 & 0 & -j62{\cdot}8 & 0 \\ 0 & 0 & 15+j94{\cdot}2 & 0 & 0 & -j94{\cdot}2 \\ -j31{\cdot}4 & 0 & 0 & 25+j62{\cdot}8 & 0 & 0 \\ 0 & -j62{\cdot}8 & 0 & 0 & 25+j125{\cdot}6 & 0 \\ 0 & 0 & -j94{\cdot}2 & 0 & 0 & 25+j188{\cdot}4 \end{bmatrix} \Omega$$

To calculate the column matrix of loop-currents, column matrices $\boldsymbol{BZ}\mathbf{I}_g$ and $\boldsymbol{B}\mathbf{U}_g$ are also needed:

$$\boldsymbol{BZ}\mathbf{I}_g = \begin{bmatrix} \boldsymbol{0} \\ \boldsymbol{Z}_{55}\boldsymbol{I}_{g5} \end{bmatrix} = \begin{bmatrix} 0 \\ 0 \\ 0 \\ 200e^{j50^o} \\ 100e^{-j30^o} \\ 0 \end{bmatrix} \mathrm{V},$$

$$\boldsymbol{B}\mathbf{U}_g = \begin{bmatrix} -\mathbf{U}_{g1} \\ \boldsymbol{0} \end{bmatrix} = \begin{bmatrix} -150 \\ 0 \\ -50^{j30^o} \\ 0 \\ 0 \\ 0 \end{bmatrix} \mathrm{V}.$$

Thus according to (2.96) the equation relating to loop-currents is:

$$\begin{bmatrix} 15+j31{\cdot}4 & 0 & 0 & -j31{\cdot}4 & 0 & 0 \\ 0 & 15+j62{\cdot}8 & 0 & 0 & -j62{\cdot}8 & 0 \\ 0 & 0 & 15+j94{\cdot}2 & 0 & 0 & -j94{\cdot}2 \\ -j31{\cdot}4 & 0 & 0 & 25+j62{\cdot}8 & 0 & 0 \\ 0 & -j62{\cdot}8 & 0 & 0 & 25+j125{\cdot}6 & 0 \\ 0 & 0 & -j94{\cdot}2 & 0 & 0 & 25+j188{\cdot}4 \end{bmatrix} \begin{bmatrix} \mathbf{J}_1 \\ \mathbf{J}_2 \end{bmatrix} =$$

$$= \begin{bmatrix} 150 \\ 0 \\ 50e^{j30^{\circ}} \\ -200e^{j50^{\circ}} \\ -100e^{-j30^{\circ}} \\ 0 \end{bmatrix}.$$

This yields the following set of matrix equations:

$$\begin{bmatrix} 15+j31{\cdot}4 & 0 & 0 \\ 0 & 15+j62{\cdot}8 & 0 \\ 0 & 0 & 15+j94{\cdot}2 \end{bmatrix} \mathbf{J}_1 - \begin{bmatrix} j31{\cdot}4 & 0 & 0 \\ 0 & j62{\cdot}8 & 0 \\ 0 & 0 & j94{\cdot}2 \end{bmatrix} \mathbf{J}_2 =$$

$$= \begin{bmatrix} 150 \\ 0 \\ 50e^{j30^{\circ}} \end{bmatrix}$$

$$-\begin{bmatrix} j31{\cdot}4 & 0 & 0 \\ 0 & j62{\cdot}8 & 0 \\ 0 & 0 & j94{\cdot}2 \end{bmatrix} \mathbf{J}_1 + \begin{bmatrix} 25+j62{\cdot}8 & 0 & 0 \\ 0 & 25+j125{\cdot}6 & 0 \\ 0 & 0 & 25+j188{\cdot}4 \end{bmatrix} \mathbf{J}_2 =$$

$$= \begin{bmatrix} 200e^{j50^{\circ}} \\ 100e^{-j30^{\circ}} \\ 0 \end{bmatrix}$$

From these:

$$\mathbf{J}_1=\begin{bmatrix}1{\cdot}15+j5{\cdot}40\\ -0{\cdot}305+j1{\cdot}22\\ -0{\cdot}728+j0{\cdot}529\end{bmatrix}\text{A}=\begin{bmatrix}5{\cdot}52e^{j78^\circ}\\ 1{\cdot}26e^{j104^\circ}\\ 0{\cdot}900e^{j144^\circ}\end{bmatrix}\text{A},$$

$$\mathbf{J}_2=\begin{bmatrix}-3{\cdot}73-j9{\cdot}62\\ 0{\cdot}015-j1{\cdot}29\\ 0{\cdot}894-j1{\cdot}078\end{bmatrix}\text{A}=-\begin{bmatrix}10{\cdot}3e^{j69^\circ}\\ 1{\cdot}30e^{j91^\circ}\\ 1{\cdot}40e^{j130^\circ}\end{bmatrix}\text{A},$$

i.e. the loop-currents expressed as functions of time are:

$$j_1(t)=[5{\cdot}52\cos(\omega t+78^\circ)+1{\cdot}26\cos(2\omega t+104^\circ)+0{\cdot}900\cos(3\omega t+144^\circ)]\ \text{A},$$

$$j_2(t)=-[10{\cdot}3\cos(\omega t+69^\circ)+1{\cdot}30\cos(2\omega t+91^\circ)+1{\cdot}40\cos(3\omega t+130^\circ)]\ \text{A}.$$

These are at the same time the time-functions of edge currents, since

$$i_{z1}(t)=i_{i2}(t)=j_1(t),$$

$$i_{z3}(t)=i_{z4}(t)=-j_2(t).$$

4. Let us now apply the method of loop-currents to determine the branch-currents of the network shown in Fig. 2.26 if the direct current source I_g is

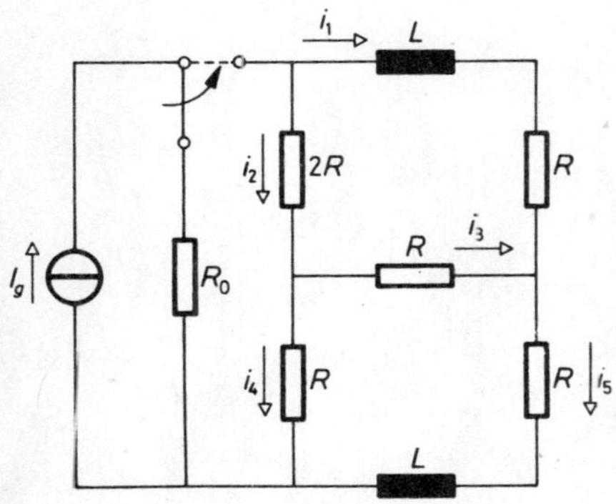

Fig. 2.26

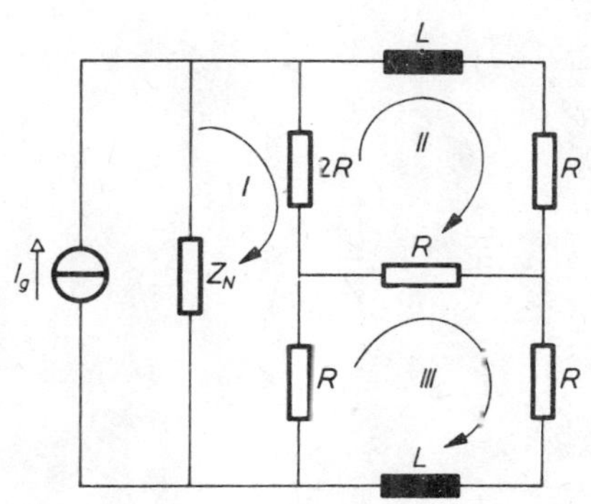

Fig. 2.27

disconnected from the resistance R_0 and switched on to the rest of the network at the instant $t=0$ by switching over the switch shown in the figure.

Considering the current-source together with an added impedance Z_N as one edge, we choose the set of loops shown in Fig. 2.27. Thus the equation relating to loop-currents (see (2.99)) is:

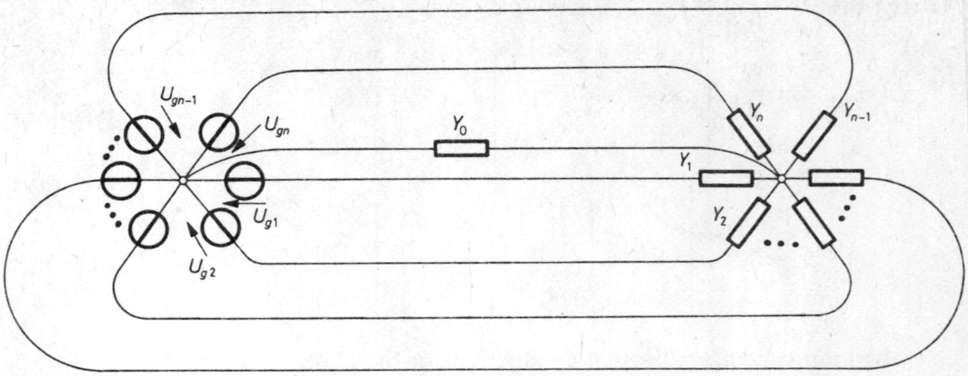

Fig. 2.28

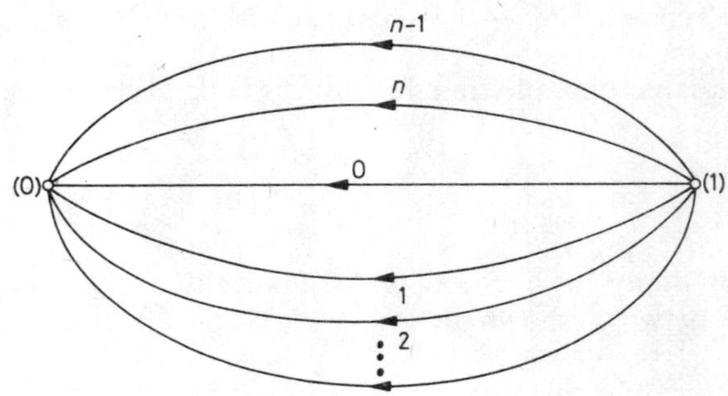

Fig. 2.29

$$\begin{bmatrix} Z_N+3R & -2R & -R \\ -2R & 4R+sL & -R \\ -R & -R & 3R+sL \end{bmatrix} \mathbf{J}(s) = \begin{bmatrix} Z_N I_g/s \\ 0 \\ 0 \end{bmatrix}.$$

If $Z_N \to \infty$:

$$\mathbf{J}(s) = \begin{bmatrix} I_g/s \\ J_2(s) \\ J_3(s) \end{bmatrix},$$

and the equation corresponding to (2.101) is:

$$\begin{bmatrix} -2R \\ -R \end{bmatrix} I_g/s + \begin{bmatrix} 4R+sL & -R \\ -R & 3R+sL \end{bmatrix} \begin{bmatrix} J_2(s) \\ J_3(s) \end{bmatrix} = \begin{bmatrix} 0 \\ 0 \end{bmatrix}.$$

From this

$$\begin{bmatrix} J_2(s) \\ J_3(s) \end{bmatrix} = \frac{I_g}{s\left(s^2 + s\dfrac{7R}{L} + \dfrac{11R^2}{L^2}\right)} \begin{bmatrix} s\dfrac{2R}{L} + \dfrac{7R^2}{L^2} \\ s\dfrac{R}{L} + \dfrac{6R^2}{L^2} \end{bmatrix}.$$

The roots of the denominator are:

$$s_1 = 0$$

$$s_2 = -\frac{7R}{2L} + \sqrt{\left(\frac{7R}{2L}\right)^2 - \frac{11R^2}{L^2}} \approx -2{\cdot}4\frac{R}{L}$$

$$s_3 \approx -4{\cdot}6\frac{R}{L}.$$

Thus

$$\mathbf{J}_E(t) = \begin{bmatrix} j_2(t) \\ j_3(t) \end{bmatrix} = 1(t)I_g \begin{bmatrix} 0{\cdot}63 & -0{\cdot}42 & -0{\cdot}22 \\ 0{\cdot}54 & -0{\cdot}68 & 0{\cdot}14 \end{bmatrix} \begin{bmatrix} 1 \\ e^{-2\cdot4Rt/L} \\ e^{-4\cdot6Rt/L} \end{bmatrix},$$

and

$$\mathbf{i}(t) = \boldsymbol{B}^+\mathbf{J}(t) = 1(t)I_g \begin{bmatrix} 1 & 0 & 0 \\ 0{\cdot}63 & -0{\cdot}42 & -0{\cdot}22 \\ 0{\cdot}37 & 0{\cdot}42 & 0{\cdot}22 \\ -0{\cdot}09 & -0{\cdot}26 & 0{\cdot}36 \\ 0{\cdot}46 & 0{\cdot}68 & -0{\cdot}14 \\ 0{\cdot}54 & -0{\cdot}68 & 0{\cdot}14 \end{bmatrix} \begin{bmatrix} 1 \\ e^{-2\cdot4Rt/L} \\ e^{-4\cdot6Rt/L} \end{bmatrix}$$

which gives the branch-currents as functions of time.

5. In the network shown in Fig. 2.28 an n-phase, Y-connected generator is connected to an n-phase, Y-connected load. The source-voltages in the phases of the generator are $U_{g1}, U_{g2}, \ldots, U_{gn}$ in this order, while the admittances of the load are $Y_1, Y_2, \ldots, Y_n$ in the same order. The two neutrals are connected by an admittance Y_0. The phase currents of the load will be determined with the aid of node-voltages.

Let us consider the source-voltage and admittance of each phase as one branch thus the graph shown in Fig. 2.29 is derived. Let us choose the neutral of the generator as the node with zero potential. Thus the incidence matrix with the orientation of the branches taken into account is:

$$\boldsymbol{A} = [1 \ 1 \ 1 \ \ldots \ 1].$$

The admittance-matrix of the network is:

$$\mathbf{Y} \doteq \langle Y_0 \;\; Y_1 \;\; Y_2 \;\; \dots \;\; Y_n \rangle,$$

i.e.

$$Y_A = \mathbf{A}\mathbf{Y}\mathbf{A}^+ = \sum_{k=0}^{n} Y_k,$$

and further

$$\mathbf{A}\mathbf{Y}\mathbf{U}_g = \sum_{k=1}^{n} Y_k U_{gk},$$

$$\mathbf{I}_g = \mathbf{0}.$$

From this according to (2.122):

$$\Phi_1 = U_0 = Y_A^{-1}\mathbf{A}\mathbf{Y}\mathbf{U}_g = \frac{\sum_{k=1}^{n} Y_k U_{gk}}{\sum_{k=0}^{n} Y_k}.$$

The result so obtained is known as Millmann's theorem. The phase voltages of the load are:

$$U_k = U_{gk} - U_0,$$

and the phase currents:

$$I_k = Y_k U_k = Y_k (U_{gk} - U_0), \quad (k = 1, 2, \dots, n).$$

6. In the network shown in Fig. 2.30 the excitation voltage is a periodic sequence of square pulses (Fig. 2.31). To determine the capacitor voltage the Fourier series of $u_g(t)$ is needed:

$$u_g(t) = \frac{4U_0}{\pi}\left[\cos\omega t - \frac{1}{3}\cos 3\omega t + \frac{1}{5}\cos 5\omega t - \frac{1}{7}\cos 7\omega t + \dots\right] =$$

$$= \frac{4U_0}{\pi}\,\mathrm{Re}\left[e^{j\omega t} - \frac{1}{3}e^{j3\omega t} + \frac{1}{5}e^{j5\omega t} - \frac{1}{7}e^{j7\omega t} + \dots\right] =$$

$$= \frac{4U_0}{\pi}\,\mathrm{Re}\sum_{\nu=0}^{\infty} \frac{(-1)^{2\nu+1}}{2\nu+1} e^{j(2\nu+1)\omega t}$$

The method of cutset-voltages is used for the analysis. The branch representing the capacitor is chosen as a tree-branch (Fig. 2.32). In the set of cutsets generated by the tree the voltage of the capacitor equals the cutset-voltage of the cutset containing this branch. The calculation will be carried out independently for each harmonic, and the results so obtained will be added to give the actual capacitor-voltage.

The admittance-matrix of the network for the k-th harmonic of angular frequency $k\omega$ is:

$$\mathbf{Y}_k = \langle G_1 \;\; G_2 \;\; G_3 \;\; jk\omega C \rangle.$$

The matrix of the set of cutsets chosen is:

$$\boldsymbol{Q}=\begin{bmatrix} -1 & 1 & 1 & 0 \\ 0 & -1 & 0 & 1 \end{bmatrix}.$$

The cutset-admittance matrix for the k-th harmonic is:

$$\boldsymbol{Y}_{Qk}=\begin{bmatrix} G_1+G_2+G_3 & -G_2 \\ -G_2 & G_2+jk\omega C \end{bmatrix}.$$

The additional matrices necessary for the analysis are:

$$\boldsymbol{YU}_{gk}=\begin{bmatrix} -G_1 U_{gk} \\ 0 \\ 0 \\ 0 \end{bmatrix}, \quad \boldsymbol{QYU}_{gk}=\begin{bmatrix} G_1 U_{gk} \\ 0 \end{bmatrix}, \quad \mathbf{I}_{gk}=0,$$

where

$$U_{gk}=U_0(-1)^k e^{jk\omega t}/k\pi, \quad k=2\nu+1, \quad \nu=0,1,2,\ldots$$

Thus the equation relating to cutset-voltages for the k-th harmonic is:

$$\mathbf{V}_{Qk}=\begin{bmatrix} U_{3k} \\ U_{4k} \end{bmatrix}=\begin{bmatrix} G_1+G_2+G_3 & -G_2 \\ -G_2 & G_2+jk\omega C \end{bmatrix}^{-1}\begin{bmatrix} G_1 U_{gk} \\ 0 \end{bmatrix}=$$

$$=\frac{G_1 U_{gk}}{(G_1+G_2+G_3)(G_2+jk\omega C)-G_2^2}\begin{bmatrix} G_2+jk\omega C \\ G_2 \end{bmatrix}.$$

From this the complex peak value of the k-th harmonic of the capacitor voltage is:

$$U_{4k}=\frac{G_1 G_2}{(G_1+G_2+G_3)(G_2+jk\omega C)-G_2^2}U_{gk}.$$

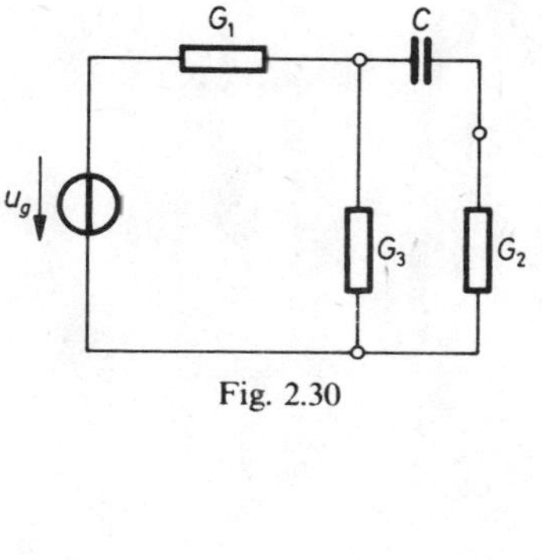

Fig. 2.30

Introducing the following notation:

$$W_k = \frac{G_1 G_2}{\sqrt{(G_1+G_3)^2 G_2^2 + k^2\omega^2 C^2 (G_1+G_2+G_3)^2}},$$

$$\psi_k = -\arctan \frac{k\omega C(G_1+G_2+G_3)}{(G_1+G_3)G_2},$$

we obtain

$$U_{4k} = W_k e^{j\psi_k} U_{gk},$$

and the time-function of the k-th harmonic of the capacitor voltage is:

$$U_{4k} = \frac{4U_0}{k\pi}(-1)^k \; W_k \mathrm{Re}\, e^{j(k\omega t+\psi_k)},$$

so that the capacitor-voltage as a function of time for the periodic square-pulse excitation is given by:

$$u_4(t) = \frac{4U_0}{\pi} \mathrm{Re} \sum_{\nu=0}^{\infty} \frac{(-1)^{2\nu+1}}{2\nu+1} W_{2\nu+1}\, e^{j((2\nu+1)\omega t+\psi_{2\nu+1})}$$

7. The branch-voltages of the network shown in Fig. 2.33 will be calculated with the aid of the method of cutset-voltages. Each source of the network gives a signal varying sinusoidally with angular frequency ω.

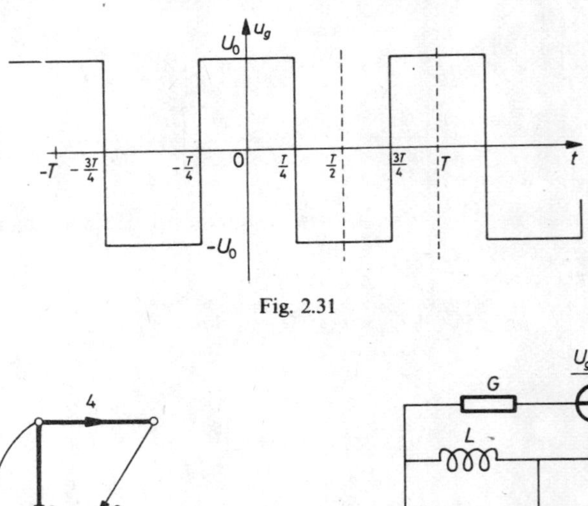

Fig. 2.31

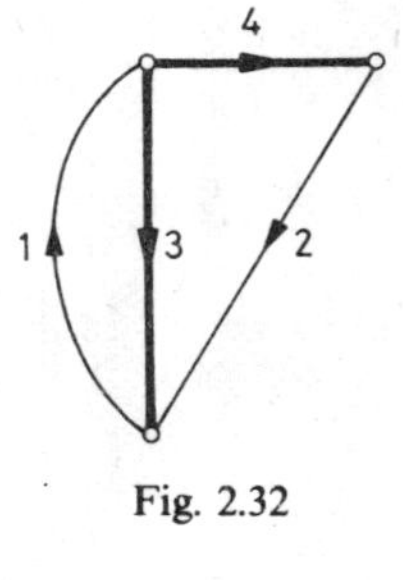

Fig. 2.32

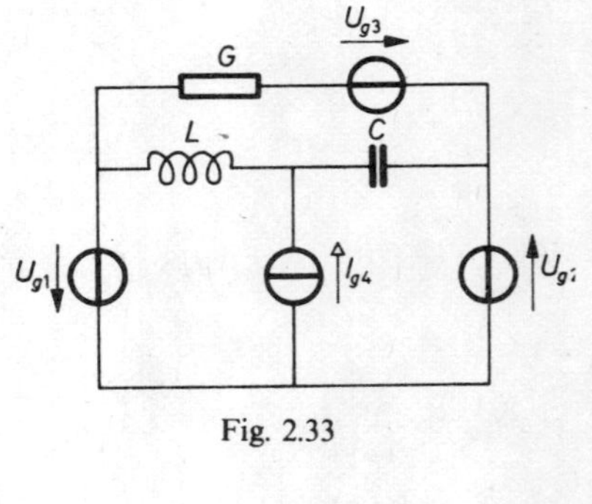

Fig. 2.33

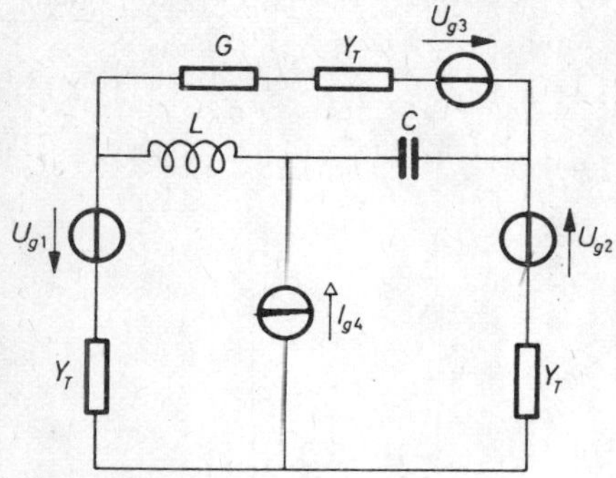

Fig. 2.34

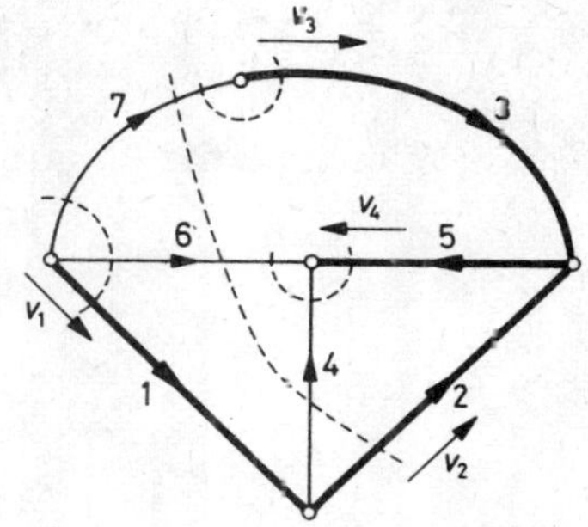

Fig. 2.35

Since the network contains voltage-sources, i.e. zero impedance branches, the equation of cutset-voltages is written for the network after modification in the manner explained previously (Fig. 2.34). Choosing the fundamental set of cutsets indicated in Fig. 2.35:

$$\left[\begin{array}{ccc|c} Y_T+G+1/j\omega L & G+1/j\omega L & -G & 1/j\omega L \\ G+1/j\omega L & Y_T+G+1/j\omega L & -G & 1/j\omega L \\ -G & -G & Y_T+G & 0 \\ \hline 1/j\omega L & 1/j\omega L & 0 & j\omega C+1/j\omega L \end{array}\right] \mathbf{V}_Q =$$

$$= \left[\begin{array}{c} Y_T U_{g1} \\ Y_T U_{g2} \\ Y_T U_{g3} \\ \hline 0 \end{array}\right] - \left[\begin{array}{c} 0 \\ I_{g4} \\ 0 \\ \hline I_{g4} \end{array}\right],$$

where the partitioning in accordance with (2.112) has been indicated. If $Y_T \to \infty$:

$$\mathbf{V}_Q = \begin{bmatrix} U_{g1} \\ U_{g2} \\ U_{g3} \\ V_4 \end{bmatrix},$$

and thus from the former equation:

$$(j\omega C + 1/j\omega L)V_4 = -I_{g4} - 1/j\omega L(U_{g1} + U_{g2}),$$

i.e. the column matrix of branch-voltages is:

$$\mathbf{U} = \begin{bmatrix} U_{g1} \\ U_{g2} \\ U_{g3} \\ -(j\omega L I_{g4} + U_{g1} + \omega^2 LC U_{g2})/(1-\omega^2 LC) \\ -(j\omega L I_{g4} + U_{g1} + U_{g2})/(1-\omega^2 LC) \\ -[j\omega L I_{g4} + \omega^2 LC(U_{g1} + U_{g2})]/(1-\omega^2 LC) \\ U_{g1} + U_{g2} - U_{g3} \end{bmatrix}.$$

CHAPTER 3

METHODS FOR THE DETERMINATION OF NETWORK CHARACTERISTICS

The matrix equations written for networks with the aid of graph theory concepts are suitable for the determination of several quantities characterizing the network. Thus, methods are available for calculating the immitance matrices of n-terminal elements, n-ports, $m \times n$-terminal elements, the hybrid-parameter matrix of $2 \times n$-terminal elements, as well as for obtaining transfer functions or transfer-function matrices. Such methods [3, 16, 24, 43, 44] will be discussed in this chapter.

Calculation of n-terminal characteristics

The number of linearly independent voltages between the terminals of an n-terminal element is $n-1$ in accordance with the loop-equation. Let these be chosen as shown in Fig. 3.1., and be denoted by $U_1, U_2, \ldots, U_{n-1}$. On application of the cutset equation $n-1$ terminal-currents are seen to be linearly independent. Therefore current I_0 can be expressed in terms of $I_1, I_2, \ldots, I_{n-1}$.

We wish to find relationships between these $n-1$ currents. Since the n-terminal element examined is linear, equations describing the relations between currents and voltages are also linear.

Let the above-mentioned voltages and currents of the n-terminal element form column matrices

$$\mathbf{U}=\begin{bmatrix} U_1 \\ U_2 \\ \vdots \\ U_{n-1} \end{bmatrix} \quad \text{and} \quad \mathbf{I}=\begin{bmatrix} I_1 \\ I_2 \\ \vdots \\ I_{n-1} \end{bmatrix}. \tag{3.1}$$

If the element contains no sources, its voltages and currents are related by

$$\mathbf{U}=\boldsymbol{Z}_1\mathbf{I} \quad \text{as well as} \quad \mathbf{I}=\boldsymbol{Y}_1\mathbf{U}, \tag{3.2}$$

where $\boldsymbol{Z}_1$ is the open-circuit impedance matrix while $\boldsymbol{Y}_1$ is the short-circuit admittance matrix of the n-terminal element. If the inverse of $\boldsymbol{Z}_1$ exists, then $\boldsymbol{Y}_1 = \boldsymbol{Z}_1^{-1}$.

For n-terminal elements containing sources the relations

$$\mathbf{U} = \boldsymbol{Z}_1 \mathbf{I} + \mathbf{U}_0 \qquad \text{and} \qquad \mathbf{I} = \boldsymbol{Y}_1 \mathbf{U} + \mathbf{I}_0 \tag{3.3}$$

hold between voltages and currents, where $\mathbf{U}_0$ and $\mathbf{I}_0$ are, respectively, the column matrices of open-circuit voltages and short-circuit currents. The order and reference directions of the elements in $\mathbf{U}_0$ and $\mathbf{I}_0$ coincide with those of the elements in $\mathbf{U}$ and $\mathbf{I}$.

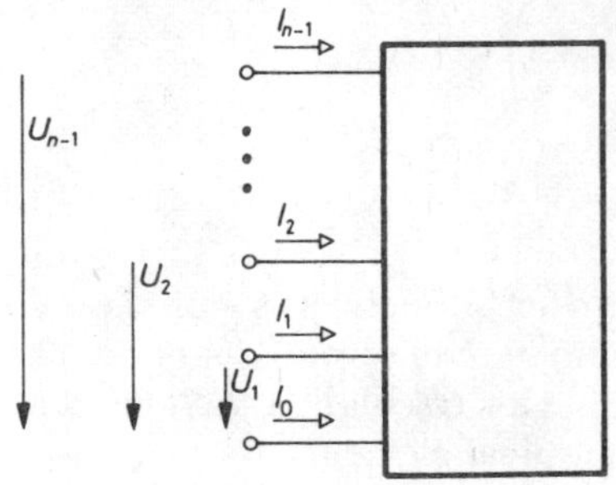

Fig. 3.1

In the following a method will be presented for the determination of the n-terminal characteristics $\boldsymbol{Z}_1$, $\boldsymbol{Y}_1$, $\mathbf{U}_0$ and $\mathbf{I}_0$. Only n-terminal elements having connected graphs will be considered.

The determination of open-circuit impedance and short-circuit admittance matrices

To calculate the open-circuit impedance matrix let the linear and passive n-terminal element be terminated by voltage-sources as shown in Fig. 3.2. In the graph of the network so obtained, let the edges associated with the voltage-sources be denoted by $1, 2, \ldots, n-1$ with reference directions identical with the currents of the branches. The order numbers and orientations of the remaining edges in the graph are arbitrary.

Let a tree of the graph be chosen with those edges corresponding to the terminations (sources) being chords. Since at least one path exists between any two terminals of the element, the existence of such a tree is ensured by the fact that any tree in the graph of the n-terminal element without the terminations is also a tree in the graph of the network terminated with sources in the manner described above, with the edges associated with the terminations being chords.

A fundamental set of loops can be based upon the tree chosen. Let the first $n-1$ loops contain the edges $1, 2, \ldots, n-1$ in that order with their orientations with respect to these edges coinciding with those of the edges. The order numbers and orientations of further loops are arbitrary. Let this set of loops be characterized by basis loop matrix $\boldsymbol{B}$. From the impedance matrix $\boldsymbol{Z}$ and loop matrix $\boldsymbol{B}$ the loop-impedance matrix $\boldsymbol{Z}_B$ of the network can be calculated. On application of the method of loop-currents:

$$-\boldsymbol{B}\mathbf{U}_g = \boldsymbol{Z}_B\mathbf{J}\,, \tag{3.4}$$

where

$$-\boldsymbol{B}\mathbf{U}_g = \begin{bmatrix} \mathbf{U} \\ \mathbf{0} \end{bmatrix} \tag{3.5}$$

and

$$\mathbf{J} = \begin{bmatrix} \mathbf{I} \\ \mathbf{J}_e \end{bmatrix}. \tag{3.6}$$

$\mathbf{J}_e$ is the column matrix of the loop-currents not flowing through the terminating edges. The number of elements in the zero submatrix in (3.5) equals that in $\mathbf{J}_e$. The order of $\boldsymbol{Z}_B$ is at least $n-1$. If the order of $\boldsymbol{Z}_B$ is exactly $n-1$, then each loop of the network contains one of the terminating sources. In this case the $\mathbf{0}$ in (3.5) and $\mathbf{J}_e$ in (3.6) does not appear, and (3.4) yields $\mathbf{U}=\boldsymbol{Z}_B\mathbf{I}$, i.e. $\boldsymbol{Z}_1=\boldsymbol{Z}_B$ is the open-circuit impedance matrix. If the order of $\boldsymbol{Z}_B$ is greater than $n-1$, it may be partitioned as follows:

$$\boldsymbol{Z}_B = \begin{bmatrix} \boldsymbol{z}_{11} & \boldsymbol{z}_{12} \\ \boldsymbol{z}_{21} & \boldsymbol{z}_{22} \end{bmatrix}, \tag{3.7}$$

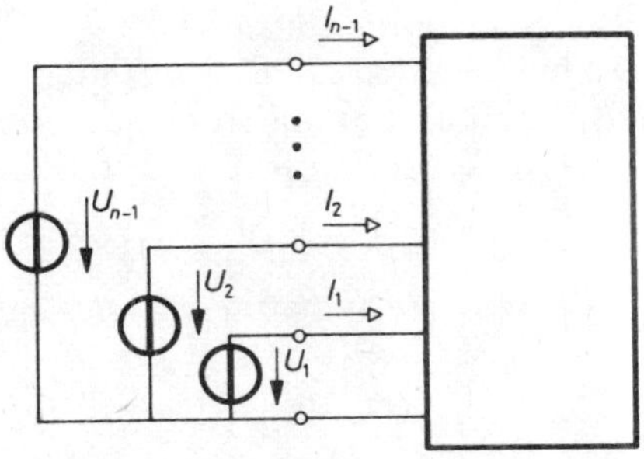

Fig. 3.2

where $\boldsymbol{z}_{11}$ is a square matrix of order $n-1$. Accordingly the number of rows in $\boldsymbol{z}_{12}$ and of columns in $\boldsymbol{z}_{21}$ is $n-1$. Let (3.5), (3.6) and (3.7) be substituted into (3.4):

$$\begin{bmatrix} \mathbf{U} \\ \mathbf{0} \end{bmatrix} = \begin{bmatrix} \boldsymbol{z}_{11} & \boldsymbol{z}_{12} \\ \boldsymbol{z}_{21} & \boldsymbol{z}_{22} \end{bmatrix} \begin{bmatrix} \mathbf{I} \\ \mathbf{J}_e \end{bmatrix}. \tag{3.8}$$

From this:

$$\mathbf{U} = \boldsymbol{z}_{11}\mathbf{I} + \boldsymbol{z}_{12}\mathbf{J}_e\,, \tag{3.9}$$

$$\mathbf{0} = \boldsymbol{z}_{21}\mathbf{I} + \boldsymbol{z}_{22}\mathbf{J}_e\,. \tag{3.10}$$

Provided that the inverse of z_{22} exists $\mathbf{J}_e$ may be obtained from (3.10) and substituted into (3.9):

$$\mathbf{U}=(z_{11}-z_{12}z_{22}^{-1}z_{21})\mathbf{I}, \tag{3.11}$$

i.e.

$$Z_1=z_{11}-z_{12}z_{22}^{-1}z_{21} \tag{3.12}$$

is the open-circuit impedance matrix of the n-terminal element.

An n-terminal element characterized by an open-circuit impedance matrix Z_1 may be replaced by a "star-connected" n-terminal element with its branches formed by the impedances on the main diagonal of Z_1 and with mutual impedances between the branches equal to corresponding impedances off the main diagonal.

Eq. (3.2) shows that, apart from degenerate cases, short-circuit admittance matrix Y_1 may be obtained as the inverse of Z_1. The following method enables Y_1 to be determined directly.

Let current-sources be connected to the linear, passive n-terminal element without sources as shown in Fig. 3.3. The numbering and choice of orientations in the network thus terminated is similar to those in the case examined above. Let a tree of the graph be chosen with edges 1, 2, . . ., $n-1$ being tree-branches. Apart from degenerate cases of no practical interest such a tree exists, since the terminations do not form loops. A fundamental set of cutsets is associated with this tree. Let its cutsets be numbered so as to have the first $n-1$ cutsets contain edges 1, 2, . . ., $n-1$, with orientations opposite to those of the edges. The order numbers and orientations of further cutsets are arbitrary.

The cutset-matrix of the fundamental set of cutsets thus chosen is denoted by Q. From the admittance matrix Y and cutset-matrix Q the cutset admittance matrix Y_Q of the network can be written. According to Eq. (2.108) of chapter 2:

$$-Q\mathbf{I}_g=Y_Q\mathbf{V}_Q, \tag{3.13}$$

since the network is excited by the terminating current-sources only. Here

$$-Q\mathbf{I}_g=\begin{bmatrix}\mathbf{I}\\ \mathbf{0}\end{bmatrix}, \tag{3.14}$$

and

$$\mathbf{V}_Q=\begin{bmatrix}\mathbf{U}\\ \mathbf{V}_e\end{bmatrix}. \tag{3.15}$$

$\mathbf{V}_e$ is formed by the cutset-voltages of cutsets not containing the terminations, while $\mathbf{0}$ is a zero column matrix with the same number of elements as $\mathbf{V}_e$. Y_Q is a square matrix of order at least $n-1$. Therefore, it can be partitioned as follows:

$$Y_Q=\begin{bmatrix}y_{11} & y_{12}\\ y_{21} & y_{22}\end{bmatrix}, \tag{3.16}$$

where y_{11} is a square matrix of order $n-1$. If the order of Y_Q is exactly $n-1$, then $\mathbf{I}=Y_Q\mathbf{U}$, and thus $Y_1=Y_Q$ is the short-circuit admittance matrix sought.

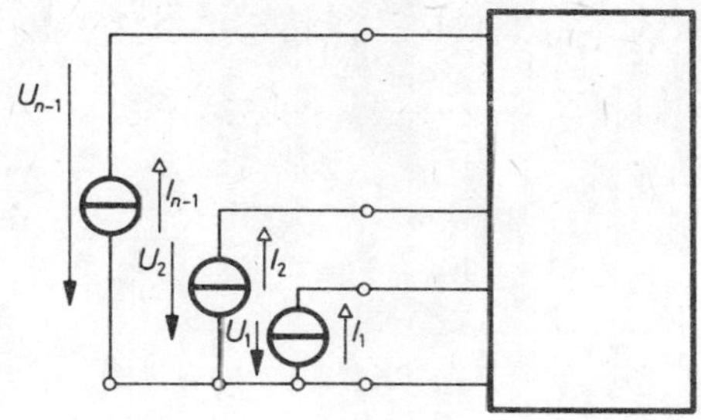

Fig. 3.3

Substituting (3.14), (3.15) and (3.16) into (3.13):

$$\begin{bmatrix} \mathbf{I} \\ \mathbf{0} \end{bmatrix} = \begin{bmatrix} y_{11} & y_{12} \\ y_{21} & y_{22} \end{bmatrix} \begin{bmatrix} \mathbf{U} \\ \mathbf{V}_e \end{bmatrix}. \tag{3.17}$$

From this:

$$\mathbf{I} = y_{11}\mathbf{U} + y_{12}\mathbf{V}_e, \tag{3.18}$$

$$\mathbf{0} = y_{21}\mathbf{U} + y_{22}\mathbf{V}_e. \tag{3.19}$$

Obtaining $\mathbf{V}_e$ from (3.19) and substituting it into (3.18):

$$\mathbf{I} = (y_{11} - y_{12}y_{22}^{-1}y_{21})\mathbf{U}, \tag{3.20}$$

provided that the inverse of y_{22} exists.

On comparison with (3.2) it is seen that

$$Y_1 = y_{11} - y_{12}y_{22}^{-1}y_{21} \tag{3.21}$$

is the short-circuit admittance matrix.

Characteristics of *n*-terminal elements containing current-sources and voltage-sources

The short-circuit admittance and open-circuit impedance matrices of *n*-terminal elements containing sources may be determined by the application of the above method to the deactivated network. Deactivation means the replacement of voltage-sources by short-circuits and current-sources by open circuits in the network, i.e. source-voltages and source-currents reduced to zero.

To calculate the column matrix $\mathbf{I}_0$ formed by the short-circuit currents of the *n*-terminal element containing sources, let the element be terminated by short-circuits (Fig. 3.4). The numbering and orientation of edges, as well as the selection of the tree and the fundamental set of loops in the graph of the element thus terminated is chosen as in the previous section. In this network let $\boldsymbol{Z}$ denote the impedance matrix, $\boldsymbol{B}$ the loop-matrix of the fundamental set of loops, and $\boldsymbol{Z}_B$ the loop-

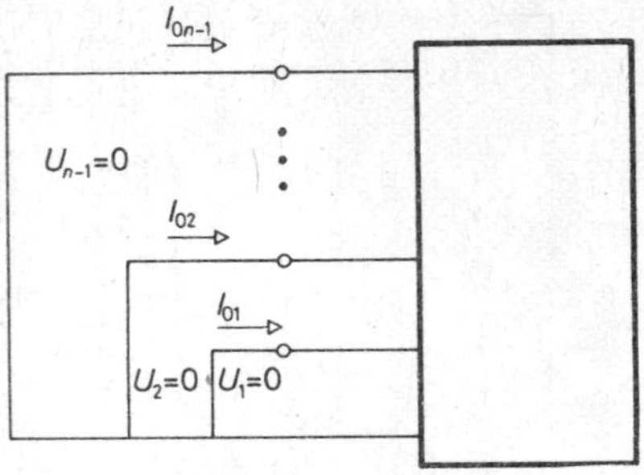

Fig. 3.4

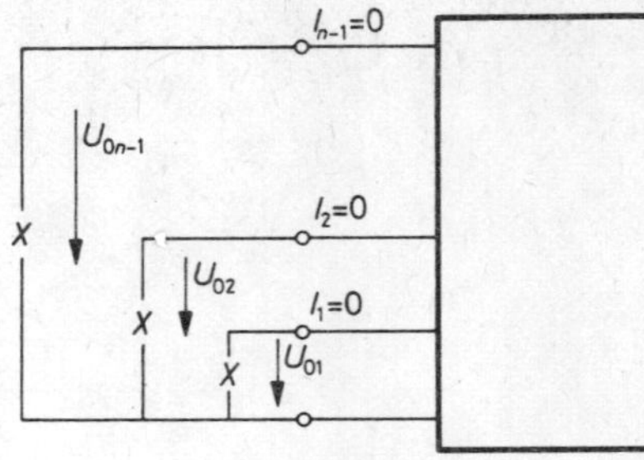

Fig. 3.5

impedance matrix obtained from these. $\boldsymbol{Z}$, $\boldsymbol{B}$ and $\boldsymbol{Z}_B$ coincide with the corresponding matrices used for the calculation of $\boldsymbol{Z}_1$. Loop-currents are obtained from

$$\boldsymbol{B}(\boldsymbol{Z}\mathbf{I}_g - \mathbf{U}_g) = \boldsymbol{Z}_B \mathbf{J}_0 \tag{3.22}$$

(see (2.96)). The first $n-1$ elements of column matrix $\mathbf{J}_0$ equal the short-circuit currents:

$$\mathbf{J}_0 = \begin{bmatrix} \mathbf{I}_0 \\ \mathbf{J}_e \end{bmatrix} = \boldsymbol{Z}_B^{-1} \boldsymbol{B}(\boldsymbol{Z}\mathbf{I}_g - \mathbf{U}_g). \tag{3.23}$$

The column matrix $\mathbf{U}_0$ of open-circuit voltages may be obtained from the n-terminal element terminated by open-circuits (Fig. 3.5). The graph and the fundamental set of cutsets of this network are similar to those used for the calculation of $\boldsymbol{Y}_1$, i.e. $\boldsymbol{Q}$ and $\boldsymbol{Y}_Q$ are equal to the matrices calculated there. The equation of the cutset-voltages is (see (2.108)):

$$\boldsymbol{Q}(\boldsymbol{Y}\mathbf{U}_g - \mathbf{I}_g) = \boldsymbol{Y}_Q \mathbf{V}_0. \tag{3.24}$$

$\mathbf{V}_0$ can be obtained from this, and its first $n-1$ elements yield column matrix $\mathbf{U}_0$:

$$\mathbf{V}_0 = \begin{bmatrix} \mathbf{U}_0 \\ \mathbf{V}_e \end{bmatrix} = \boldsymbol{Y}_Q^{-1} \boldsymbol{Q}(\boldsymbol{Y}\mathbf{U}_g - \mathbf{I}_g). \tag{3.25}$$

Thus the matrices characterizing the n-terminal element containing sources have been determined.

$m \times n$-terminal element

An $m \times n$-terminal network has mn terminals, with each group of n terminals forming one of m connection-points (Fig. 3.6). Each of these connection-points can be terminated by an n-terminal element, i.e. the sum of currents of the terminals corresponding to one connection-point equals zero, the maximum number of linearly independent currents at one connection-point being $n-1$. Let at least one path exist between any two terminals of the same connection-point in the graph of the network. At each connection-point the maximum number of linearly independent voltages is $n-1$, as for n-terminal elements. The reference directions of voltages and currents at connection-points are chosen as shown in Fig. 3.6.

Let the independent voltages and currents at connection-points form column-matrices:

$$\mathbf{U}_1 = \begin{bmatrix} U_{11} \\ U_{12} \\ \vdots \\ U_{1n-1} \end{bmatrix}; \quad \mathbf{U}_2 = \begin{bmatrix} U_{21} \\ U_{22} \\ \vdots \\ U_{2n-1} \end{bmatrix}; \quad \ldots \quad \mathbf{U}_m = \begin{bmatrix} U_{m1} \\ U_{m2} \\ \vdots \\ U_{mn-1} \end{bmatrix},$$

$$\mathbf{I}_1 = \begin{bmatrix} I_{11} \\ I_{12} \\ \vdots \\ I_{1n-1} \end{bmatrix}; \quad \mathbf{I}_2 = \begin{bmatrix} I_{21} \\ I_{22} \\ \vdots \\ I_{2n-1} \end{bmatrix}; \quad \ldots \quad \mathbf{I}_m = \begin{bmatrix} I_{m1} \\ I_{m2} \\ \vdots \\ I_{mn-1} \end{bmatrix}. \tag{3.26}$$

The analysis of $m \times n$-terminal elements involves finding relationships between $\mathbf{U}_1, \mathbf{U}_2, \ldots, \mathbf{U}_m, \mathbf{I}_1, \mathbf{I}_2, \ldots, \mathbf{I}_m$. Two $m \times n$-terminal elements are considered to be equivalent, if the relationships between the above quantities are described by identical equations. In $m \times n$-terminal elements voltages may exist between terminals corresponding to different connection-points, but these voltages are not relevant to the discussion of $m \times n$-terminal elements.

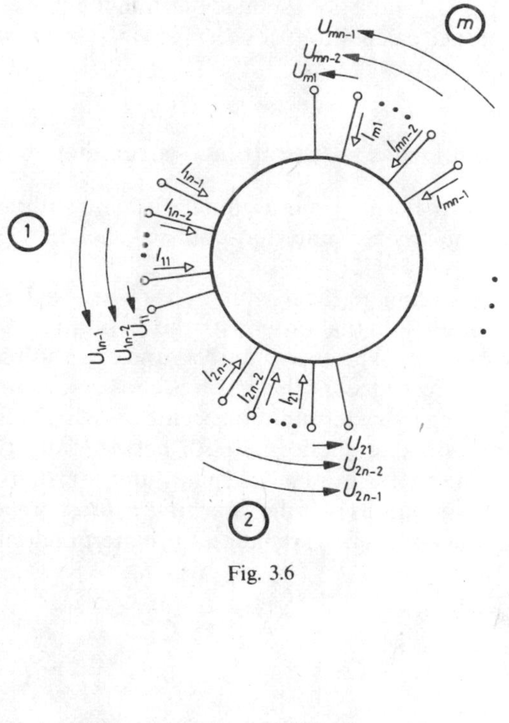

Fig. 3.6

Let hypermatrices

$$\mathbf{U}=\begin{bmatrix}\mathbf{U}_1\\ \mathbf{U}_2\\ \vdots\\ \mathbf{U}_m\end{bmatrix} \quad \text{and} \quad \mathbf{I}=\begin{bmatrix}\mathbf{I}_1\\ \mathbf{I}_2\\ \vdots\\ \mathbf{I}_m\end{bmatrix} \tag{3.27}$$

be formed by the column-matrices of (3.26).

For an $m \times n$-terminal element without sources relationships

$$\mathbf{U}=\boldsymbol{Z}_p\mathbf{I} \quad \text{and} \quad \mathbf{I}=\boldsymbol{Y}_p\mathbf{U} \tag{3.28}$$

hold, where $\boldsymbol{Z}_p$ is the impedance- and $\boldsymbol{Y}_p=\boldsymbol{Z}_p^{-1}$ the admittance-parameter matrix. These characterize the $m \times n$-terminal element independently from any excitations connected to the network.

The relations between the connection-point voltages and currents of linear $m \times n$-terminal elements containing sources are described by the equations

$$\mathbf{U}=\boldsymbol{Z}_p\boldsymbol{I}+\mathbf{U}_0 \quad \text{and} \quad \mathbf{I}=\boldsymbol{Y}_p\mathbf{U}+\mathbf{I}_0\,. \tag{3.29}$$

Here $\mathbf{U}_0$ is the column matrix of the open-circuit voltages, while $\mathbf{I}_0$ that of the short-circuit currents of the $m \times n$-terminal element.

The following section presents a method for the determination of the characteristics $\boldsymbol{Z}_p$, $\boldsymbol{Y}_p$, $\mathbf{U}_0$ and $\mathbf{I}_0$ of the $m \times n$-terminal element.

The applications of such calculations concerning $m \times n$-terminal elements include analysis of multiphase networks.

The impedance- and admittance-parameter matrices

Let us first consider a linear $m \times n$-terminal element without sources, having voltage-sources connected to its connection-points as shown in Fig. 3.7. Let a graph be associated with this network.

Let the edges corresponding to the terminations be denoted by $1, 2, \ldots, n-1, n, n+1, \ldots, 2n-2, \ldots m(n-1)$ in that order with the orientations coinciding with the directions of branch-currents. The order numbers and orientations of the remaining edges are arbitrary. A forest (tree) of the graph is chosen for our calculations with those edges corresponding to the terminations being chords. Such forest (tree) exists, since on the presumption that there is a path between any two terminals of a connection-point of the $m \times n$-terminal element, any forest in the graph of the element without the terminations is at the same time a forest of the network with the terminations added, the edges corresponding to the terminations naturally being chords.

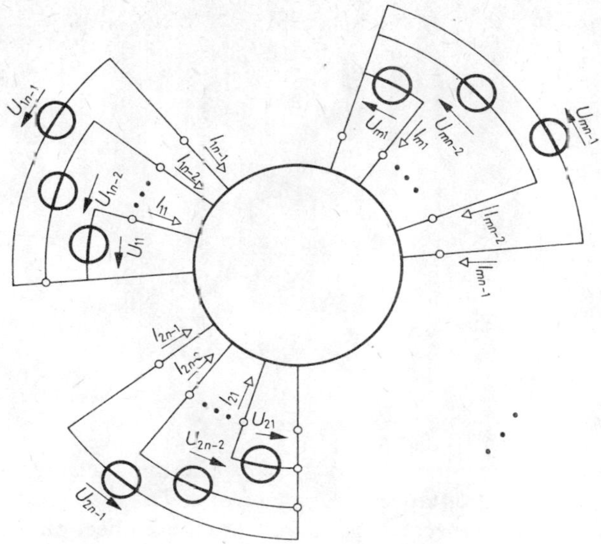

Fig. 3.7

A fundamental set of loops is based upon the forest so chosen in the manner previously described. In this set of loops the first $m(n-1)$ loops contain the edges 1, 2, . . ., $m(n-1)$ with their orientations along these edges coinciding with those of the edges. The numbering and orientation of the remaining loops is arbitrary. Let this set of loops be characterized by loop-matrix $\boldsymbol{B}$. From this and the edge-impedance matrix $\boldsymbol{Z}$ the loop-impedance matrix $\boldsymbol{Z}_B$ of the network can be calculated.

Applying the method of loop-currents:

$$-\boldsymbol{B}\mathbf{U}_g = \boldsymbol{Z}_B\mathbf{J}. \tag{3.30}$$

Here

$$-\boldsymbol{B}\mathbf{U}_g = \begin{bmatrix} \mathbf{U}_1 \\ \mathbf{U}_2 \\ \vdots \\ \mathbf{U}_m \\ \mathbf{0} \end{bmatrix} = \begin{bmatrix} \mathbf{U} \\ \mathbf{0} \end{bmatrix} \quad \text{and} \quad \mathbf{J} = \begin{bmatrix} \mathbf{I}_1 \\ \mathbf{I}_2 \\ \vdots \\ \mathbf{I}_m \\ \mathbf{J}_e \end{bmatrix} = \begin{bmatrix} \mathbf{I} \\ \mathbf{J}_e \end{bmatrix}, \tag{3.31}$$

where relations (3.27) have been used, and $\mathbf{J}_e$ is the column matrix of loop-currents not flowing through the terminations. Let $\boldsymbol{Z}_B$ be partitioned:

$$\boldsymbol{Z}_B = \begin{bmatrix} \boldsymbol{z}_{11} & \boldsymbol{z}_{12} \\ \boldsymbol{z}_{21} & \boldsymbol{z}_{22} \end{bmatrix}, \tag{3.32}$$

8

where z_{11} is a square matrix of order $m(n-1)$. Let us substitute (3.31) and (3.32) into (3.30):

$$\begin{bmatrix} \mathbf{U} \\ \mathbf{0} \end{bmatrix} = \begin{bmatrix} z_{11} & z_{12} \\ z_{21} & z_{22} \end{bmatrix} \begin{bmatrix} \mathbf{I} \\ \mathbf{J}_e \end{bmatrix}. \tag{3.33}$$

Hence

$$\mathbf{U} = z_{11}\mathbf{I} + z_{12}\mathbf{J}_e, \tag{3.34}$$

$$\mathbf{0} = z_{21}\mathbf{I} + z_{22}\mathbf{J}_e, \tag{3.35}$$

i.e., provided that z_{22} is non-singular,

$$\mathbf{U} = (z_{11} - z_{12}z_{22}^{-1}z_{21})\mathbf{I}, \tag{3.36}$$

and thus

$$Z_p = z_{11} - z_{12}z_{22}^{-1}z_{21} \tag{3.37}$$

is the impedance-parameter matrix of the $m \times n$-terminal element.

To determine the admittance-parameters of a linear $m \times n$-terminal element without sources, let us connect current-sources to the connection points as shown in Fig. 3.8. Let the edges corresponding to the terminations in the graph of the network so terminated be denoted by $1, 2, \ldots, n-1, n, n+1, \ldots, 2n-2, \ldots, m(n-1)$ in that order with orientations equal to those of the current-source currents. The order numbers and orientations of the remaining edges are arbitrary. Let a forest (tree) of

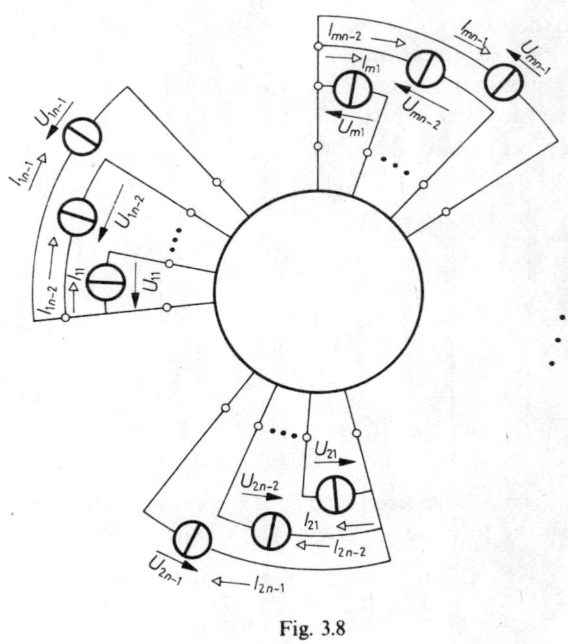

Fig. 3.8

the graph be chosen such that the edges corresponding to the terminations are tree-branches. This method can be applied only if such a forest exists, i.e. the edges associated with the terminations do not form loops. A fundamental set of cutsets belongs to the forest thus chosen. Let the cutsets in this set be so numbered that the first $m(n-1)$ cutsets contain edges $1, 2, \ldots, m(n-1)$, with orientations on these edges opposite to those of the edges, the order numbers and orientations of the remaining cutsets being arbitrary.

The cutset-matrix of the fundamental set of cutsets so chosen is denoted by $\boldsymbol{Q}$. Given $\boldsymbol{Q}$ and the admittance-matrix $\boldsymbol{Y}$ the cutset-admittance matrix $\boldsymbol{Y}_Q$ of the network can be derived. According to the equation of cutset-voltages:

$$-\boldsymbol{Q}\mathbf{I}_g = \boldsymbol{Y}_Q \mathbf{V}_Q, \tag{3.38}$$

taking into account that the network is excited at the terminations only. Thus:

$$-\boldsymbol{Q}\mathbf{I}_g = \begin{bmatrix} \mathbf{I}_1 \\ \mathbf{I}_2 \\ \vdots \\ \mathbf{I}_m \\ \mathbf{0} \end{bmatrix} = \begin{bmatrix} \mathbf{I} \\ \mathbf{0} \end{bmatrix} \quad \text{and} \quad \mathbf{V}_Q = \begin{bmatrix} \mathbf{U}_1 \\ \mathbf{U}_2 \\ \vdots \\ \mathbf{U}_m \\ \mathbf{V}_e \end{bmatrix} = \begin{bmatrix} \mathbf{U} \\ \mathbf{V}_e \end{bmatrix}, \tag{3.39}$$

where $\mathbf{V}_e$ is formed by the cutset-voltages of cutsets not containing the terminations, and $\mathbf{0}$ is a zero matrix with its number of elements identical to that of $\mathbf{V}_e$. Let $\boldsymbol{Y}_Q$ be partitioned as follows:

$$\boldsymbol{Y}_Q = \begin{bmatrix} y_{11} & y_{12} \\ y_{21} & y_{22} \end{bmatrix}, \tag{3.40}$$

where y_{11} is a square matrix of order $m(n-1)$. Substituting (3.39) and (3.40) into (3.38):

$$\begin{bmatrix} \mathbf{I} \\ \mathbf{0} \end{bmatrix} = \begin{bmatrix} y_{11} & y_{12} \\ y_{21} & y_{22} \end{bmatrix} \begin{bmatrix} \mathbf{U} \\ \mathbf{V}_e \end{bmatrix}. \tag{3.41}$$

Hence

$$\mathbf{I} = y_{11}\mathbf{U} + y_{12}\mathbf{V}_e, \tag{3.42}$$

$$\mathbf{0} = y_{21}\mathbf{U} + y_{22}\mathbf{V}_e. \tag{3.43}$$

Obtaining $\mathbf{V}_e$ from the latter equation and substituting into the former (provided that y_{22} is non-singular):

$$\mathbf{I} = (y_{11} - y_{12}y_{22}^{-1}y_{21})\mathbf{U}, \tag{3.44}$$

i.e.

$$\boldsymbol{Y}_p = y_{11} - y_{12}y_{22}^{-1}y_{21} \tag{3.45}$$

is the admittance-parameter matrix of the $m \times n$-terminal element.

Calculation of characteristics of $m \times n$-terminal elements containing current-sources or voltage-sources

The admittance- and impedance-parameter matrices of $m \times n$-terminal elements containing current-sources or voltage-sources are determined by the application of one of the previously presented methods to the deactivated network.

To calculate the column matrix $\mathbf{I}_0$ of the short-circuit currents of an $m \times n$-terminal element containing sources, the terminations at each connection-point should be short-circuits (Fig. 3.9). The graph of this network is the same as the graph used to determine $\boldsymbol{Z}_p$. Let the same fundamental set of loops be used as for the calculation of $\boldsymbol{Z}_p$. The edge-impedance matrix $\boldsymbol{Z}$ and loop matrix $\boldsymbol{B}$ are identical to those of the previous calculation, so $\boldsymbol{Z}_B$ is also similar. The equation of loop-currents:

$$\boldsymbol{B}(\boldsymbol{Z}\mathbf{I}_g - \mathbf{U}_g) = \boldsymbol{Z}_B \mathbf{J}_0 \tag{3.46}$$

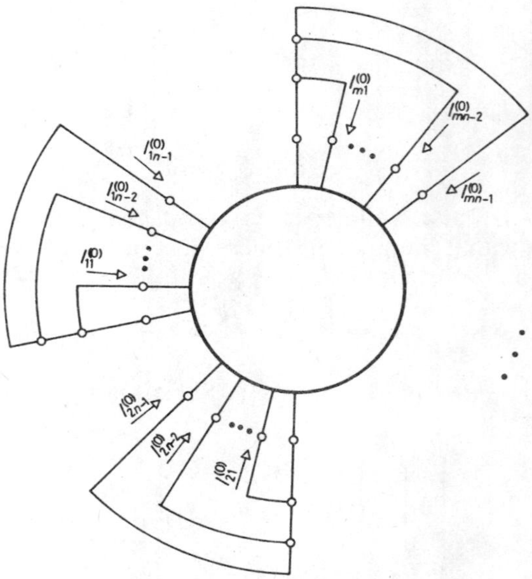

Fig. 3.9

differs from (3.30) owing to the difference in the networks, and $\mathbf{U}_g$ appearing here is also different from the column matrix of voltage-sources in (3.30). $\mathbf{J}_0$ may be obtained from (3.46) and its first $m(n-1)$ elements form the column matrix $\mathbf{I}_0$ of the short-circuit currents:

$$\mathbf{J}_0 = \begin{bmatrix} \mathbf{I}_0 \\ \mathbf{J}_e \end{bmatrix} = \boldsymbol{Z}_B^{-1} \boldsymbol{B}(\boldsymbol{Z}\mathbf{I}_g - \mathbf{U}_g). \tag{3.47}$$

The column matrix $\mathbf{U}_0$ of the open-circuit voltages is determined from the network terminated by open-circuits (Fig. 3.10). The graph of this network and the fundamental set of cutsets used for the calculation are identical to those involved in the calculation of $\boldsymbol{Y}_p$. Accordingly $\boldsymbol{Q}$ and $\boldsymbol{Y}_Q$ are also the same as the corresponding

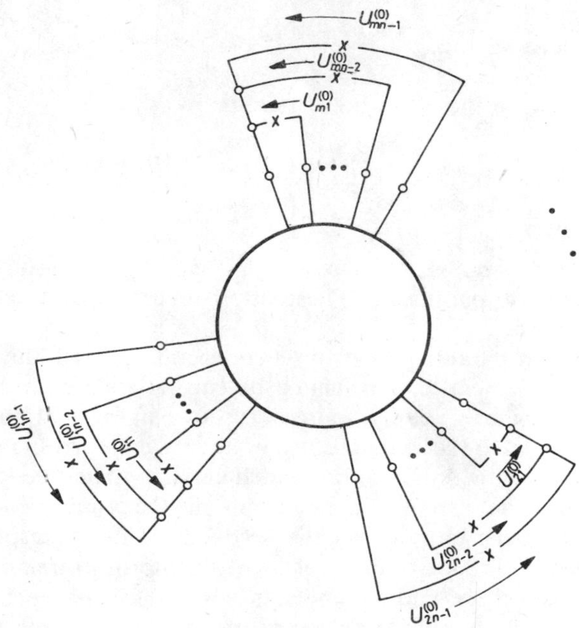

Fig. 3.10

matrices in the previous calculation. The equation of cutset-voltages for this case is:

$$\boldsymbol{Q}(\boldsymbol{Y}\mathbf{U}_g-\mathbf{I}_g)=\boldsymbol{Y}_Q\mathbf{V}_0\,, \tag{3.48}$$

where however $-\boldsymbol{Q}\mathbf{I}_g$ is not the same column matrix as in (3.39). $\mathbf{V}_0$ can be expressed from (3.48):

$$\mathbf{V}_0=\begin{bmatrix}\mathbf{U}_0\\ \mathbf{V}_e\end{bmatrix}=\boldsymbol{Y}_Q^{-1}\boldsymbol{Q}(\boldsymbol{Y}\mathbf{U}_g-\mathbf{I}_g), \tag{3.49}$$

and the column matrix $\mathbf{U}_0$ of the open-circuit voltages is formed by the first $m(n-1)$ elements of $\mathbf{V}_0$.

Calculation of hybrid-parameters of $2 \times n$-terminal elements

To determine the hybrid-parameters of $2 \times n$-terminal elements without sources the column matrices defined in (3.26) will be denoted for $m=2$ as follows:

$$\mathbf{U}_p=\mathbf{U}_1; \quad \mathbf{U}_s=\mathbf{U}_2; \quad \mathbf{I}_p=\mathbf{I}_1; \quad \mathbf{I}_s=\mathbf{I}_2. \tag{3.50}$$

The relationship between these can be written as

$$\begin{bmatrix}\mathbf{U}_p\\ \mathbf{I}_s\end{bmatrix}=\boldsymbol{H}\begin{bmatrix}\mathbf{I}_p\\ \mathbf{U}_s\end{bmatrix}=\begin{bmatrix}\boldsymbol{h}_{11} & \boldsymbol{h}_{12}\\ \boldsymbol{h}_{21} & \boldsymbol{h}_{22}\end{bmatrix}\begin{bmatrix}\mathbf{I}_p\\ \mathbf{U}_s\end{bmatrix}, \tag{3.51}$$

where $\boldsymbol{H}$ is the hybrid-parameter matrix, and $\boldsymbol{h}_{11}$, $\boldsymbol{h}_{12}$, $\boldsymbol{h}_{21}$, $\boldsymbol{h}_{22}$, square blocks of order $n-1$, are the hybrid-parameters. These latter are scalar quantities for $n=2$, i.e. in the case of two-ports.

To obtain the hybrid-parameter matrix, let connection-point 1, the primary side of the $2 \times n$-terminal element, be terminated by current-sources and connection-point 2, the secondary side by voltage-sources, as shown in Fig. 3.11. In the graph of the network so obtained the orientations of the edges corresponding to the terminations coincide with those of the branch-currents. The tree (forest) of the graph is chosen so that the current-sources terminating the primary side correspond to chords, and the edges terminating the secondary side correspond to tree-branches. The edges of the $2 \times n$-terminal element are chords or tree-branches with the admittance of each chord and the impedance of each tree-branch being finite. Thus the branches of the terminated network can be classified into four groups:

1. branches terminating the primary side (chords);
2. branches of the $2 \times n$-terminal element corresponding to chords;
3. branches of the $2 \times n$-terminal element corresponding to tree-branches;
4. branches terminating the secondary side (tree-branches).

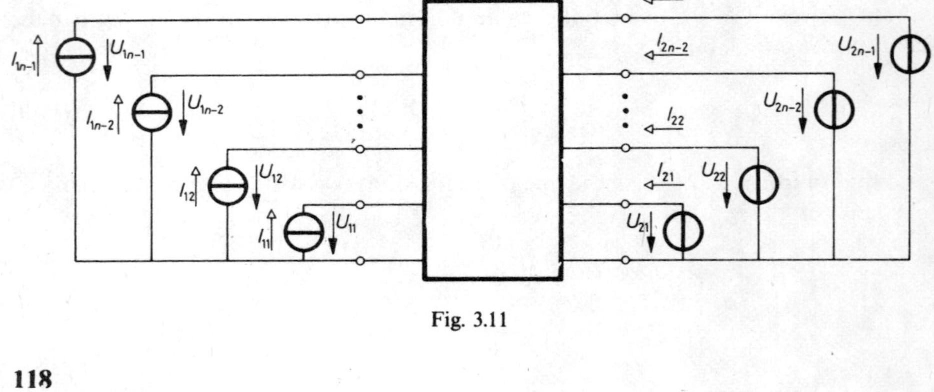

Fig. 3.11

The branches are numbered in the order of the groups, and the loop- and cutset-equations of the network are written with the aid of matrices partitioned in accordance with the classification:

$$\boldsymbol{B}\mathbf{U} = \begin{bmatrix} \mathbf{1} & \mathbf{0} & \boldsymbol{F}_{11} & \boldsymbol{F}_{21} \\ \mathbf{0} & \mathbf{1} & \boldsymbol{F}_{21} & \boldsymbol{F}_{22} \end{bmatrix} \begin{bmatrix} \mathbf{U}_1 \\ \mathbf{U}_2 \\ \mathbf{U}_3 \\ \mathbf{U}_4 \end{bmatrix} = \mathbf{0}, \tag{3.52}$$

$$\boldsymbol{Q}\mathbf{I} = \begin{bmatrix} -\boldsymbol{F}_{11}^{+} & -\boldsymbol{F}_{21}^{+} & \mathbf{1} & \mathbf{0} \\ -\boldsymbol{F}_{12}^{+} & -\boldsymbol{F}_{22}^{+} & \mathbf{0} & \mathbf{1} \end{bmatrix} \begin{bmatrix} \mathbf{I}_1 \\ \mathbf{I}_2 \\ \mathbf{I}_3 \\ \mathbf{I}_4 \end{bmatrix} = \mathbf{0}, \tag{3.53}$$

where $\mathbf{U}_1 = -\mathbf{U}_p$, $\mathbf{U}_4 = -\mathbf{U}_s$, $\mathbf{I}_1 = \mathbf{I}_p$, $\mathbf{I}_4 = \mathbf{I}_s$.

Taking these into account, from (3.52) and (3.53):

$$\mathbf{U}_p = \boldsymbol{F}_{11}\mathbf{U}_3 - \boldsymbol{F}_{12}\mathbf{U}_s, \tag{3.54}$$

$$\mathbf{U}_2 = -\boldsymbol{F}_{21}\mathbf{U}_3 + \boldsymbol{F}_{22}\mathbf{U}_s, \tag{3.55}$$

$$\mathbf{I}_3 = \boldsymbol{F}_{11}^{+}\mathbf{I}_p + \boldsymbol{F}_{21}^{+}\mathbf{I}_2, \tag{3.56}$$

$$\mathbf{I}_s = \boldsymbol{F}_{12}^{+}\mathbf{I}_p + \boldsymbol{F}_{22}^{+}\mathbf{I}_2, \tag{3.57}$$

If there is no coupling between branches in groups 2. and 3.:

$$\mathbf{I}_2 = \boldsymbol{Y}_2\mathbf{U}_2; \quad \mathbf{U}_3 = \boldsymbol{Z}_3\mathbf{I}_3, \tag{3.58}$$

where $\boldsymbol{Y}_2$ is the admittance-matrix of the branches in group 2., while $\boldsymbol{Z}_3$ is the impedance-matrix of the branches in group 3. Substituting these into (3.55) and (3.56):

$$\mathbf{I}_2 = -\boldsymbol{Y}_2\boldsymbol{F}_{21}\mathbf{U}_3 + \boldsymbol{Y}_2\boldsymbol{F}_{22}\mathbf{U}_s, \tag{3.59}$$

$$\mathbf{U}_3 = \boldsymbol{Z}_3\boldsymbol{F}_{11}^{+}\mathbf{I}_p + \boldsymbol{Z}_3\boldsymbol{F}_{21}^{+}\mathbf{I}_2. \tag{3.60}$$

$\mathbf{I}_2$ and $\mathbf{U}_3$ can hence be expressed in terms of $\mathbf{I}_p$ and $\mathbf{U}_s$. Substituting these into (3.54) and (3.57) the following is obtained:

$$\begin{aligned} \mathbf{U}_p = {} & \boldsymbol{F}_{11}\boldsymbol{Z}_3[\boldsymbol{F}_{11}^{+} - \boldsymbol{F}_{21}^{+}(\mathbf{1} + \boldsymbol{Y}_2\boldsymbol{F}_{21}\boldsymbol{Z}_3\boldsymbol{F}_{21}^{+})^{-1}\,\boldsymbol{Y}_2\boldsymbol{F}_{21}\boldsymbol{Z}_3\boldsymbol{F}_{11}^{+}]\mathbf{I}_p + \\ & + [\boldsymbol{F}_{11}\boldsymbol{Z}_3\boldsymbol{F}_{21}^{+}(\mathbf{1} + \boldsymbol{Y}_2\boldsymbol{F}_{21}\boldsymbol{Z}_3\boldsymbol{F}_{21}^{+})^{-1}\,\boldsymbol{Y}_2\boldsymbol{F}_{22} - \boldsymbol{F}_{12}]\mathbf{U}_s, \end{aligned} \tag{3.61}$$

$$\begin{aligned} \mathbf{I}_s = {} & [\boldsymbol{F}_{12}^{+} - \boldsymbol{F}_{22}^{+}(\mathbf{1} + \boldsymbol{Y}_2\boldsymbol{F}_{21}\boldsymbol{Z}_3\boldsymbol{F}_{21}^{+})^{-1}\,\boldsymbol{Y}_2\boldsymbol{F}_{21}\boldsymbol{Z}_3\boldsymbol{F}_{11}^{+}]\mathbf{I}_p + \\ & + \boldsymbol{F}_{22}^{+}(\mathbf{1} + \boldsymbol{Y}_2\boldsymbol{F}_{21}\boldsymbol{Z}_3\boldsymbol{F}_{21}^{+})^{-1}\,\boldsymbol{Y}_2\boldsymbol{F}_{22}\mathbf{U}_s. \end{aligned} \tag{3.62}$$

By comparison with (3.51) these can be seen to be the hybrid-parameter equations.

Transfer-function matrix

Applying specific source-voltage or source-current excitations to the connection points of a linear n-terminal element without sources, causes responses (voltages or currents) to appear on the terminals of the n-terminal element. The relationship between the excitations and certain selected responses is to be determined. The responses sought are solely voltages or currents.

The column matrix of excitations is denoted by $\mathbf{U}_1$ or $\mathbf{I}_1$ and that of the responses by $\mathbf{U}_2$ or $\mathbf{I}_2$. Thus the transfer-impedance matrix $\boldsymbol{Z}_t$, transfer-admittance matrix $\boldsymbol{Y}_t$, the voltage-transfer matrix $\boldsymbol{W}_U$ and the current-transfer matrix $\boldsymbol{W}_I$ are defined by the following relations:

$$\mathbf{U}_2 = \boldsymbol{Z}_t \mathbf{I}_1; \quad \mathbf{I}_2 = \boldsymbol{Y}_t \mathbf{U}_1; \quad \mathbf{U}_2 = \boldsymbol{W}_U \mathbf{U}_1; \quad \mathbf{I}_2 = \boldsymbol{W}_I \mathbf{I}_1 . \tag{3.63}$$

The number of elements in the column matrices of excitations and responses is not necessarily equal. If the number of elements is identical the transfer matrix is square and in general the number of columns equals the number of elements in the column matrix of excitations and the number of rows those in the column matrix of responses.

In case of one excitation and one response, the transfer-function matrix has one element, a scalar quantity, which is called the transfer function (or, occasionally, transfer-coefficient).

In the following the branches with the excitations will be referred to as primary branches and those with the responses as secondary branches.

The application of the loop-impedance matrix

First, the loop-impedance matrix will be used in our analysis.

Suppose $n-1$ voltage-sources are connected to the input terminals as shown in Fig. 3.2, and let the graph of the network so terminated be drawn. The edges of the graph are numbered, the primary edges being assigned order numbers $1, 2, \ldots, n-1 = p$ (p is the number of primary edges), and the secondary edges order numbers $p+1$, $p+2$, ..., $p+s$ (s is the number of secondary edges), all with orientations coinciding with those of edge-currents. The order numbers and orientations of the remaining edges are arbitrary. A tree (forest) is chosen with the primary and secondary edges being chords. The method presented below is valid provided that the above choice of a tree is possible.

In the fundamental set of loops specified by the tree (forest) chosen as above the numbering of the loops is arranged so as to have the first p loops contain the primary edges and loops $p+1, p+2, \ldots, p+s$ the secondary edges in the order of the edge-numbering. The orientations of the loops on these edges coincide with those of the edges. Let the basis loop-matrix $\boldsymbol{B}$ of the fundamental set of loops and the branch-impedance matrix $\boldsymbol{Z}$ of the network be written. From these the loop-impedance matrix Z_B can be determined. Applying the method of loop-currents:

$$-\boldsymbol{B}\mathbf{U}_g = Z_B\mathbf{J}\,. \tag{3.64}$$

The first p elements of the column matrix of loop-currents yields $\mathbf{I}_1$, while the following s elements $\mathbf{I}_2$. Thus $\mathbf{J}$ can be written in the form

$$\mathbf{J} = \begin{bmatrix} \mathbf{I}_1 \\ \mathbf{I}_2 \\ \mathbf{J}_e \end{bmatrix}. \tag{3.65}$$

The column matrix $\mathbf{U}_1$ is formed by the first p elements of the column matrix $-\boldsymbol{B}\mathbf{U}_g$ and zeros for the remaining elements. Let this column matrix be partitioned as follows:

$$-\boldsymbol{B}\mathbf{U}_g = \begin{bmatrix} \mathbf{U}_1 \\ \mathbf{0} \\ \mathbf{0} \end{bmatrix}, \tag{3.66}$$

where the number of elements in the first $\mathbf{0}$ column matrix is s. Thus (3.64) can be written as

$$\begin{bmatrix} \mathbf{U}_1 \\ \mathbf{0} \\ \mathbf{0} \end{bmatrix} = Z_B \begin{bmatrix} \mathbf{I}_1 \\ \mathbf{I}_2 \\ \mathbf{J}_e \end{bmatrix}. \tag{3.67}$$

Partitioning Z_B^{-1}:

$$Z_B^{-1} = \begin{bmatrix} y_{11} & y_{12} & y_{13} \\ y_{21} & y_{22} & y_{23} \\ y_{31} & y_{32} & y_{33} \end{bmatrix}, \tag{3.68}$$

where y_{11} and y_{22} are square matrices of orders p and s respectively. Thus according to (3.67):

$$\begin{bmatrix} \mathbf{I}_1 \\ \mathbf{I}_2 \\ \mathbf{J}_e \end{bmatrix} = \begin{bmatrix} y_{11} & y_{12} & y_{13} \\ y_{21} & y_{22} & y_{23} \\ y_{31} & y_{32} & y_{33} \end{bmatrix} \begin{bmatrix} \mathbf{U}_1 \\ \mathbf{0} \\ \mathbf{0} \end{bmatrix}. \tag{3.69}$$

Hence:

$$\mathbf{I}_1 = y_{11}\mathbf{U}_1 = Y_1\mathbf{U}_1 \quad \text{and} \quad \mathbf{I}_2 = y_{21}\mathbf{U}_1 = Y_t\mathbf{U}_1\,, \tag{3.70}$$

i.e. y_{11} is the input and y_{21} the transfer admittance matrix. If the relationship

$$\mathbf{I}_2 = Y_2\mathbf{U}_2 \tag{3.71}$$

holds between the currents and voltages of the secondary side, then (3.70) yields:

$$\boldsymbol{Y}_2 \mathbf{U}_2 = \boldsymbol{y}_{21} \mathbf{U}_1 , \tag{3.72}$$

and thus

$$\mathbf{U}_2 = \boldsymbol{Y}_2^{-1} \boldsymbol{y}_{21} \mathbf{U}_1 = \boldsymbol{W}_U \mathbf{U}_1 , \tag{3.73}$$

i.e.

$$\boldsymbol{W}_U = \boldsymbol{Y}_2^{-1} \boldsymbol{y}_{21} \tag{3.74}$$

is the voltage-transfer matrix.

The application of the cutset-admittance matrix

The cutset-admittance matrix can also be used for the determination of transfer matrices. In this case $p = n - 1$ current-sources are connected to the input terminals as shown in Fig. 3.3. In the graph of the network thus derived the edges are numbered to have the first p of them to correspond to primary edges, and the edges $p+1, p+2, \ldots, p+s$ to secondary edges with orientations coinciding with those of edge-currents. A tree of the network is chosen with primary and secondary edges being tree-branches. In the following it is assumed that such choice is possible.

The numbering of edges in the set of cutsets generated by the tree thus chosen is carried out to have the first $p+s$ cutsets contain the tree-branches of the same order numbers with orientations opposite to these edges. The order numbers and orientations of further cutsets are arbitrary. The cutset-admittance matrix $\boldsymbol{Y}_Q$ can be determined with the aid of the basis cutset-matrix $\boldsymbol{Q}$ of the fundamental set of cutsets, and the branch-admittance matrix $\boldsymbol{Y}$ of the network. According to Eq. (2.108) of chapter 2:

$$-\boldsymbol{Q}\mathbf{I}_g = \boldsymbol{Y}_Q \mathbf{V}_Q . \tag{3.75}$$

The first p elements of $\mathbf{V}_Q$ form the column matrix $\mathbf{U}_1$ and its further s elements $\mathbf{U}_2$. Thus:

$$\mathbf{V}_Q = \begin{bmatrix} \mathbf{U}_1 \\ \mathbf{U}_2 \\ \mathbf{V}_e \end{bmatrix} . \tag{3.76}$$

The column matrix $-\boldsymbol{Q}\mathbf{I}_g$ is accordingly partitioned:

$$-\boldsymbol{Q}\mathbf{I}_g = \begin{bmatrix} \mathbf{I}_1 \\ \mathbf{0} \\ \mathbf{0} \end{bmatrix} , \tag{3.77}$$

where the number of elements in $\mathbf{I}_1$ is p, and that in the first $\mathbf{0}$ is s. Substituting (3.76) and (3.77) into (3.75):

$$\begin{bmatrix} \mathbf{I}_1 \\ \mathbf{0} \\ \mathbf{0} \end{bmatrix} = Y_Q \begin{bmatrix} \mathbf{U}_1 \\ \mathbf{U}_2 \\ \mathbf{V}_e \end{bmatrix}, \tag{3.78}$$

The column matrix of cutset-voltages can be expressed from here. To this end Y_Q^{-1} is partitioned:

$$Y_Q^{-1} = \begin{bmatrix} z_{11} & z_{12} & z_{13} \\ z_{21} & z_{22} & z_{23} \\ z_{31} & z_{32} & z_{33} \end{bmatrix}, \tag{3.79}$$

where z_{11} and z_{22} are square matrices of orders p and s respectively. Thus from (3.78):

$$\begin{bmatrix} \mathbf{U}_1 \\ \mathbf{U}_2 \\ \mathbf{V}_e \end{bmatrix} = \begin{bmatrix} z_{11} & z_{12} & z_{13} \\ z_{21} & z_{22} & z_{23} \\ z_{31} & z_{32} & z_{33} \end{bmatrix} \begin{bmatrix} \mathbf{I}_1 \\ \mathbf{0} \\ \mathbf{0} \end{bmatrix}, \tag{3.80}$$

and so

$$\mathbf{U}_1 = z_{11}\mathbf{I}_1 = Z_1\mathbf{I}_1; \quad \mathbf{U}_2 = z_{21}\mathbf{I}_1 = Z_t\mathbf{I}_1, \tag{3.81}$$

i.e. z_{11} is the input and z_{21} is the transfer impedance matrix. If

$$\mathbf{U}_2 = Z_2\mathbf{I}_2, \tag{3.82}$$

then according to (3.81):

$$Z_2\mathbf{I}_2 = Z_t\mathbf{I}_1, \tag{3.83}$$

i.e.

$$\mathbf{I}_2 = Z_2^{-1}Z_t\mathbf{I}_1, \tag{3.84}$$

or

$$W_I = Z_2^{-1}Z_t = Y_2 Z_t \tag{3.85}$$

is the current transfer matrix.

Thus, the transfer matrices defined in (3.63) have now all been determined.

Examples

In the following section the methods presented for the determination of network characteristics will be illustrated by a few examples.

1. To calculate the open-circuit impedance matrix of the three-terminal network shown in Fig. 3.12, let us terminate the network by voltage-sources as shown in Fig. 3.13. In the graph of the network thus obtained (Fig. 3.14, a) a tree is chosen (Fig.

3.14, b) with the voltage-sources corresponding to chords. The matrix of the fundamental set of loops generated by the chosen tree is:

$$B=\begin{bmatrix} 1 & 0 & 0 & 1 & -1 & 0 & 1 & 0 \\ 0 & 1 & 1 & 0 & -1 & 1 & 1 & 0 \\ 0 & 0 & 0 & 0 & 0 & -1 & -1 & 1 \end{bmatrix}.$$

The impedance-matrix of the network is:

$$Z=\langle 0\ 0\ Z_1\ Z_1\ Z_1\ Z_2\ Z_2\ Z_2 \rangle.$$

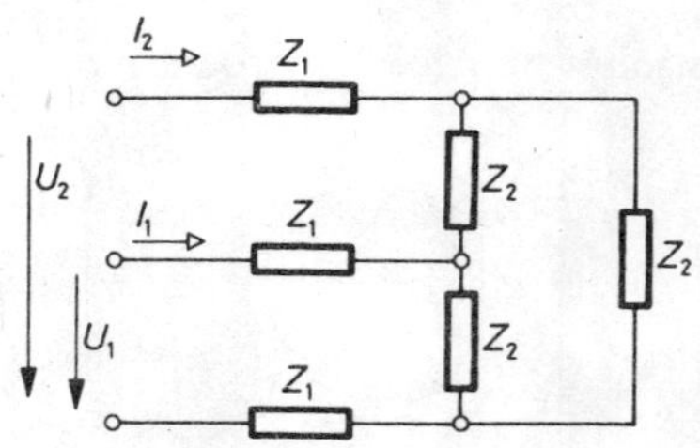

Fig. 3.12

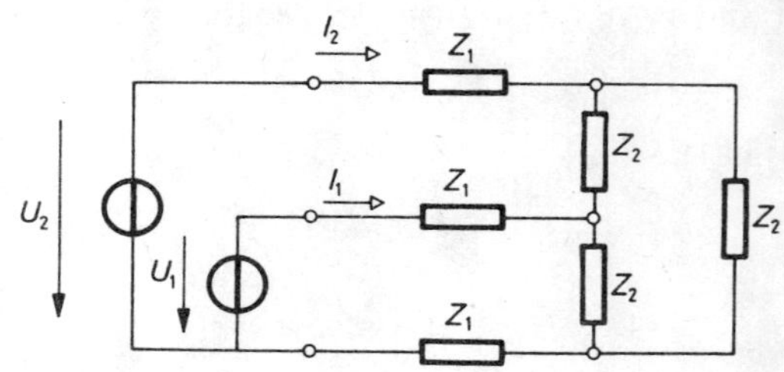

Fig. 3.13

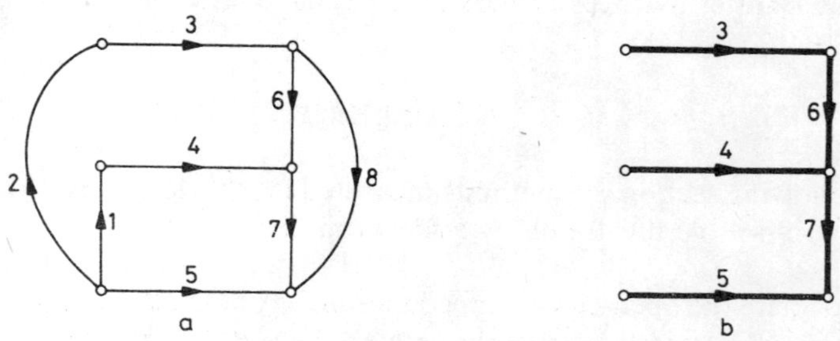

Fig. 3.14

The loop-impedance matrix from these is:

$$
Z_B = \left[\begin{array}{cc|c} 2Z_1+Z_2 & Z_1+Z_2 & -Z_2 \\ Z_1+Z_2 & 2(Z_1+Z_2) & -2Z_2 \\ \hline -Z_2 & -2Z_2 & 3Z_2 \end{array}\right],
$$

with the partitioning according to (3.7) indicated. (3.12) yields:

$$
Z_1 = \begin{bmatrix} 2Z_1+Z_2 & Z_1+Z_2 \\ Z_1+Z_2 & 2(Z_1+Z_2) \end{bmatrix} - \begin{bmatrix} Z_2 \\ 2Z_2 \end{bmatrix} [Z_2 2Z_2] \frac{1}{3Z_2} = \frac{3Z_1+Z_2}{3} \begin{bmatrix} 2 & 1 \\ 1 & 2 \end{bmatrix},
$$

which is the open-circuit impedance matrix of the three-terminal element.

2. The four-terminal element shown in Fig. 3.15 consists of self-impedances and mutual impedances. To calculate its open-circuit impedance matrix it is terminated by voltage-sources as shown in Fig. 3.16. A tree with the voltage-sources being chords is generated by edges 4, 5, 6 in the graph of the network (Fig. 3.17). The

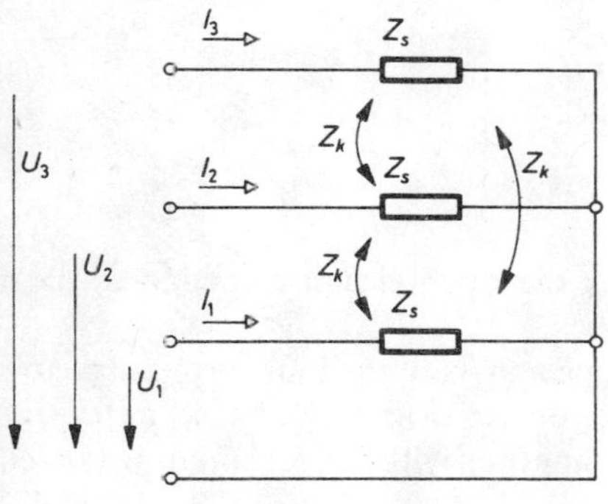

Fig. 3.15

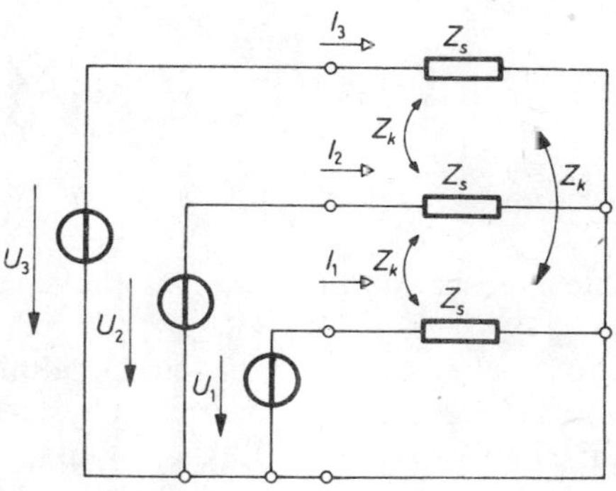

Fig. 3.16

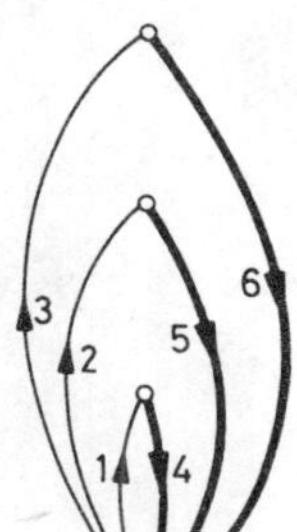

Fig. 3.17

matrix of the fundamental set of loops assigned to it is:

$$B=\begin{bmatrix} 1 & 0 & 0 & 1 & 0 & 0 \\ 0 & 1 & 0 & 0 & 1 & 0 \\ 0 & 0 & 1 & 0 & 0 & 1 \end{bmatrix}.$$

The impedance-matrix of the network is:

$$Z=\begin{bmatrix} 0 & 0 & 0 & 0 & 0 & 0 \\ 0 & 0 & 0 & 0 & 0 & 0 \\ 0 & 0 & 0 & 0 & 0 & 0 \\ 0 & 0 & 0 & Z_s & Z_k & Z_k \\ 0 & 0 & 0 & Z_k & Z_s & Z_k \\ 0 & 0 & 0 & Z_k & Z_k & Z_s \end{bmatrix}.$$

Thus the loop-impedance matrix is:

$$Z_B=\begin{bmatrix} Z_s & Z_k & Z_k \\ Z_k & Z_s & Z_k \\ Z_k & Z_k & Z_s \end{bmatrix},$$

which, being of third order, is the same as the open-circuit impedance matrix.

3. To determine the short-circuit admittance matrix of the four-terminal element shown in Fig. 3.18 it is terminated by current-sources (Fig. 3.19). A tree in the graph of the network (Fig. 3.20, a) is chosen for the analysis with the terminating current-

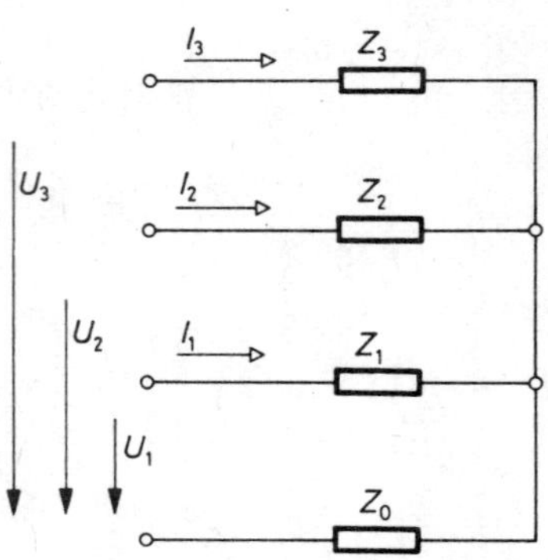

Fig. 3.18

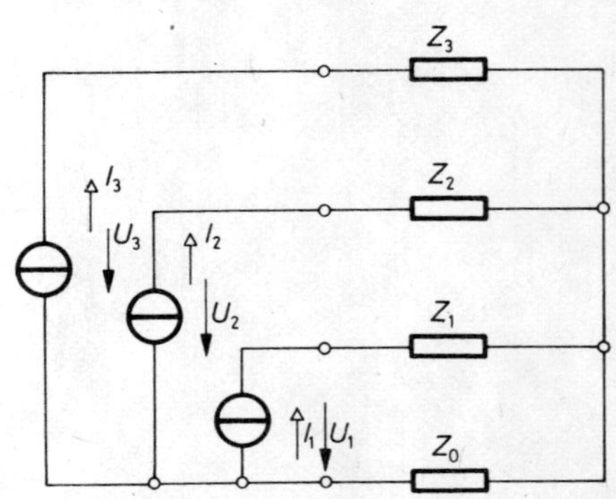

Fig. 3.19

sources being tree-branches (Fig. 3.20, b). The matrix of the fundamental set of cutsets generated by this tree is:

$$\boldsymbol{Q}=\begin{bmatrix} -1 & 0 & 0 & 0 & 1 & 0 & 0 \\ 0 & -1 & 0 & 0 & 0 & 1 & 0 \\ 0 & 0 & -1 & 0 & 0 & 0 & 1 \\ 0 & 0 & 0 & -1 & -1 & -1 & -1 \end{bmatrix}.$$

The admittance-matrix of the network is:

$$\boldsymbol{Y}=\langle 0\ 0\ 0\ Y_0\ Y_1\ Y_2\ Y_3\rangle,$$

where the notation $Y_i=1/Z_i$ ($i=0, 1, 2, 3$) has been employed. Thus the cutset-admittance matrix is:

$$\boldsymbol{Y}_Q=\left[\begin{array}{ccc|c} Y_1 & 0 & 0 & -Y_1 \\ 0 & Y_2 & 0 & -Y_2 \\ 0 & 0 & Y_3 & -Y_3 \\ \hline -Y_1 & -Y_2 & -Y_2 & Y_0+Y_1+Y_2+Y_3 \end{array}\right].$$

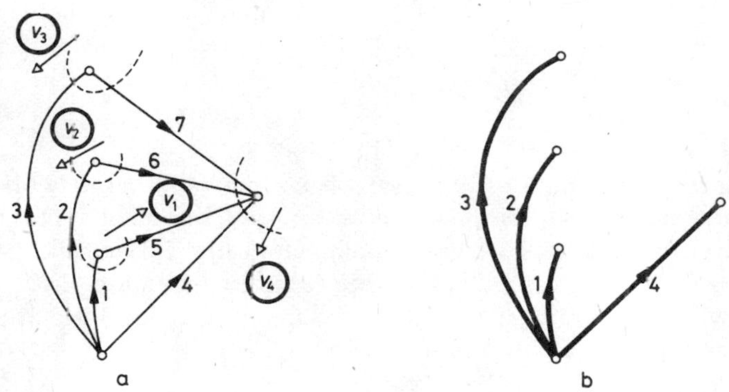

Fig. 3.20

The partitioning of (3.16) has been indicated by dashed lines. So according to (3.21)

$$\boldsymbol{Y}_1=\begin{bmatrix} Y_1 & 0 & 0 \\ 0 & Y_2 & 0 \\ 0 & 0 & Y_3 \end{bmatrix}-\frac{1}{Y_0+Y_1+Y_2+Y_3}\begin{bmatrix} Y_1 \\ Y_2 \\ Y_3 \end{bmatrix}[Y_1\ Y_2\ Y_3]=$$

$$= \frac{1}{Y_0+Y_1+Y_2+Y_3}\begin{bmatrix} Y_1(Y_0+Y_2+Y_3) & -Y_1Y_2 & -Y_1Y_3 \\ -Y_1Y_2 & Y_2(Y_0+Y_1+Y_3) & -Y_2Y_3 \\ -Y_1Y_3 & -Y_2Y_3 & Y_3(Y_0+Y_1+Y_2) \end{bmatrix}$$

is the short-circuit admittance matrix of the four-terminal element.

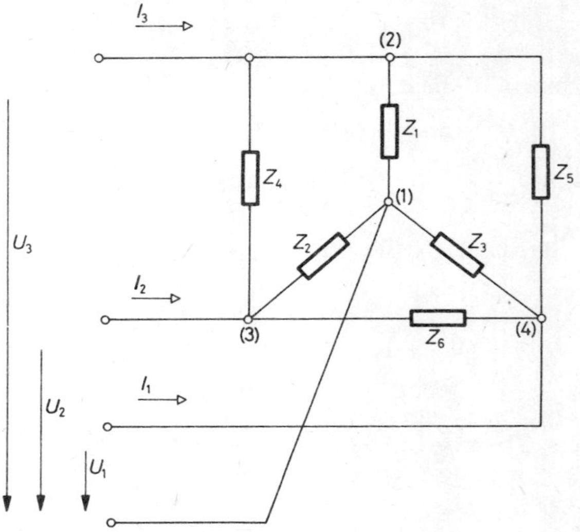

Fig. 3.21

4. The short-circuit admittance matrix of the four-terminal network shown in Fig. 3.21 is now calculated. To this end the network is terminated by current-sources in accordance with Fig. 3.22. Its graph is shown in Fig. 3.23, a. The set of cutsets generated by the tree in Fig. 3.23, b is used for the calculation. Since the cutset-admittance matrix is of order three, it is the same as the short-circuit admittance matrix. Using the notation $Y_i = 1/Z_i$ $(i = 1, 2, \ldots, 6)$:

$$\boldsymbol{Y}_Q = \boldsymbol{Y}_1 = \begin{bmatrix} Y_3+Y_5+Y_6 & -Y_6 & -Y_5 \\ -Y_6 & Y_2+Y_4+Y_6 & -Y_4 \\ -Y_5 & -Y_4 & Y_1+Y_4+Y_5 \end{bmatrix}.$$

5. Let us now determine the open-circuit impedance matrix $\boldsymbol{Z}_1$ and the column matrix $\mathbf{U}_0$ of open-circuit voltages characterizing the four-terminal element shown in Fig. 3.24.

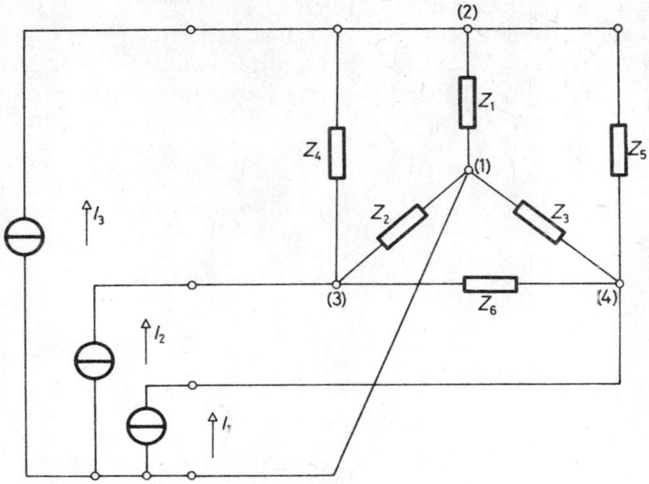

Fig. 3.22

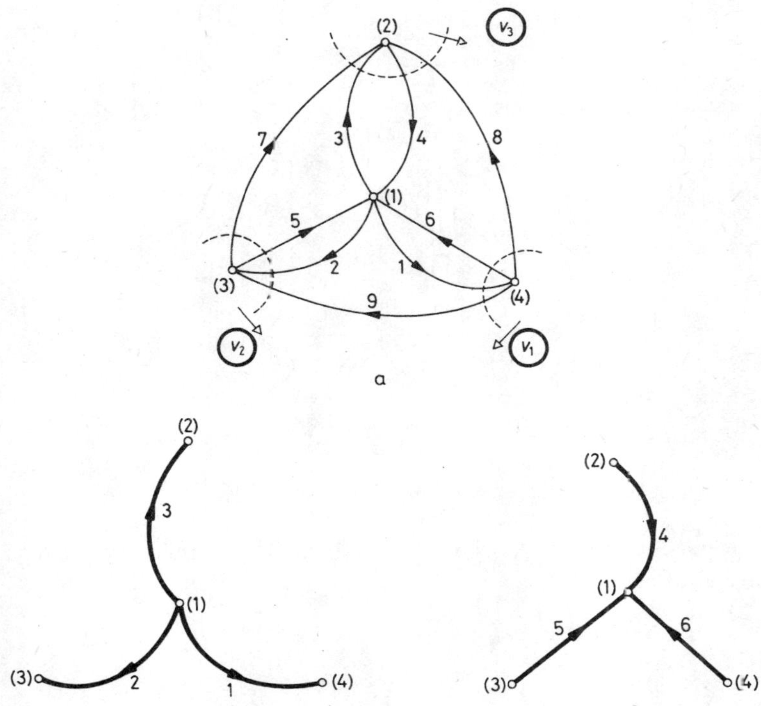

Fig. 3.23

To calculate the open-circuit impedance matrix the deactivated element is terminated by voltage-sources (Fig. 3.25). In the graph drawn in Fig. 3.23 a tree is chosen with the terminating voltage-sources corresponding to chords. Thus edges 4, 5, 6 are tree-branches (Fig. 3.23, c). The loop-impedance matrix of the network is:

$$Z_B = \left[\begin{array}{ccc|ccc} Z_1 & 0 & 0 & 0 & -Z_1 & -Z_1 \\ 0 & Z_1 & 0 & -Z_1 & 0 & Z_1 \\ 0 & 0 & Z_1 & Z_1 & Z_1 & 0 \\ \hline 0 & -Z_1 & Z_1 & 2Z_1+Z_2 & Z_1 & -Z_1 \\ -Z_1 & 0 & Z_1 & Z_1 & 2Z_1+Z_2 & Z_1 \\ -Z_1 & Z_1 & 0 & -Z_1 & Z_1 & 2Z_1+Z_2 \end{array}\right],$$

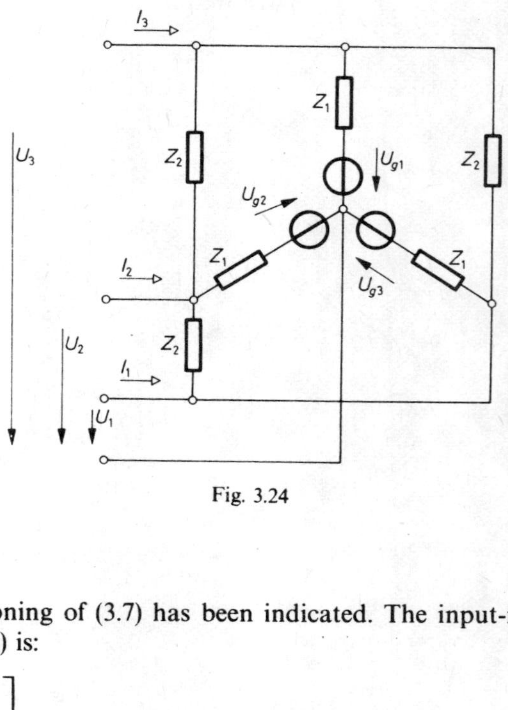

Fig. 3.24

where the partitioning of (3.7) has been indicated. The input-impedance matrix according to (3.12) is:

$$Z_1 = \begin{bmatrix} Z_1 & 0 & 0 \\ 0 & Z_1 & 0 \\ 0 & 0 & Z_1 \end{bmatrix} -$$

$$-\frac{Z_1^2}{Z_2(3Z_1+Z_2)}\begin{bmatrix}0 & -1 & -1\\ -1 & 0 & 1\\ 1 & 1 & 0\end{bmatrix}\begin{bmatrix}Z_1+Z_2 & -Z_1 & Z_1\\ -Z_1 & Z_1+Z_2 & -Z_1\\ Z_1 & -Z_1 & Z_1+Z_2\end{bmatrix}\begin{bmatrix}0 & -1 & 1\\ -1 & 0 & 1\\ -1 & 1 & 0\end{bmatrix}=$$

$$=\frac{Z_1}{3Z_1+Z_2}\begin{bmatrix}Z_1+Z_2 & Z_1 & Z_1\\ Z_1 & Z_1+Z_2 & Z_1\\ Z_1 & Z_1 & Z_1+Z_2\end{bmatrix}.$$

To determine the column matrix of open-circuit voltages the element is terminated by open-circuits (Fig. 3.26). For this network, with $Y_1=1/Z_1$ and $Y_2=1/Z_2$:

$$\boldsymbol{Y}=\langle 0\ 0\ 0\ Y_1\ Y_1\ Y_1\ Y_2\ Y_2\ Y_2\rangle.$$

In this case the short-circuit admittance matrix is the same as the cutset-admittance matrix, i.e. $\boldsymbol{Y}_Q^{-1}=\boldsymbol{Z}_1$, and thus the equation of cutset-voltages is:

$$\mathbf{U}_0=\begin{bmatrix}U_{10}\\ U_{20}\\ U_{30}\end{bmatrix}=\frac{Z_1}{3Z_1+Z_2}\begin{bmatrix}Z_1+Z_2 & Z_1 & Z_1\\ Z_1 & Z_1+Z_2 & Z_1\\ Z_1 & Z_1 & Z_1+Z_2\end{bmatrix}\begin{bmatrix}Y_1U_{g1}\\ Y_1U_{g2}\\ Y_1U_{g3}\end{bmatrix}=$$

$$=\frac{1}{3Z_1+Z_2}\begin{bmatrix}U_{g1}Z_2+(U_{g1}+U_{g2}+U_{g3})Z_1\\ U_{g2}Z_2+(U_{g1}+U_{g2}+U_{g3})Z_1\\ U_{g3}Z_2+(U_{g1}+U_{g2}+U_{g3})Z_1\end{bmatrix},$$

and this yields the column matrix of open-circuit voltages.

6. The three-terminal element shown in Fig. 3.27 can be characterized by its short-circuit admittance matrix and the column matrix of its short-circuit currents. To determine the short-circuit admittance matrix the deactivated element is terminated by current-sources (Fig. 3.28). In the graph of the network thus obtained edges 1, 2 corresponding to the terminating current-sources are chosen tree-branches (Fig. 3.29). The cutset-admittance matrix of the fundamental set of cutsets generated by the tree thus chosen is

$$\boldsymbol{Y}_Q=\begin{bmatrix}Y_1+Y_2 & -Y_2\\ -Y_2 & Y_2+Y_3\end{bmatrix},$$

where

$$Y_i=1/Z_i\quad (i=1,2,3)$$

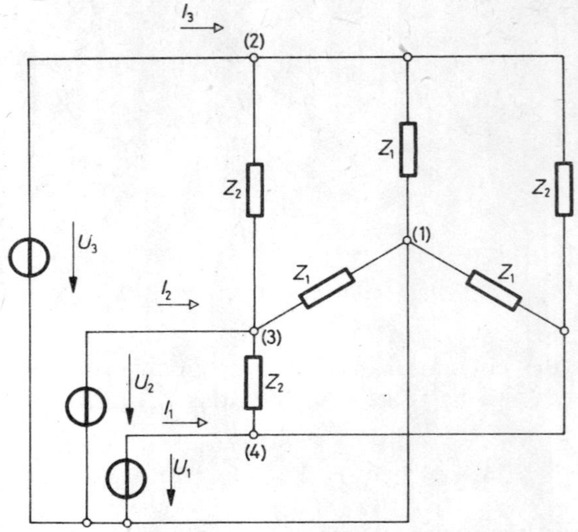

Fig. 3.25

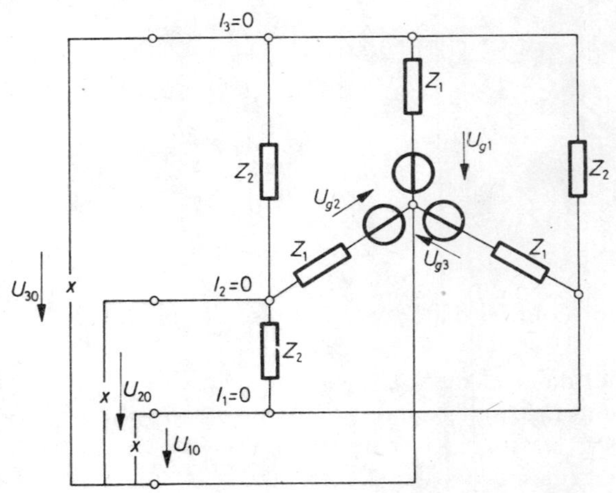

Fig. 3.26

and this is the same as the short-circuit admittance matrix. To calculate the short-circuit currents the equation of loop-currents is written for the network terminated by short-circuits (Fig. 3.30). A tree with edges 1 and 2 as chords is chosen for this, e.g. the one consisting of edges 3 and 5 (Fig. 3.29). The equation of loop-currents is:

$$\begin{bmatrix} I_{10} \\ I_{20} \\ J_{III} \end{bmatrix} = \begin{bmatrix} Z_1 & 0 & -Z_1 \\ 0 & Z_3 & Z_3 \\ -Z_1 & Z_3 & Z_1+Z_2+Z_3 \end{bmatrix}^{-1} \begin{bmatrix} -U_{g1} \\ 0 \\ U_{g1}+U_{g2} \end{bmatrix} =$$

$$= \begin{bmatrix} Y_1+Y_2 & -Y_2 & Y_2 \\ -Y_2 & Y_2+Y_3 & -Y_2 \\ Y_2 & -Y_2 & Y_2 \end{bmatrix} \begin{bmatrix} -U_{g1} \\ 0 \\ U_{g1}+U_{g2} \end{bmatrix},$$

i.e.

$$\mathbf{I}_0 = \begin{bmatrix} I_{10} \\ I_{20} \end{bmatrix} = \begin{bmatrix} -Y_1 U_{g1} + Y_2 U_{g2} \\ -Y_2 U_{g2} \end{bmatrix}.$$

7. To determine the impedance-parameter matrix of a Π-section (Fig. 3.31) voltage-sources are connected to the two-port (Fig. 3.32). The loop-impedance

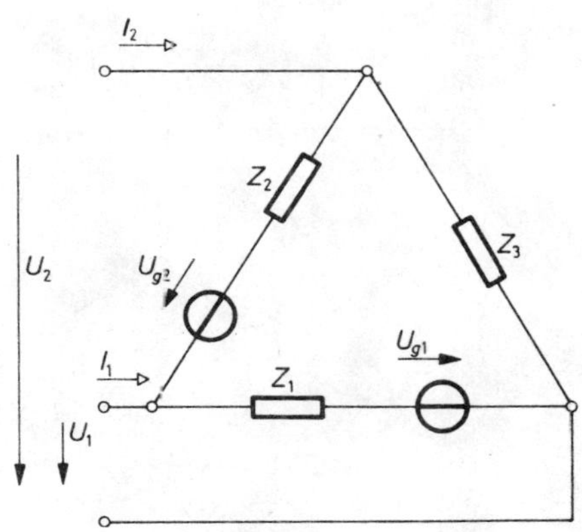

Fig. 3.27

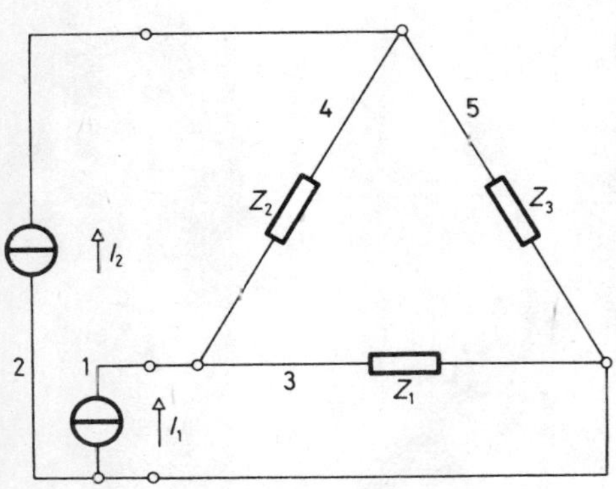

Fig. 3.28

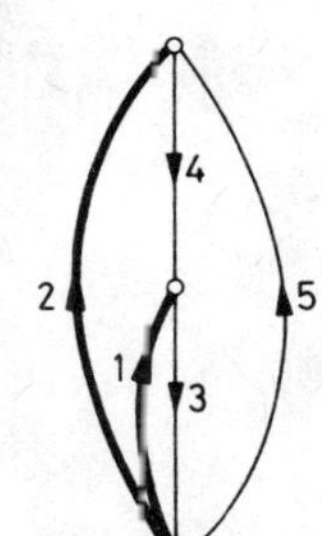

Fig. 3.29

matrix of the fundamental set of loops generated by the tree in Fig. 3.33 is:

$$\boldsymbol{Z}_B = \left[\begin{array}{cc|c} Z_1 & 0 & -Z_1 \\ 0 & Z_3 & Z_3 \\ \hline -Z_1 & Z_3 & Z_1+Z_2+Z_3 \end{array}\right].$$

Dashed lines indicate the partitioning of (3.32). Now according to (3.37):

$$\boldsymbol{Z}_p = \begin{bmatrix} Z_1 & 0 \\ 0 & Z_3 \end{bmatrix} - \frac{1}{Z_1+Z_2+Z_3}\begin{bmatrix} -Z_1 \\ Z_3 \end{bmatrix}[-Z_1 \ \ Z_3] =$$

$$= \frac{1}{Z_1+Z_2+Z_3}\begin{bmatrix} Z_1(Z_2+Z_3) & Z_1Z_3 \\ Z_1Z_3 & (Z_1+Z_2)Z_3 \end{bmatrix}$$

is the impedance-parameter matrix.

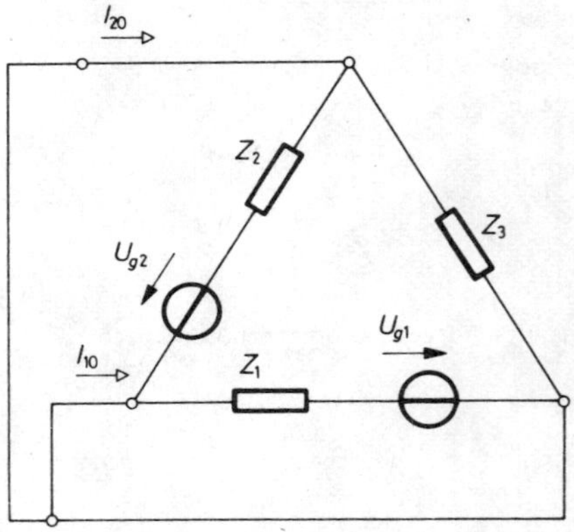

Fig. 3.30

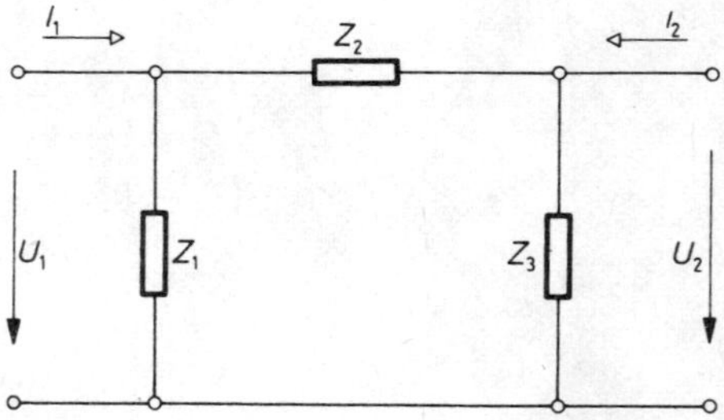

Fig. 3.31

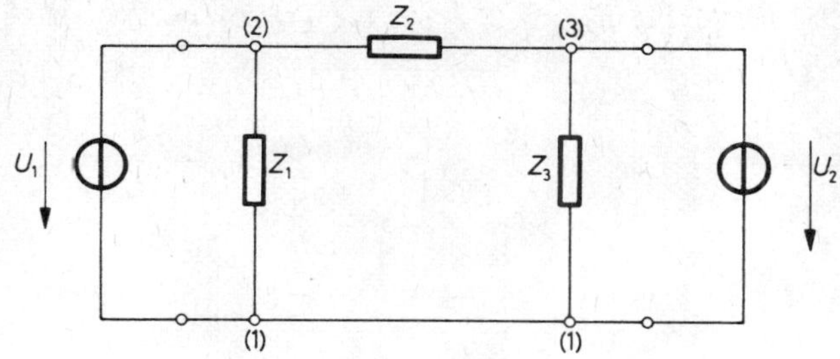

Fig. 3.32

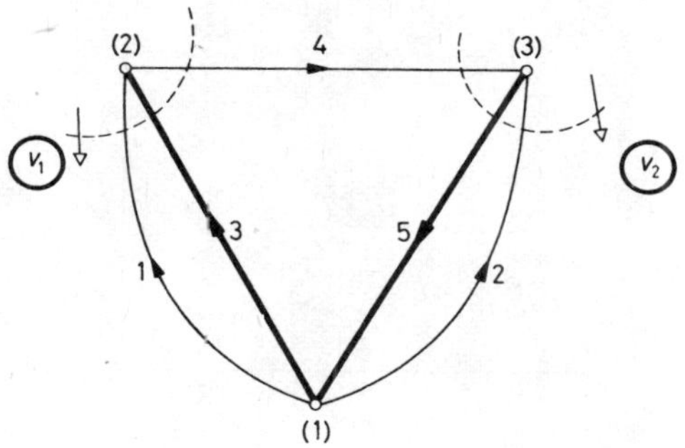

Fig. 3.33

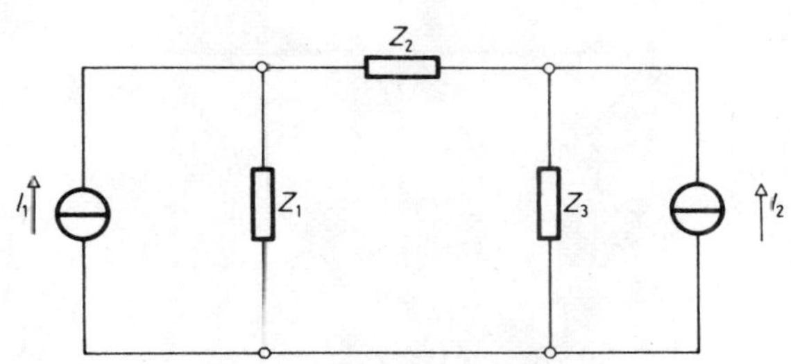

Fig. 3.34

To determine the admittance-parameter matrix, the Π-section is terminated by current-sources (Fig. 3.34). Edges 1 and 2 are chosen as tree-branches for the calculation (Fig. 3.33). The cutset-admittance matrix of the fundamental set of cutsets generated by this tree is at the same time the admittance-parameter matrix of the Π-section:

$$\boldsymbol{Y}_p = \boldsymbol{Y}_Q = \begin{bmatrix} Y_1 + Y_2 & -Y_2 \\ -Y_2 & Y_2 + Y_3 \end{bmatrix},$$

where the notation $Y_i = 1/Z_i$ $(i = 1, 2, 3)$ has been employed.

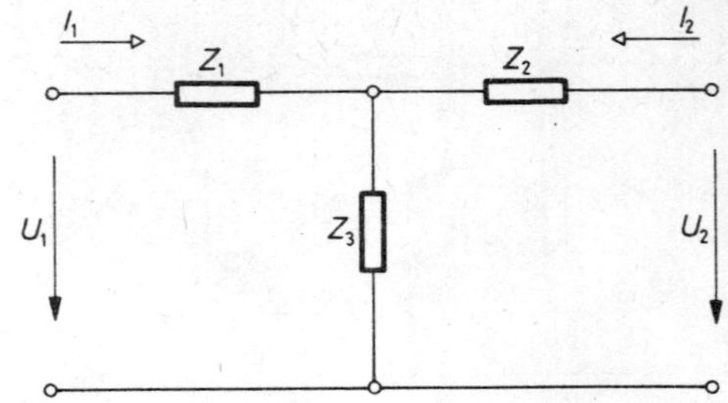

Fig. 3.35

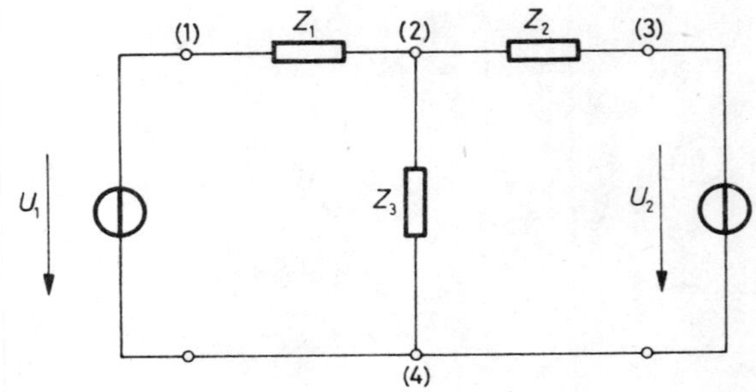

Fig. 3.36

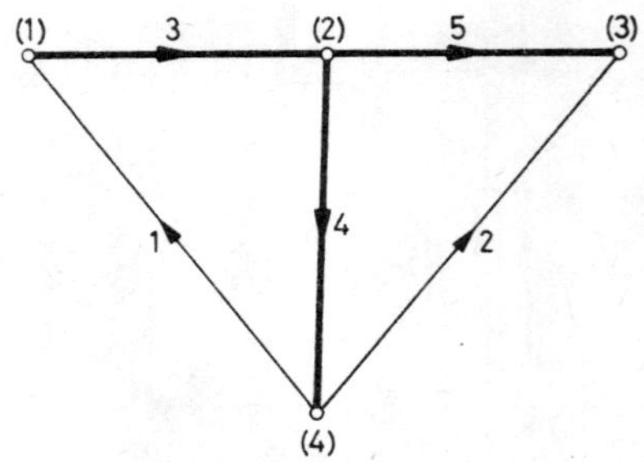

Fig. 3.37

8. The impedance-parameter matrix of a T-section (Fig. 3.35) is the same as the loop-impedance matrix of the two-port terminated by voltage-sources (Fig. 3.36), provided that the fundamental set of loops generated by the tree indicated in Fig. 3.37 is chosen, i.e.:

$$\mathbf{Z}_p = \mathbf{Z}_B = \begin{bmatrix} Z_1 + Z_3 & Z_3 \\ Z_3 & Z_2 + Z_3 \end{bmatrix}.$$

To calculate the admittance-parameter matrix, the T-section is terminated by current-sources (Fig. 3.38). The cutset-admittance matrix of the fundamental set of cutsets generated by the tree consisting of edges 1, 2 and 4 (Fig. 3.37) is:

$$Y_Q = \left[\begin{array}{cc|c} Y_1 & 0 & -Y_1 \\ 0 & Y_2 & -Y_2 \\ \hline -Y_1 & -Y_2 & Y_1+Y_2+Y_3 \end{array}\right],$$

where the partitioning of (3.40) has been indicated. By (3.45) this yields:

$$Y_p = \begin{bmatrix} Y_1 & 0 \\ 0 & Y_2 \end{bmatrix} - \frac{1}{Y_1+Y_2+Y_3}\begin{bmatrix} Y_1 \\ Y_2 \end{bmatrix}[Y_1 \ \ Y_2] =$$

$$= \frac{1}{Y_1+Y_2+Y_3}\begin{bmatrix} Y_1(Y_2+Y_3) & -Y_1 Y_2 \\ -Y_1 Y_2 & Y_2(Y_1+Y_3) \end{bmatrix}$$

the admittance-parameter matrix of the T-section.

9. To calculate the impedance-parameter matrix of the lattice-section shown in Fig. 3.39 it is terminated by voltage-sources (Fig. 3.40). In the graph of the network

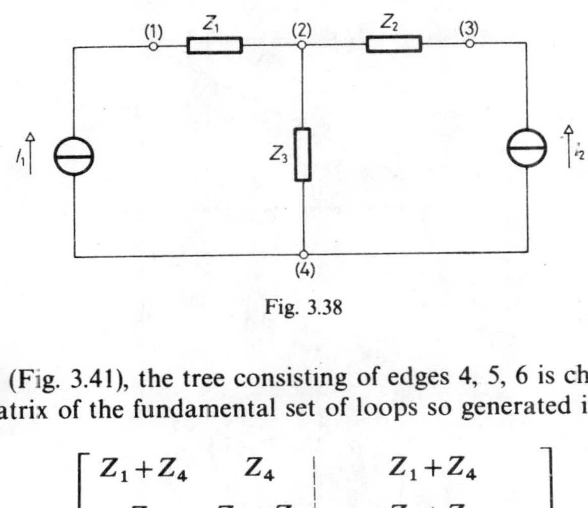

Fig. 3.38

thus obtained (Fig. 3.41), the tree consisting of edges 4, 5, 6 is chosen. The loop-impedance matrix of the fundamental set of loops so generated is:

$$Z_B = \left[\begin{array}{cc|c} Z_1+Z_4 & Z_4 & Z_1+Z_4 \\ Z_4 & Z_2+Z_4 & Z_2+Z_4 \\ \hline Z_1+Z_4 & Z_2+Z_4 & Z_1+Z_2+Z_3+Z_4 \end{array}\right],$$

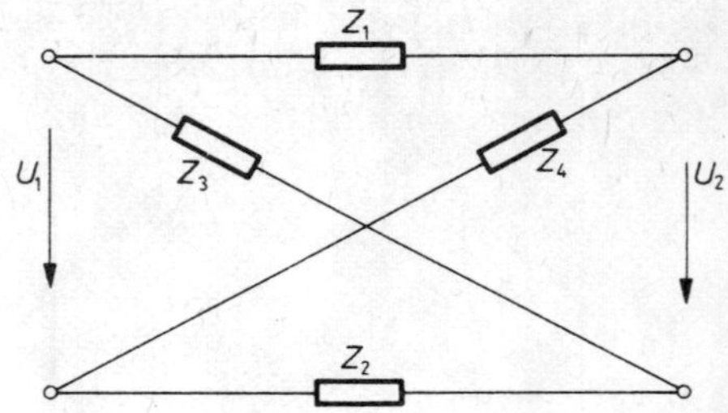

Fig. 3.39

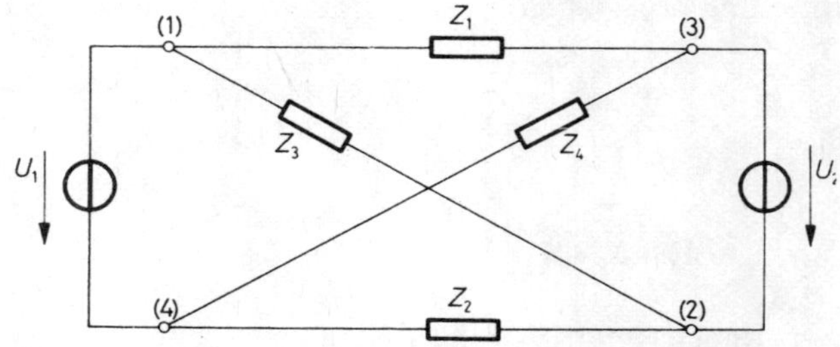

Fig. 3.40

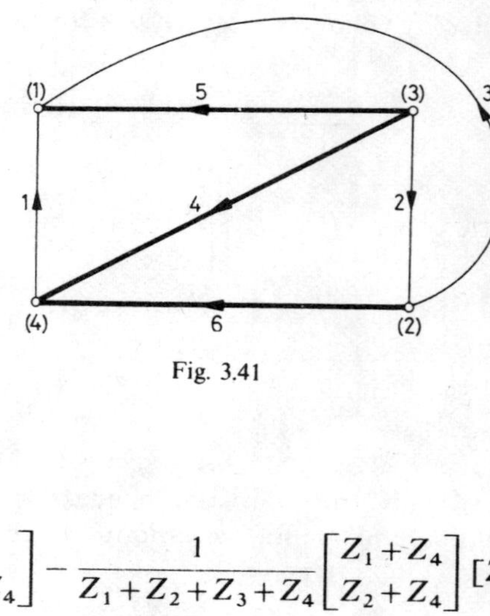

Fig. 3.41

and hence

$$\boldsymbol{Z}_p = \begin{bmatrix} Z_1+Z_4 & Z_4 \\ Z_4 & Z_2+Z_4 \end{bmatrix} - \frac{1}{Z_1+Z_2+Z_3+Z_4} \begin{bmatrix} Z_1+Z_4 \\ Z_2+Z_4 \end{bmatrix} [Z_1+Z_4 \;\; Z_2+Z_4] =$$

$$= \frac{1}{Z_1+Z_2+Z_3+Z_4} \begin{bmatrix} (Z_1+Z_4)(Z_2+Z_3) & Z_3Z_4-Z_1Z_2 \\ Z_3Z_4-Z_1Z_2 & (Z_1+Z_3)(Z_2+Z_4) \end{bmatrix}$$

is the impedance-parameter matrix of the two-port.

10. To calculate the admittance-parameters of the bridged T-section shown in Fig. 3.42 the network is terminated by current-sources (Fig. 3.43). In the graph of the terminated network (Fig. 3.44, a) the tree in Fig. 3.44, b is chosen. The fundamental set of cutsets generated by this tree has been indicated in Fig. 3.44, a. Its cutset-admittance matrix is:

$$\mathbf{Y}_Q = \left[\begin{array}{cc|c} Y_1+Y_4 & -Y_4 & -Y_1 \\ -Y_4 & Y_3+Y_4 & -Y_3 \\ \hline -Y_1 & -Y_3 & Y_1+Y_2+Y_3 \end{array}\right].$$

From here according to (3.45):

$$\mathbf{Y}_p = \begin{bmatrix} Y_1+Y_4 & -Y_4 \\ -Y_4 & Y_3+Y_4 \end{bmatrix} - \frac{1}{Y_1+Y_2+Y_3}\begin{bmatrix} Y_1 \\ Y_3 \end{bmatrix}[Y_1 \quad Y_3] =$$

$$= \frac{1}{Y_1+Y_2+Y_3}\begin{bmatrix} (Y_1+Y_4)(Y_1+Y_2+Y_3)-Y_1^2 & -(Y_1+Y_2+Y_3)Y_4-Y_1Y_3 \\ -(Y_1+Y_2+Y_3)Y_4-Y_1Y_3 & (Y_3+Y_4)(Y_1+Y_2+Y_3)-Y_3^2 \end{bmatrix}.$$

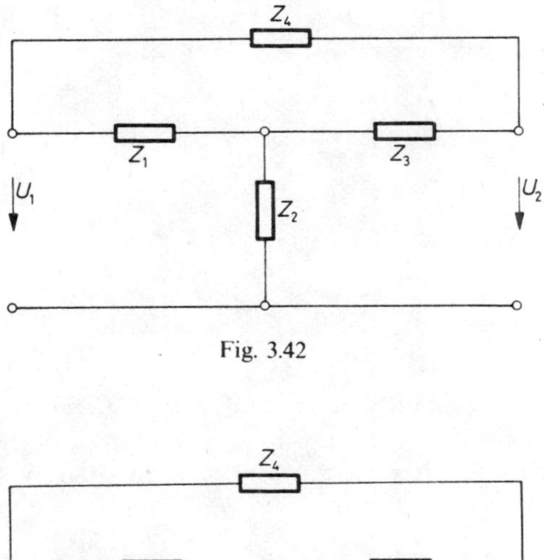

Fig. 3.42

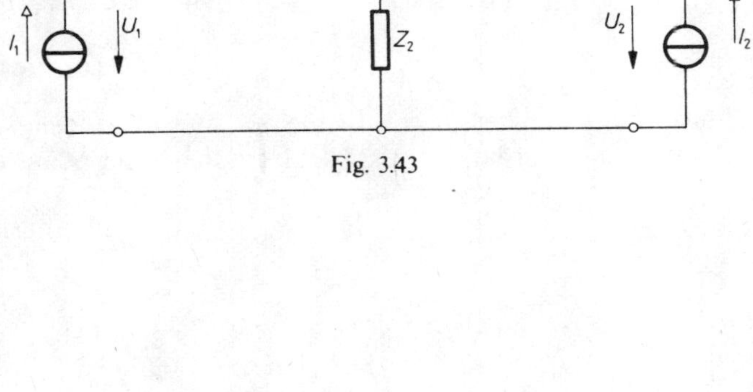

Fig. 3.43

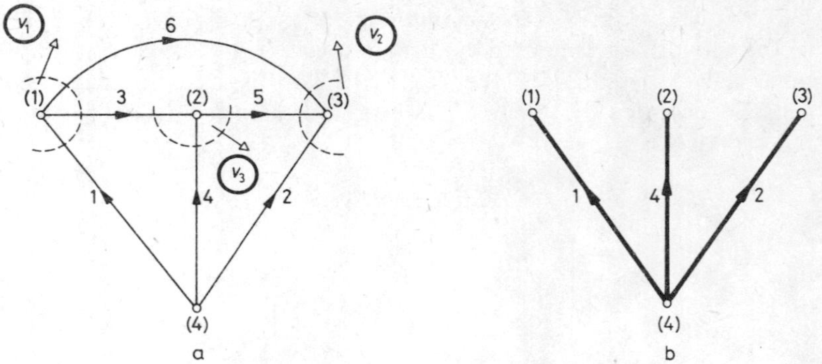

Fig. 3.44

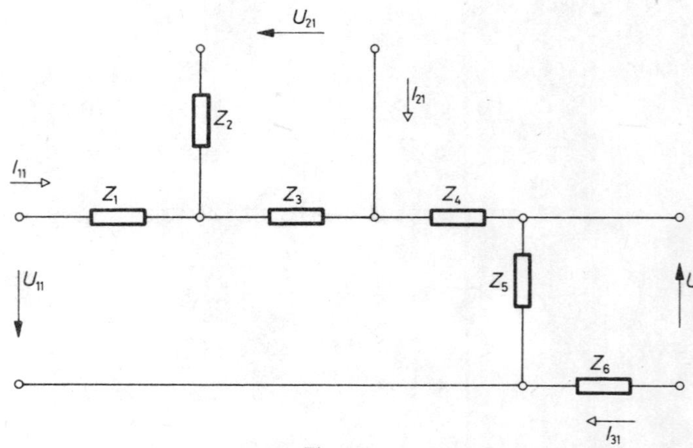

Fig. 3.45

11. The impedance-parameter matrix of the 3 × 2-terminal element (3-port) shown in Fig. 3.45 is given by the loop-impedance matrix of the fundamental set of loops chosen as explained when the network is terminated by voltage-sources (Figs 3.46, 3.47):

$$\boldsymbol{Z}_B = \boldsymbol{Z}_p = \begin{bmatrix} Z_1+Z_3+Z_4+Z_5 & -Z_3 & -Z_5 \\ -Z_3 & Z_2+Z_3 & 0 \\ -Z_5 & 0 & Z_5+Z_6 \end{bmatrix}.$$

12. The admittance-parameter matrix of the 2 × 3-terminal element drawn in Fig. 3.48 is calculated in accordance with (3.45) from the cutset-admittance matrix

$$
\boldsymbol{Y}_Q = \left[\begin{array}{cccc|c}
Y_1 & 0 & 0 & 0 & 0 \\
0 & Y_2 & 0 & 0 & -Y_1 \\
0 & 0 & Y_3 & 0 & 0 \\
0 & 0 & 0 & Y_3 & -Y_3 \\
\hline
0 & -Y_1 & 0 & -Y_3 & Y_1+Y_2+Y_3
\end{array}\right]
$$

of the set of cutsets indicated in Fig. 3.50 in the graph of the network terminated by current-sources (Fig. 3.49). Thus

$$
\boldsymbol{Y}_p = \frac{1}{Y_1+Y_2+Y_3}\begin{bmatrix}
Y_1(Y_1+Y_2+Y_3) & 0 & 0 & 0 \\
0 & Y_1(Y_2+Y_3) & 0 & -Y_1 Y_3 \\
0 & 0 & Y_3(Y_1+Y_2+Y_3) & 0 \\
0 & -Y_1 Y_3 & 0 & Y_3(Y_1+Y_2)
\end{bmatrix}
$$

is the admittance-parameter matrix.

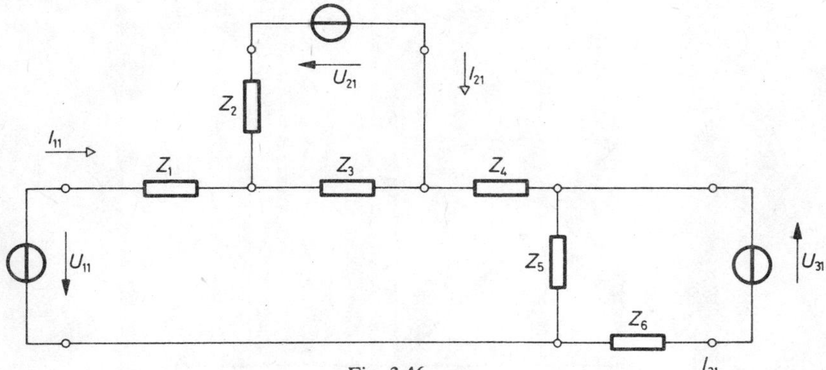

Fig. 3.46

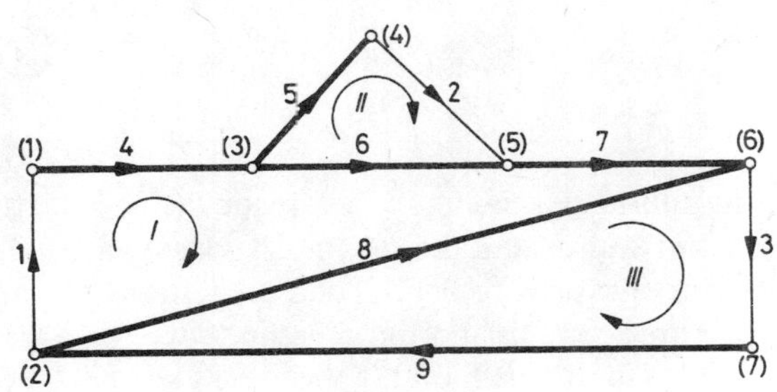

Fig. 3.47

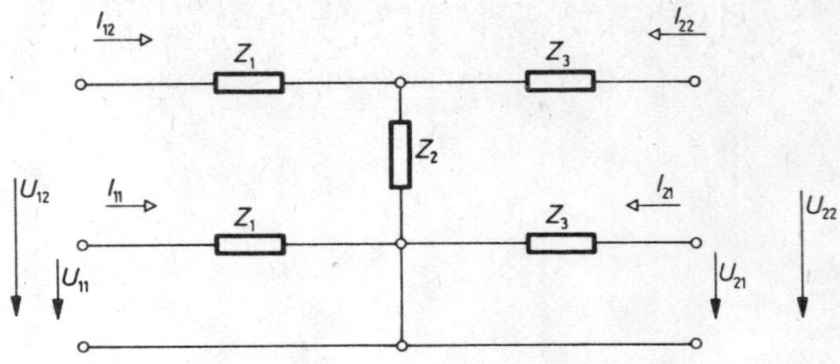

Fig. 3.48

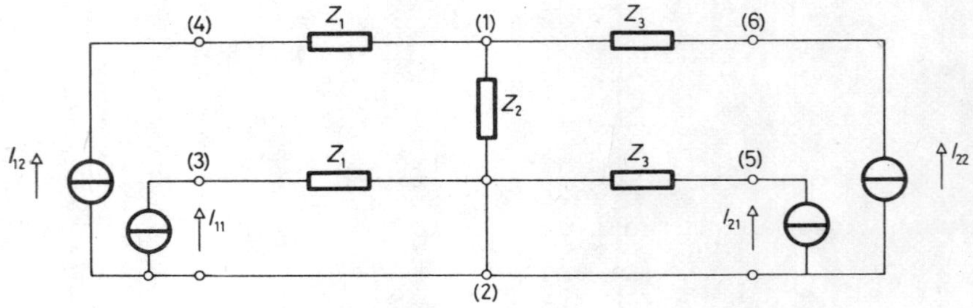

Fig. 3.49

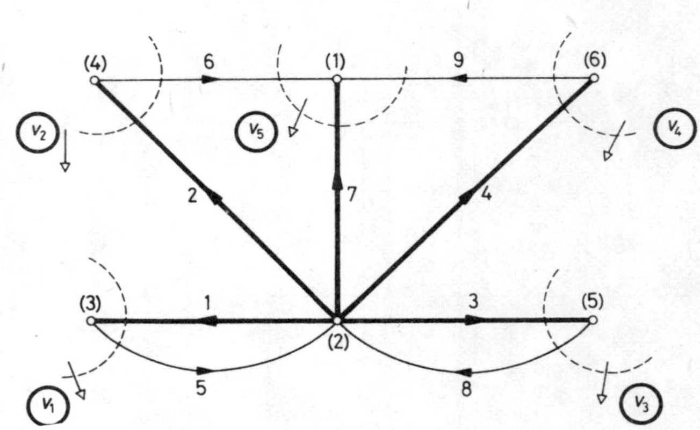

Fig. 3.50

13. To determine the hybrid-parameters of a T-section (Fig. 3.35) the primary side is terminated by a current-source and the secondary side by a voltage-source (Fig. 3.51). A tree of the network graph (Fig. 3.52) is chosen with the terminating voltage-source represented by a tree-branch and the current-source by a chord. The order number of the primary termination is 1, that of the other chord is 2. Edges 3, 4 are the tree-branches of the T-section, and 5 corresponds to the secondary termination.

The locp-matrix is

$$B=\left[\begin{array}{c|c|cc|c} 1 & 0 & 1 & -1 & 0 \\ \hline 0 & 1 & 0 & -1 & 1 \end{array}\right],$$

where the partitioning of (3.52) has been indicated. The immitance mat-ices necessary for the analysis are:

$$Y_2=1/Z_2=Y_2; \quad Z_3=\langle Z_1 \ \ Z_3\rangle.$$

From these and (3.61), (3.62):

$$h_{11}=[1 \ \ -1]\begin{bmatrix} Z_1 & 0 \\ 0 & Z_3 \end{bmatrix}\left\{\begin{bmatrix} 1 \\ -1 \end{bmatrix}-\begin{bmatrix} 0 \\ -1 \end{bmatrix}\left(1+\right.\right.$$

$$\left.\left.+Y_2[0 \ \ -1]\begin{bmatrix} Z_1 & 0 \\ 0 & Z_3 \end{bmatrix}\begin{bmatrix} 0 \\ 1 \end{bmatrix}\right)^{-1} Y_2[0 \ \ -1]\begin{bmatrix} Z_1 & 0 \\ 0 & Z_3 \end{bmatrix}\begin{bmatrix} -1 \\ -1 \end{bmatrix}\right\}=$$

$$=Z_1+\frac{Z_2 Z_3}{Z_2+Z_3},$$

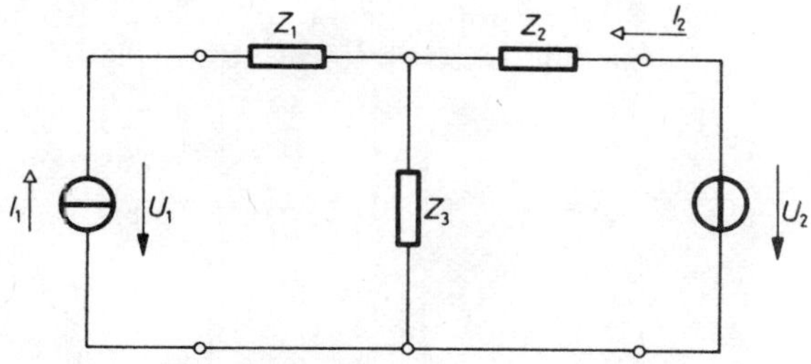

Fig. 3.51

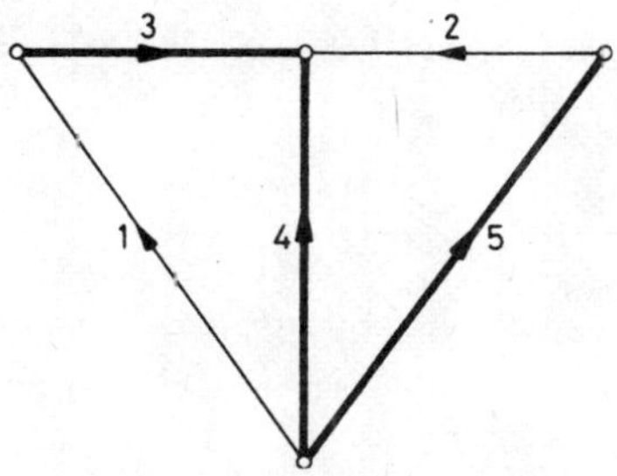

Fig. 3.52

$$h_{12}=[1\ \ -1]\begin{bmatrix} Z_1 & 0 \\ 0 & Z_3 \end{bmatrix}\begin{bmatrix} 0 \\ -1 \end{bmatrix}\left(1+Y_2[0\ \ -1]\begin{bmatrix} Z_1 & 0 \\ 0 & Z_3 \end{bmatrix}\begin{bmatrix} 0 \\ -1 \end{bmatrix}\right)^{-1} Y_2-0=$$

$$=\frac{Z_3}{Z_2+Z_3},$$

$$h_{21}=-\left(1+Y_2[0\ \ -1]\begin{bmatrix} Z_1 & 0 \\ 0 & Z_3 \end{bmatrix}\begin{bmatrix} 0 \\ -1 \end{bmatrix}\right)^{-1} Y_2[0\ \ -1]\begin{bmatrix} Z_1 & 0 \\ 0 & Z_3 \end{bmatrix}\begin{bmatrix} 1 \\ -1 \end{bmatrix}=$$

$$=-\frac{Z_3}{Z_2+Z_3},$$

$$h_{22}=\left(1+Y_2[0\ \ -1]\begin{bmatrix} Z_1 & 0 \\ 0 & Z_3 \end{bmatrix}\begin{bmatrix} 0 \\ -1 \end{bmatrix}\right)^{-1} Y_2=\frac{1}{Z_2+Z_3},$$

i.e.

$$H=\frac{1}{Z_2+Z_3}\begin{bmatrix} Z_1Z_2+Z_1Z_3+Z_2Z_3 & Z_3 \\ -Z_3 & 1 \end{bmatrix}$$

is the required hybrid matrix.

14. To calculate the input- and transfer admittances of the T-section terminated by an impedance Z_2 (Fig. 3.53) the fundamental set of loops generated by the tree consisting of edges 3, 4, 5 in the network graph (Fig. 3.54) is chosen. Its loop-

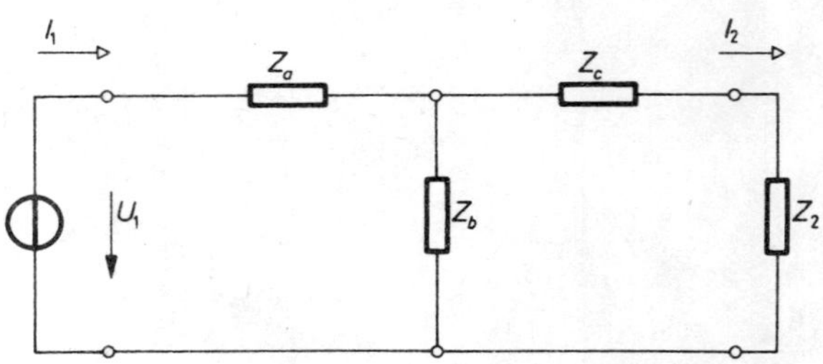

Fig. 3.53

impedance matrix is:

$$Z_B=\begin{bmatrix} Z_a+Z_b & -Z_b \\ -Z_b & Z_b+Z_c+Z_2 \end{bmatrix}.$$

Its inverse is:

$$Z_B^{-1}=\frac{1}{D}\begin{bmatrix} Z_b+Z_c+Z_2 & Z_b \\ Z_b & Z_a+Z_b \end{bmatrix},$$

where

$$D = Z_a(Z_b + Z_c + Z_2) + Z_b(Z_c + Z_2).$$

Thus according to (3.70) the input admittance

$$Y_1 = \frac{Z_b + Z_c + Z_2}{Z_a(Z_b + Z_c + Z_2) + Z_b(Z_c + Z_2)}$$

and the transfer admittance

$$Y_t = \frac{Z_b}{Z_a(Z_b + Z_c + Z_2) + Z_b(Z_c + Z_2)},$$

which are scalar quantities in this case.

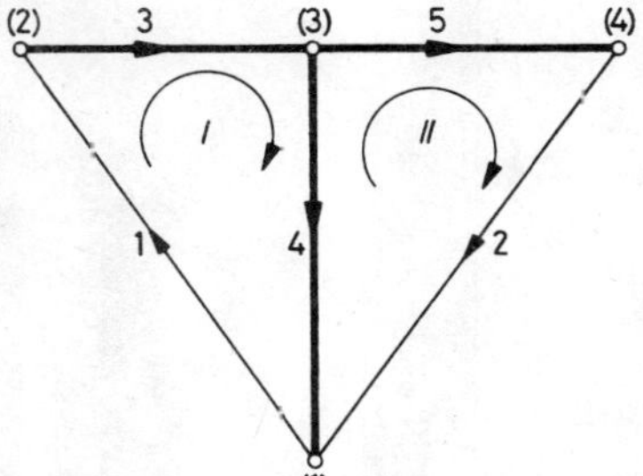

Fig. 3.54

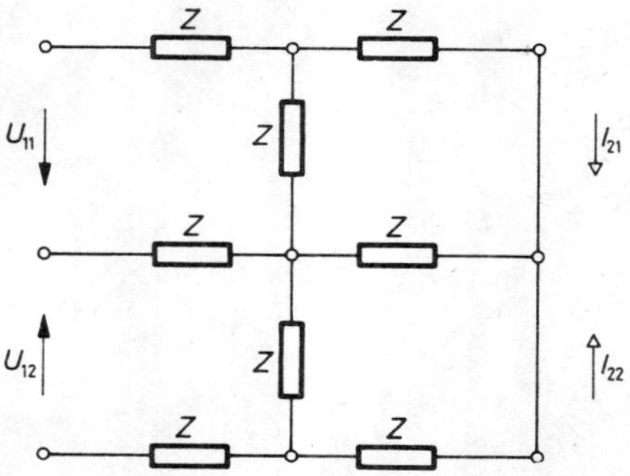

Fig. 3.55

15. To determine the transfer-admittance matrix of the 2 × 3-terminal element terminated by short-circuits shown in Fig. 3.55 the fundamental set of loops generated by the tree drawn in Fig. 3.56, b of the network graph (Fig. 3.56, a) has been chosen. The loop-impedance matrix is:

$$\boldsymbol{Z}_B = \begin{bmatrix} 3Z & Z & -Z & 0 \\ Z & 3Z & 0 & -Z \\ -Z & 0 & 3Z & Z \\ 0 & -Z & Z & 3Z \end{bmatrix}.$$

Its inverse is of the form

$$\boldsymbol{Z}_B^{-1} = \begin{bmatrix} \boldsymbol{Y}_{11} & \boldsymbol{Y}_{12} \\ \boldsymbol{Y}_{21} & \boldsymbol{Y}_{22} \end{bmatrix},$$

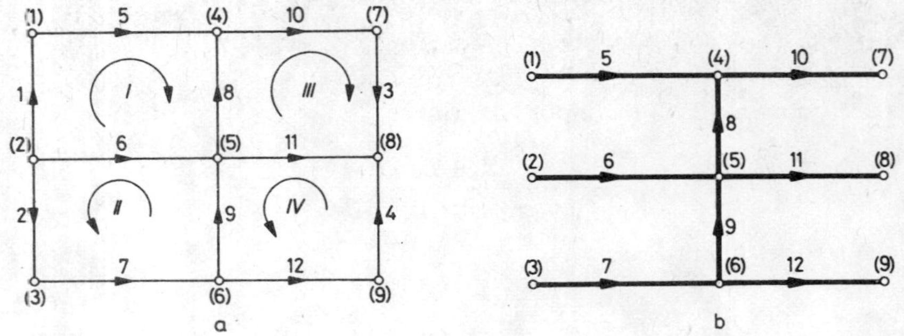

Fig. 3.56

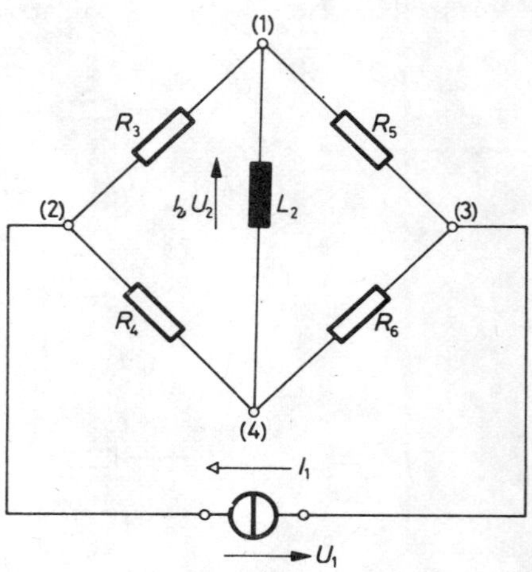

Fig. 3.57

where $\boldsymbol{Y}_{11}$, $\boldsymbol{Y}_{12}$, $\boldsymbol{Y}_{21}$, $\boldsymbol{Y}_{22}$ are square matrices of order two, and

$$\boldsymbol{Y}_{21} = \boldsymbol{Y}_t = \frac{1}{15Z}\begin{bmatrix} 3 & -2 \\ -2 & 3 \end{bmatrix}$$

is the transfer-admittance matrix.

16. The input-impedance and transfer-impedance $Z_t = U_2/I_1$ of the circuit shown in Fig. 3.57 will now be determined as a function of angular frequency. The data of the network are:

$$L_2 = 50 \text{ mH},$$

$$R_3 = 100\ \Omega, \quad R_4 = 250\ \Omega$$

$$R_5 = 400\ \Omega, \quad R_6 = 500\ \Omega.$$

For our analysis let us choose the tree consisting of edges 1, 2, 4 in the graph of the network (Fig. 3.58). The cutset-admittance matrix of the fundamental set of cutsets generated by this tree is:

$$Y_Q = \begin{bmatrix} G_5+G_6 & G_5 & G_5+G_6 \\ G_5 & \dfrac{1}{j\omega L_2}+G_3+G_5 & G_3+G_5 \\ G_5+G_6 & G_3+G_5 & G_3+G_4+G_5+G_6 \end{bmatrix},$$

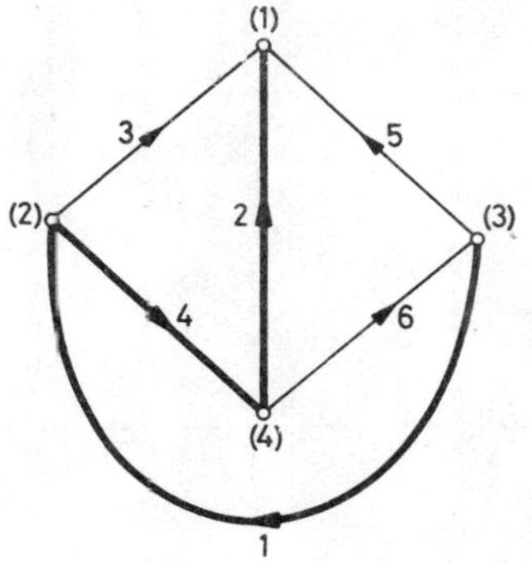

Fig. 3.58

where the notation $G_i = 1/R_i$ $(i=3, 4, 5, 6)$ has been employed. With the numerical values:

$$Y_Q = 0{\cdot}5 \cdot 10^{-3} \begin{bmatrix} 9 & 5 & 9 \\ 5 & 25-j40{,}000/\omega & 25 \\ 9 & 25 & 37 \end{bmatrix} S.$$

According to (3.79) and (3.81) the first element in the first column in Y_Q^{-1} is input-impedance Z_1, while the second element is transfer-impedance Z_t, i.e.:

$$Z_1 = \frac{300\omega - j1{\cdot}48 \cdot 10^6}{\omega - j5{\cdot}04 \cdot 10^3}\ \Omega,$$

and

$$Z_t = \frac{40\omega}{\omega - j5{\cdot}04 \cdot 10^3}\ \Omega$$

with these:

$$U_1 = Z_1 I_1 = \frac{300\omega - j1{\cdot}48 \cdot 10^6}{\omega - j5{\cdot}04 \cdot 10^3} I_1.$$

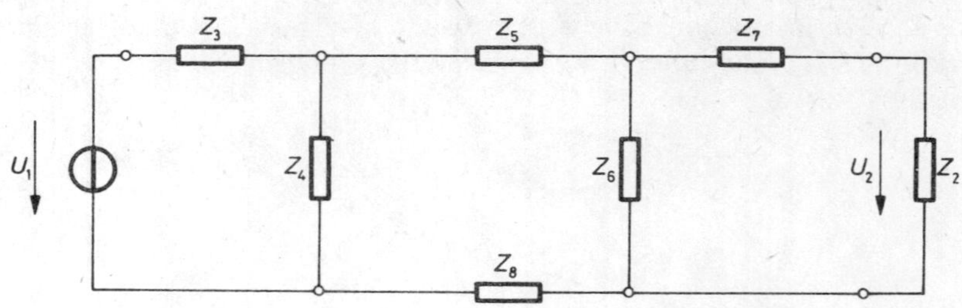

Fig. 3.59

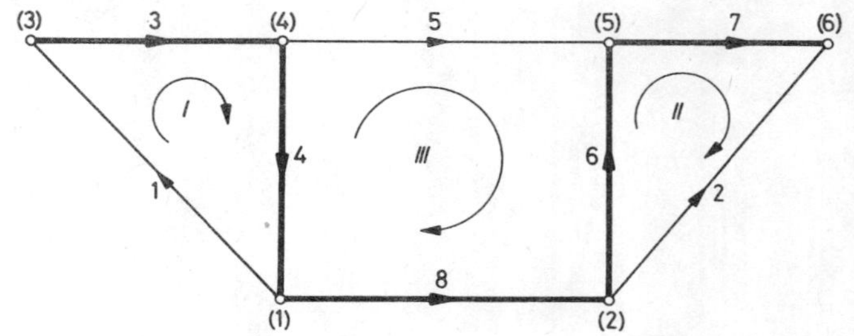

Fig. 3.60

and

$$U_2 = Z_t I_1 = \frac{40\omega}{\omega - j5{\cdot}04 \cdot 10^3} I_1 .$$

17. To determine the voltage-ratio $W_U = U_2/U_1$ of the network drawn in Fig. 3.59, the fundamental set of loops generated by the tree indicated in Fig. 3.60 is chosen. The loop-impedance matrix is:

$$Z_B = \begin{bmatrix} Z_3 + Z_4 & 0 & -Z_4 \\ 0 & Z_2 + Z_6 + Z_7 & -Z_6 \\ -Z_4 & -Z_6 & Z_4 + Z_5 + Z_6 + Z_8 \end{bmatrix} .$$

The voltage-ratio is given by the product of the first element in the second row of Z_B^{-1} and Z_2 according to (3.68) and (3.74), i.e.

$$W_U = \frac{U_2}{U_1} =$$

$$= \frac{Z_2 Z_4 Z_6}{(Z_3 + Z_4)\,[(Z_2 + Z_6 + Z_7)(Z_4 + Z_5 + Z_6 + Z_8) - Z_6^2] - Z_4^2(Z_2 + Z_6 + Z_7)}$$

is the ratio required.

CHAPTER 4

NETWORKS CONTAINING TRANSMISSION LINES

In this chapter the solution of the equations of transmission line networks is dealt with [18, 50, 51, 52, 53]. Our analysis has so far dealt only with lumped-element networks, whereas the transmission line is a distributed network which is frequently used. The transmission line consists of two parallel conductors. The resistance and inductance of the conductors as well as the conductance and capacitance between the two conductors are distributed along the transmission line. Therefore, Kirchhoff equations may be written for a section of differential length in the transmission line.

Let z denote the coordinate parallel to the axes of the conductors. The voltage $u(z)$ between the two conductors and their current $i(z)$ are functions of location because of the resistance and inductance of the conductors as well as the conductance and capacitance between them (Fig. 4.1). From Kirchhoff equations:

$$-\frac{\partial u}{\partial z} = Ri + L\frac{\partial i}{\partial t}, \tag{4.1}$$

$$-\frac{\partial i}{\partial z} = Gu + C\frac{\partial u}{\partial t}. \tag{4.2}$$

where R, L, G and C are the resistance, inductance, conductance and capacitance per unit length of the transmission line, and t denotes time.* The partial differential equations (4.1) and (4.2) are called the telegraph equations. Their steady-state solution in the case of an excitation varying sinusoidally with time is

$$u(z,t) = U^{(+)}e^{-\gamma z + j\omega t} + U^{(-)}e^{\gamma z + j\omega t}, \tag{4.3}$$

$$i(z,t) = Y_0 U^{(+)}e^{-\gamma z + j\omega t} - Y_0 U^{(-)}e^{\gamma z + j\omega t}, \tag{4.4}$$

where complex notation is employed. This steady-state solution will be considered in the following.

* For uniform, linear, time-invariant lines, the line parameters R, L, G, C are constants.

In the solution γ is the propagation coefficient, while Y_0 is the characteristic admittance of the transmission line given, respectively by

$$\gamma = \sqrt{(R+j\omega L)(G+j\omega C)}, \tag{4.5}$$

$$Y_0 = \sqrt{\frac{G+j\omega C}{R+j\omega L}}. \tag{4.6}$$

$U^{(+)}$ and $U^{(-)}$ may be determined from a knowledge of the terminations of the transmission line. In (4.3) and (4.4) damped waves propagating in directions $+z$ and

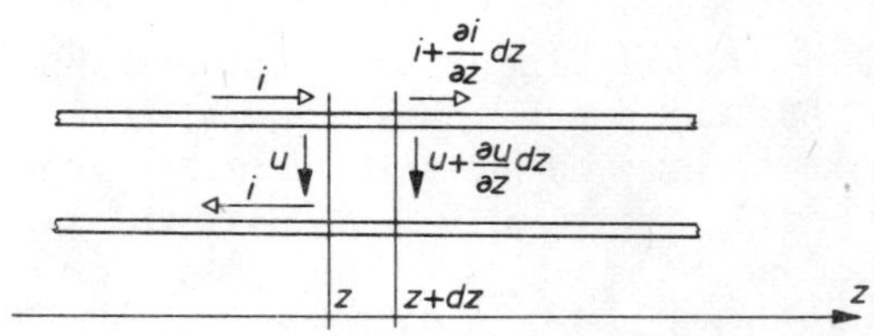

Fig. 4.1

$-z$ are described by the first and second terms respectively on the right-hand side. Thus, $U^{(+)}$ and $U^{(-)}$ are the amplitudes of the voltage waves at $z=0$ propagating in directions $+z$ and $-z$, respectively.

A transmission line section may be characterized by characteristic admittance Y_0, propagation coefficient γ and length l. The transmission line section may be considered as a two-port from the point of view of network theory, with its two-port parameters expressible with the aid of Y_0 and γl from (4.3) and (4.4).

The connection of transmission line sections with lumped-element two-terminal elements and two-ports forms a transmission line network. Two transmission line sections are connected by joining each conductor of one transmission line section to one conductor of the other section (Fig. 4.2). Such connection points of sections will be called vertices, while the common points of conductors of two or more transmission line section are nodes. There are two nodes at each vertex of a transmission line network. The analysis of transmission line networks will be first considered for the case of lumped-element two-terminal elements connected to nodes at vertices as shown in Fig. 4.2.

In the course of network analysis the admittance parameters of the transmission line sections will be used. The method presented is also applicable to the analysis of networks containing lumped-element two-ports characterized by admittance parameters instead of transmission line sections. Although the fact that a transmission line section is a symmetrical two-port is utilized, the method may be generalized to apply to asymmetrical two-ports. These however will not be discussed.

In addition to their use for electrical power distribution, transmission line networks are also used for various connections between radio transmitters, receivers and aerials. (The supply of aerial systems, the interconnection of several simultaneously operating transmitters sharing one aerial, diplexers connecting a transmitter with the aerial or ones used to connect the sound and video transmitters for television broadcasting, etc. may be realized by the employment of transmission line networks.)

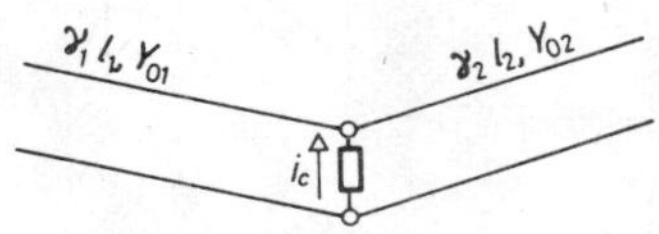

Fig. 4.2

Parallel transmission lines are coupled due to the electric and magnetic fields generated in their surroundings, thus constituting a coupled transmission line system. The equations previously written for transmission lines may be generalized to coupled transmission line systems. The theory of parallel wires running horizontally above the earth may be reduced to the theory of systems consisting of coupled conductor pairs.

The analysis of models of multiphase power distributing transmission line networks is similarly discussed. The network consists of sections forming a system of coupled transmission lines arranged above the earth. n-terminal elements may connect to the connection points of the sections.

In the course of our discussion relations are derived, applicable to both power distribution and to telecommunication transmission line networks. In the case of telecommunication applications the calculation of certain matrices (impedance, admittance, scattering matrix) is usually the objective. In the case of power distribution the primary task is to determine the voltages, currents and powers in multiphase transmission line networks.

Transmission line networks

A graph is derived for transmission line networks [51] by associating each transmission line section with an edge of the graph. Thus, for example, the graph of the transmission line system drawn in Fig. 4.3, a is shown in Fig. 4.3, b. The connection points of transmission line sections correspond to the vertices of the graph.

Besides transmission lines, impedances and sources may connect to vertices. In our calculations the two-terminal elements connected to vertices are represented by the equivalent circuit shown in Fig. 4.4. In the course of the calculations the source-voltages and source-currents of the sources and the impedances connected to the vertices are assumed to be known. If only an impedance is connected to a vertex, the

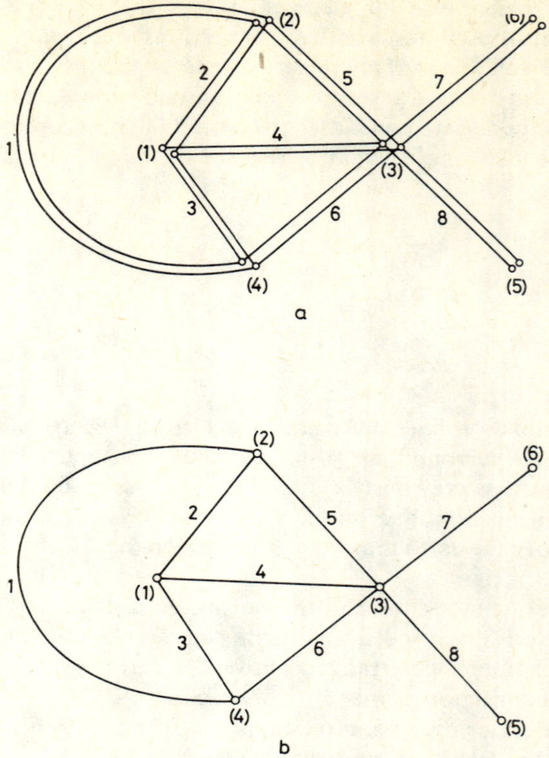

Fig. 4.3

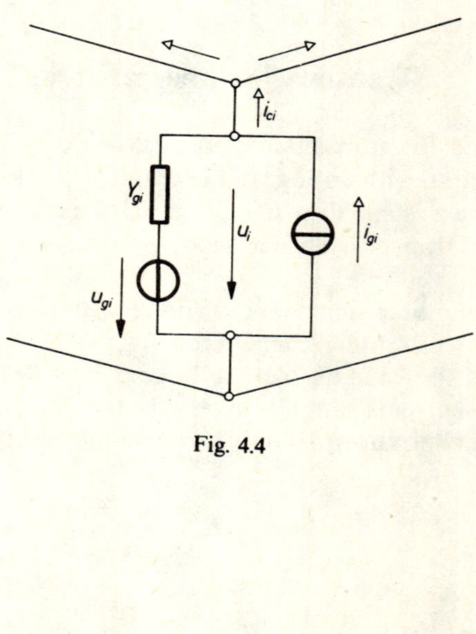

Fig. 4.4

relevant source-voltage and source-current are zero. Similarly if only a voltage-source is connected to the vertex, the source-current is zero, etc. Furthermore, the characteristic admittances, propagation coefficients and lengths of transmission line sections are assumed to be known. Our objective is to determine the voltages and currents at the ends of transmission line sections. Then, by means of (4.3) and (4.4) the voltage and current of any transmission line section at any location can be calculated.

The graph of transmission line networks, as opposed to graphs of lumped element networks discussed previously, frequently contains end-edges. Therefore the incidence matrix is used to characterize the graph.

To write the equations of the network the edges of the graph, and thus the transmission line sections, should be assigned orientations. These can be arbitrarily chosen. Thus a possible choice of orientation of the edges in the graph in Fig. 4.3. b is shown in Fig. 4.5. The termination of vertices in deactivated networks may be extreme, that is, either short-circuit or open-circuit. If both short-circuits and open-circuits appear among the terminations of vertices in the deactivated network, vertices are numbered as explained later. Otherwise the order of numbering of the vertices is arbitrary.

In the following the graph's complete (non-basis) incidence matrix $\boldsymbol{A}_t$ indicating orientations and $\boldsymbol{A}_{t0}$ disregarding orientations as well as matrices $\frac{1}{2}(\boldsymbol{A}_{t0}+\boldsymbol{A}_t)$ and $\frac{1}{2}(\boldsymbol{A}_{t0}-\boldsymbol{A}_t)$ will also be employed (see Chapter 1).

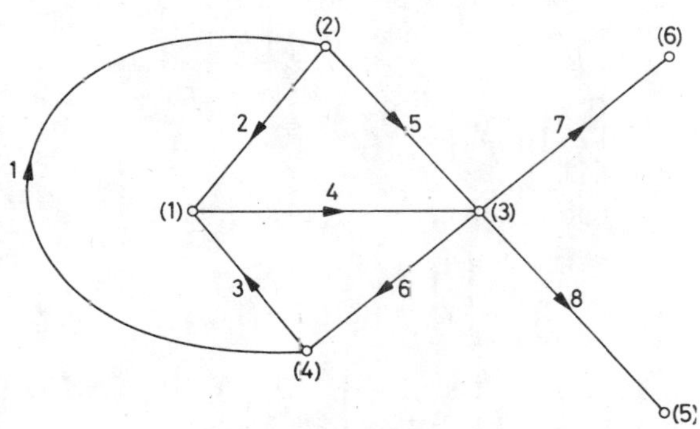

Fig. 4.5

In a network containing several transmission line sections connected to each other the ends of a particular section are usually not distinguishable. Accordingly, the reference directions of the voltages and currents at vertices are chosen to be symmetric as shown in Fig. 4.6. The reference directions of voltages at ends connecting to the same vertex are chosen to coincide. To characterize a transmission line section the relation between the two voltages (u_i, u_j) and the two currents (i_i, i_j) at the ends should be given. The admittance parameter matrix is employed to this end.

The voltage u and current i along a transmission line are usually written as a sum of two waves propagating in directions z and $-z$. At the initial point of the transmission line ($z=0$), omitting the factor $e^{j\omega t}$:

$$u(z=0)=U_1=U^{(+)}+U^{(-)}, \tag{4.7}$$

$$i(z=0)=I_1=Y_0U^{(+)}-Y_0U^{(-)}. \tag{4.8}$$

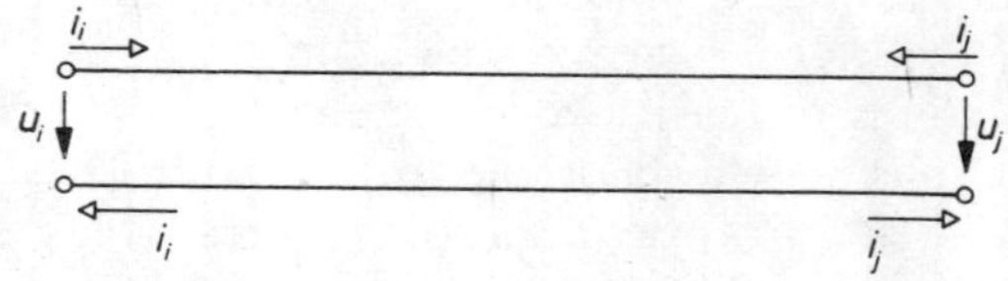

Fig. 4.6

At the end of the transmission line section of length l, observing the reference direction of I_2:

$$u(z=l)=U_2=U^{(+)}e^{-\gamma l}+U^{(-)}e^{\gamma l}, \tag{4.9}$$

$$i(z=l)=I_2=-Y_0U^{(+)}e^{-\gamma l}+Y_0U^{(-)}e^{\gamma l}. \tag{4.10}$$

The voltages and currents at the two ends of the section are thus expressed as

$$\begin{bmatrix} U_1 \\ U_2 \end{bmatrix} = \begin{bmatrix} 1 & 1 \\ e^{-\gamma l} & e^{\gamma l} \end{bmatrix} \begin{bmatrix} U^{(+)} \\ U^{(-)} \end{bmatrix}, \tag{4.11}$$

$$\begin{bmatrix} I_1 \\ I_2 \end{bmatrix} = Y_0 \begin{bmatrix} 1 & -1 \\ -e^{-\gamma l} & e^{\gamma l} \end{bmatrix} \begin{bmatrix} U^{(+)} \\ U^{(-)} \end{bmatrix}. \tag{4.12}$$

Hence the admittance parameter matrix in

$$\begin{bmatrix} I_1 \\ I_2 \end{bmatrix} = \boldsymbol{Y} \begin{bmatrix} U_1 \\ U_2 \end{bmatrix} \tag{4.13}$$

is:

$$\boldsymbol{Y} = Y_0 \begin{bmatrix} 1 & -1 \\ -e^{-\gamma l} & e^{\gamma l} \end{bmatrix} \begin{bmatrix} 1 & 1 \\ e^{-\gamma l} & e^{\gamma l} \end{bmatrix}^{-1} = \frac{Y_0}{\sinh\gamma l} \begin{bmatrix} \cosh\gamma l & -1 \\ -1 & \cosh\gamma l \end{bmatrix}. \tag{4.14}$$

Thus the admittance parameter matrix of section m with characteristic admittance Y_{0m}, propagation coefficient γ_m and length l_m in the transmission line network can be written as follows:

$$\boldsymbol{Y}_m = Y_{0m} \begin{bmatrix} \coth\gamma_m l_m & -\dfrac{1}{\sinh\gamma_m l_m} \\ -\dfrac{1}{\sinh\gamma_m l_m} & \coth\gamma_m l_m \end{bmatrix} = \begin{bmatrix} p_m & r_m \\ r_m & p_m \end{bmatrix}, \tag{4.15}$$

where

$$p_m = Y_{0m} \coth\gamma_m l_m; \quad r_m = -Y_{0m}\frac{1}{\sinh\gamma_m l_m}. \tag{4.16}$$

Let diagonal matrices

$$\boldsymbol{P} = \langle p_1 \; p_2 \; \dots \; p_b\rangle; \quad \boldsymbol{R} = \langle r_1 \; r_2 \; \dots \; r_b\rangle \tag{4.17}$$

be formed by the values p_m and r_m ($m=1, 2, \dots b$) characterizing the branches of a network consisting of b transmission line sections. In the following matrices $\boldsymbol{P}$ and $\boldsymbol{R}$ will be used to characterize the transmission line sections of the network.

Circuit equations

The Kirchhoff equations of the network will in the following be written so as to satisfy the equations relating to voltages automatically. If the number of vertices in the network is n, n independent node-equations should be written, since the Kirchhoff equations for the two nodes of a vertex coincide. The unknowns in the equations are the voltages of the vertices. Their number is also n, and therefore the voltages of the nodes may be determined from the node-equations. Knowing the voltages of the vertices the currents at the ends of transmission line sections as well as those of the impedances between the two nodes of the vertices can be calculated.

The node-equation is written for the node with the voltage of the vertex pointing away from the node. The currents flowing out of a node of a vertex or those flowing

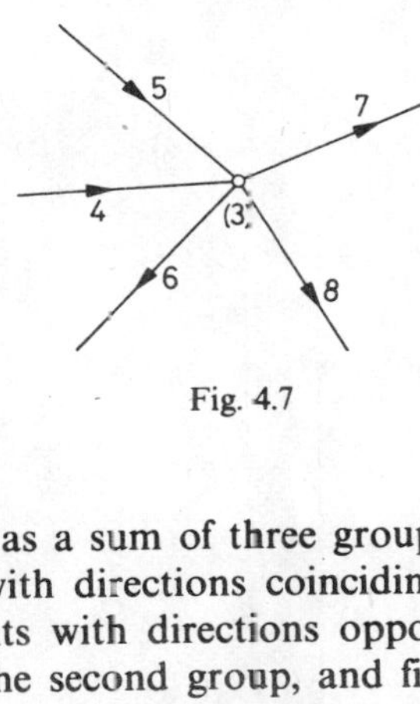

Fig. 4.7

into the node are written as a sum of three groups of currents. The first group consists of the currents with directions coinciding with the orientation of the relevant edge. The currents with directions opposite to the orientation of the relevant edge belong to the second group, and finally the currents of the two-terminal elements between the two nodes of the vertex constitute the third group.

The method is presented for the network drawn in Fig. 4.3, a. Branches 4, 5, 6, 7, 8 are incident with vertex (3) (Fig. 4.7). The directions of edges 6, 7, 8 point away from the vertex, so the sum of their currents is written in the first group. The directions of edges 4 and 5 point towards (3), so their current is written in the second group.

Finally the third group is given by the current of the two-terminal element between the nodes of the vertex.

The node-equation is written for one node of each vertex in the network.

If edge m is incident with vertices (i) and (j) with orientation pointing from (i) to (j), the current of the edge in the first group is

$$i_{mi} = p_m u_i + r_m u_j \,. \tag{4.18}$$

Similar equations may be written for each branch. The system of equation thus derived is summarized in the following matrix equation

$$\mathbf{I}' = \frac{1}{2}\boldsymbol{P}(\boldsymbol{A}_{t0} + \boldsymbol{A}_t)^+ \mathbf{U} + \frac{1}{2}\boldsymbol{R}(\boldsymbol{A}_{t0} - \boldsymbol{A}_t)^+ \mathbf{U}, \tag{4.19}$$

where $\mathbf{U}$ is the column matrix formed by the voltages of the vertices.

The column matrix formed by the voltages of the vertices in the network drawn in Fig. 4.3 is:

$$\mathbf{U} = \begin{bmatrix} u_1 \\ u_2 \\ u_3 \\ u_4 \\ u_5 \\ u_6 \end{bmatrix}.$$

Thus in accordance with (4.19) the column matrix of the currents in the first group is

$$\mathbf{I}' = \begin{bmatrix} i_{14} \\ i_{22} \\ i_{34} \\ i_{41} \\ i_{52} \\ i_{63} \\ i_{73} \\ i_{83} \end{bmatrix} = \begin{bmatrix} p_1 u_4 + r_1 u_2 \\ p_2 u_2 + r_2 u_1 \\ p_3 u_4 + r_3 u_1 \\ p_4 u_1 + r_4 u_3 \\ p_5 u_2 + r_5 u_3 \\ p_6 u_3 + r_6 u_4 \\ p_7 u_3 + r_7 u_6 \\ p_8 u_3 + r_8 u_5 \end{bmatrix}.$$

The elements of $\mathbf{I}'$ correspond to the currents written in (4.18). These currents are shown in Fig. 4.8.

The current in the second group for branch m is:

$$i_{mj} = r_m u_i + p_m u_j\,. \tag{4.20}$$

Such an equation may be written for all of the branches. This system of equations is as follows:

$$\mathbf{I}'' = \frac{1}{2}\boldsymbol{R}(\boldsymbol{A}_{t0} + \boldsymbol{A}_t)^+\mathbf{U} + \frac{1}{2}\boldsymbol{P}(\boldsymbol{A}_{t0} - \boldsymbol{A}_t)^+\mathbf{U}\,. \tag{4.21}$$

In our example:

$$\mathbf{I}'' = \begin{bmatrix} r_1u_4 + p_1u_2 \\ r_2u_2 + p_2u_1 \\ r_3u_4 + p_3u_1 \\ r_4u_1 + p_4u_3 \\ r_5u_2 + p_5u_3 \\ r_6u_3 + p_6u_4 \\ r_7u_3 + p_7u_6 \\ r_8u_3 + p_8u_5 \end{bmatrix}.$$

The currents in the second group have been indicated in Fig. 4.9.

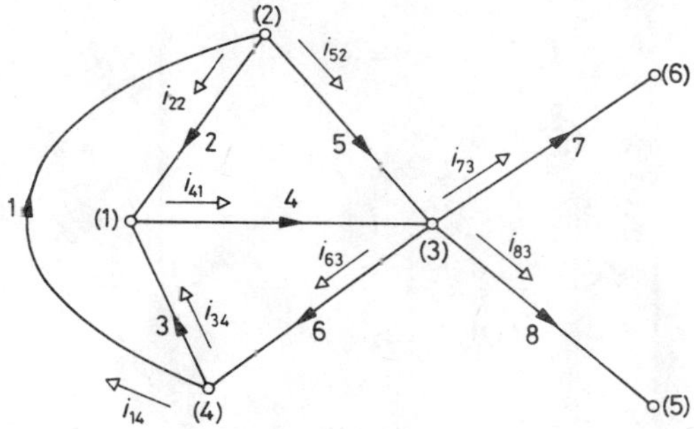

Fig. 4.8

The currents of the two-terminal elements at the vertices are yet to be determined. The case of the connection drawn in Fig. 4.4 placed between the nodes of the vertex will be examined. For vertex (i):

$$i_{ci} = Y_{gi}(u_{gi} - u_i) + i_{gi}\,. \tag{4.22}$$

Writing these for all of the vertices they are summarized in

$$\mathbf{I}_c = \boldsymbol{Y}_g(\mathbf{U}_g - \mathbf{U}) + \mathbf{I}_g, \tag{4.23}$$

where $\mathbf{I}_c$, $\mathbf{U}_g$ and $\mathbf{I}_g$ are the column matrices formed by the currents of the two-terminal elements in the vertices, the source-voltages of the voltage-sources and the source-currents of current-sources respectively, while $\boldsymbol{Y}_g$ is a diagonal matrix with the admittances of the two-terminal elements in the vertices in the main diagonal.

The currents must satisfy the node-equations. The currents written in $\mathbf{I}'$ flow away from one node of each vertex. Let $\mathbf{I}'_c$ denote the column matrix of the sum of these

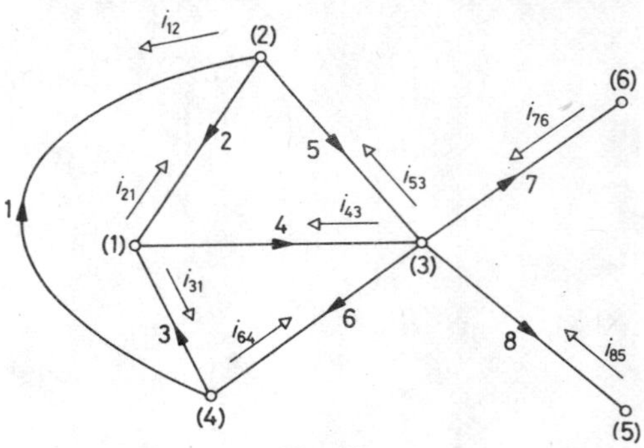

Fig. 4.9

currents at each vertex:

$$\mathbf{I}'_c = \frac{1}{2}(\boldsymbol{A}_{t0} + \boldsymbol{A}_t)\mathbf{I}'. \tag{4.24}$$

In our example the following is obtained:

$$\mathbf{I}'_c = \begin{bmatrix} p_4 u_1 + r_4 u_3 \\ p_2 u_2 + r_2 u_1 + p_5 u_2 + r_5 u_3 \\ p_6 u_3 + r_6 u_4 + p_7 u_3 + r_7 u_6 + p_8 u_3 + r_8 u_5 \\ p_1 u_4 + r_1 u_2 + p_3 u_4 + r_3 u_1 \\ 0 \\ 0 \end{bmatrix}.$$

The branch-currents forming $\mathbf{I}''$ flow away from one node of each vertex. The sums of currents flowing away from each vertex constitute column matrix $\mathbf{I}''_c$:

$$\mathbf{I}''_c = \frac{1}{2}(\boldsymbol{A}_{t0} - \boldsymbol{A}_t)\mathbf{I}''. \tag{4.25}$$

Let us write it in our example:

$$\mathbf{I}_c'' = \begin{bmatrix} r_2u_2+p_2u_1+r_3u_4+p_3u_1 \\ r_1u_4+p_1u_2 \\ r_4u_1+p_4u_3+r_5u_2+p_5u_3 \\ r_6u_3+p_6u_4 \\ r_8u_3+p_8u_5 \\ r_7u_3+p_7u_6 \end{bmatrix}.$$

The currents in $\mathbf{I}_c$ flow towards the node of each vertex having the relevant currents in $\mathbf{I}_c'$ flow outwards (Fig. 4.4). Thus the matrix form of the node-equations using (4.24), (4.25), (4.23) as well as (4.19) and (4.21), is:

$$\begin{aligned} \mathbf{I}_c' + \mathbf{I}_c'' - \mathbf{I}_c &= \frac{1}{4}(\boldsymbol{A}_{t0} + \boldsymbol{A}_t)\,[\boldsymbol{P}(\boldsymbol{A}_{t0} + \boldsymbol{A}_t)^+ + \boldsymbol{R}(\boldsymbol{A}_{t0} - \boldsymbol{A}_t)^+]\mathbf{U} + \\ &+ \frac{1}{4}(\boldsymbol{A}_{t0} - \boldsymbol{A}_t)\,[\boldsymbol{P}(\boldsymbol{A}_{t0} - \boldsymbol{A}_t)^+ + \boldsymbol{R}(\boldsymbol{A}_{t0} + \boldsymbol{A}_t)^+]\mathbf{U} - \\ &- \boldsymbol{Y}_g\mathbf{U}_g + \boldsymbol{Y}_g\mathbf{U} - \mathbf{I}_g = \mathbf{0}. \end{aligned} \tag{4.26}$$

On rearrangement:

$$(\boldsymbol{Y}_c + \boldsymbol{Y}_g)\mathbf{U} = \mathbf{I}_g + \boldsymbol{Y}_g\mathbf{U}_g, \tag{4.27}$$

where the notation

$$\boldsymbol{Y}_c = \frac{1}{2}\boldsymbol{A}_{t0}(\boldsymbol{P} - \boldsymbol{R})\boldsymbol{A}_{t0}^+ + \frac{1}{2}\boldsymbol{A}_t(\boldsymbol{P} - \boldsymbol{R})\boldsymbol{A}_t^+ \tag{4.28}$$

has been used. $\boldsymbol{Y}_c$ is called *vertex admittance matrix*. In our example $\boldsymbol{Y}_c$ is as follows:

$$\boldsymbol{Y}_c = \begin{bmatrix} p_2+p_3+p_4 & r_2 & r_4 & r_3 & 0 & 0 \\ r_2 & p_1+p_2+p_5 & r_5 & r_1 & 0 & 0 \\ r_4 & r_5 & p_4+p_5+p_6+p_7+p_8 & r_6 & r_8 & r_7 \\ r_3 & r_1 & r_6 & p_1+p_3+p_6 & 0 & 0 \\ 0 & 0 & r_8 & 0 & p_8 & 0 \\ 0 & 0 & r_7 & 0 & 0 & p_7 \end{bmatrix}.$$

This matrix is seen to be symmetric. The elements of the main diagonal are the sums of values p of branches incident with the vertex associated with the relevant row (column). The remaining elements are the values r of the branches connecting the vertices associated with the appropriate row and column. If no branch connects the two vertices the corresponding element of the matrix is 0.

From (4.27):

$$\mathbf{U}=(\boldsymbol{Y}_c+\boldsymbol{Y}_g)^{-1}(\boldsymbol{Y}_g\mathbf{U}_g+\mathbf{I}_g), \tag{4.29}$$

i.e. the voltages of the vertices have been expressed in terms of known quantities.

If vertices terminated by a short-circuit or by a source of zero internal impedance are present, with no vertex terminated by an open circuit or source of zero internal admittance, the matrix $\boldsymbol{Z}_g$ is employed instead of $\boldsymbol{Y}_g$, and thus the voltage of vertices are determined from

$$(\boldsymbol{Z}_g\boldsymbol{Y}_c+\boldsymbol{1})\mathbf{U}=\boldsymbol{Z}_g\mathbf{I}_g+\mathbf{U}_g \tag{4.30}$$

instead of (4.27).

If both short-circuits and open-circuits appear as terminations at the vertices of the deactivated network the vertices are classified into two groups. The first group is formed by the vertices with no short-circuits as termination between their nodes in the deactivated network. The vertices with short-circuit between their nodes constitute the second group. (The impedance between the nodes is zero.) The numbering of vertices is carried out observing the following classification: order numbers (1), (2), . . ., (n_1) are associated with the vertices in the first group, while

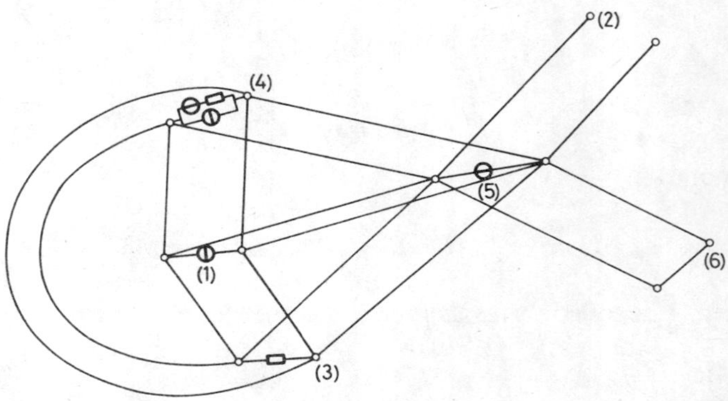

Fig. 4.10

those in the second group are (n_1+1), (n_1+2), . . ., (n). Thus, for example, in the network shown in Fig. 4.10 vertices (1), (2), (3), (4) belong to the first group, while (5), (6) belong to the second. Let $\boldsymbol{Y}_c$, $\boldsymbol{Y}_g$, $\mathbf{U}$ and $\boldsymbol{Y}_g\mathbf{U}_g+\mathbf{I}_g$ be partitioned in accordance with the classification of vertices:

$$\boldsymbol{Y}_c=\begin{bmatrix}\boldsymbol{Y}_{11} & \boldsymbol{Y}_{12}\\ \boldsymbol{Y}_{21} & \boldsymbol{Y}_{22}\end{bmatrix}, \tag{4.31}$$

$$\boldsymbol{Y}_g=\begin{bmatrix}\boldsymbol{Y}_{g1} & \boldsymbol{0}\\ \boldsymbol{0} & \boldsymbol{Y}_{g2}\end{bmatrix}, \tag{4.32}$$

$$\mathbf{U}=\begin{bmatrix}\mathbf{U}_1\\ \mathbf{U}_2\end{bmatrix}=\begin{bmatrix}\mathbf{U}_1\\ \mathbf{U}_{g2}\end{bmatrix}, \tag{4.33}$$

$$\boldsymbol{Y}_g\mathbf{U}_g+\mathbf{I}_g=\begin{bmatrix}\boldsymbol{Y}_{g1}\mathbf{U}_{g1}+\mathbf{I}_{g1}\\ \boldsymbol{Y}_{g2}\mathbf{U}_{g2}\end{bmatrix}, \tag{4.34}$$

where $\mathbf{U}_2=\mathbf{U}_{g2}$ is known and $\boldsymbol{Y}_{g2}$ does not exist. Therefore, the equations are so rearranged that $\boldsymbol{Y}_{g2}$ does not appear in the solution. With the matrices partitioned (4.27) is as follows:

$$\begin{bmatrix}\boldsymbol{Y}_{11} & \boldsymbol{Y}_{12}\\ \boldsymbol{Y}_{21} & \boldsymbol{Y}_{22}\end{bmatrix}\begin{bmatrix}\mathbf{U}_1\\ \mathbf{U}_{g2}\end{bmatrix}+\begin{bmatrix}\boldsymbol{Y}_{g1}\mathbf{U}_1\\ \boldsymbol{Y}_{g2}\mathbf{U}_{g2}\end{bmatrix}=\begin{bmatrix}\boldsymbol{Y}_{g1}\mathbf{U}_{g1}+\mathbf{I}_{g1}\\ \boldsymbol{Y}_{g2}\mathbf{U}_{g2}\end{bmatrix}. \tag{4.35}$$

Hence:

$$(\boldsymbol{Y}_{11}+\boldsymbol{Y}_{g1})\mathbf{U}_1+\boldsymbol{Y}_{12}\mathbf{U}_{g2}=\boldsymbol{Y}_{g1}\mathbf{U}_{g1}+\mathbf{I}_{g1}, \tag{4.36}$$

i.e.

$$\mathbf{U}_1=(\boldsymbol{Y}_{11}+\boldsymbol{Y}_{g1})^{-1}(\boldsymbol{Y}_{g1}\mathbf{U}_{g1}-\boldsymbol{Y}_{12}\mathbf{U}_{g2}+\mathbf{I}_{g1}) \tag{4.37}$$

are the voltages of the vertices not terminated by short-circuits in the deactivated network.

The matrices characterizing the network

The relationships between the voltages and currents at the connection points of a deactivated transmission line network can be characterized by various different matrices. Such matrices may be utilized, for example, in the analysis of passive transmission line networks connecting receiver-transmitter units operating in the UHF range with the aerial (diplexers, networks connecting community receiver aerials) or of single-phase power distribution transmission line networks.

A transmission line network may connect to other networks at certain vertices. These vertices are called *connection points*, and their number in the network is denoted by a. To determine the matrices characterizing the network the column matrices $\mathbf{U}_1$ and $\mathbf{I}_1$ are formed from the voltages and currents at the connection points. The relations

$$\mathbf{U}_1=\boldsymbol{Z}_1\mathbf{I}_1, \qquad \mathbf{I}_1=\boldsymbol{Y}_1\mathbf{U}_1 \tag{4.38}$$

may be written between them, where $\boldsymbol{Z}_1$ is the impedance, and $\boldsymbol{Y}_1$ the admittance matrix of the network, and

$$\boldsymbol{Z}_1=\boldsymbol{Y}_1^{-1} \tag{4.39}$$

To determine the scattering matrix of the network, the diagonal matrix

$$\boldsymbol{Y}_{c0}=\langle Y_{01} \;\; Y_{02} \;\; \ldots \;\; Y_{0a}\rangle \tag{4.40}$$

formed of the characteristic admittances of the branches incident with the connection points is defined. The order numbers of the characteristic admittances coincide with those of the connection points incident with the branch. If more than one branch is incident with one connection point, the sum of the characteristic admittances of these branches appears in the matrix in (4.40).

Let the column matrices $\mathbf{U}_1$ and $\mathbf{I}_1$ be decomposed as the sum of two column matrices describing an incident and a reflected wave respectively:

$$\mathbf{U}_1=\mathbf{U}^{(+)}+\mathbf{U}^{(-)}, \tag{4.41}$$

$$\mathbf{I}_1=\mathbf{I}^{(+)}+\mathbf{I}^{(-)}, \tag{4.42}$$

where

$$\mathbf{I}^{(+)}=\boldsymbol{Y}_{c0}\mathbf{U}^{(+)}, \tag{4.43}$$

$$\mathbf{I}^{(-)}=-\boldsymbol{Y}_{c0}\mathbf{U}^{(-)}, \tag{4.44}$$

i.e.

$$\mathbf{I}_1=\boldsymbol{Y}_{c0}(\mathbf{U}^{(+)}-\mathbf{U}^{(-)})=\boldsymbol{Y}_1(\mathbf{U}^{(+)}+\mathbf{U}^{(-)}). \tag{4.45}$$

Hence

$$\mathbf{U}^{(-)}=(\boldsymbol{Y}_{c0}+\boldsymbol{Y}_1)^{-1}(\boldsymbol{Y}_{c0}-\boldsymbol{Y}_1)\mathbf{U}^{(+)}. \tag{4.46}$$

The scattering matrix $\boldsymbol{S}$ is defined by

$$\mathbf{U}^{(-)}=\boldsymbol{S}\mathbf{U}^{(+)}. \tag{4.47}$$

On comparison with (4.46)

$$\boldsymbol{S}=(\boldsymbol{Y}_{c0}+\boldsymbol{Y}_1)^{-1}(\boldsymbol{Y}_{c0}-\boldsymbol{Y}_1) \tag{4.48}$$

is obtained for the scattering matrix.

In certain applications, the connection points may be classified into two groups. At the connection points in the first group the supply of energy to the network takes place. In the following these will be called primary connection points. The connection points in the second group, hereinafter called secondary, are connected to consumers. Let $\mathbf{U}_p$ and $\mathbf{I}_p$ denote the column matrices formed by the voltages and current at the primary connection points, and $\mathbf{U}_s$ and $\mathbf{I}_s$ the same at secondary connection points. If the primary connection points are assigned the first order numbers:

$$\mathbf{U}_1=\begin{bmatrix}\mathbf{U}_p\\ \mathbf{U}_s\end{bmatrix};\qquad \mathbf{I}_1=\begin{bmatrix}\mathbf{I}_p\\ \mathbf{I}_s\end{bmatrix}. \tag{4.49}$$

If the finite and non-zero impedances of the consumers at the secondary connection points are known:

$$\mathbf{U}_s=\boldsymbol{Z}_s\mathbf{I}_s;\qquad \mathbf{I}_s=\boldsymbol{Y}_s\mathbf{U}_s, \tag{4.50}$$

where $\boldsymbol{Z}_s$ is the diagonal matrix formed by the impedances of the consumers, and $\boldsymbol{Y}_s=\boldsymbol{Z}_s^{-1}$.

In the following a method is presented for the determination of matrices characterizing the relationships between $\mathbf{U}_1, \mathbf{I}_1, \mathbf{U}^{(+)}, \mathbf{U}^{(-)}, \mathbf{U}_p, \mathbf{I}_p, \mathbf{U}_s, \mathbf{I}_s$. These are $\boldsymbol{Z}_1$ and $\boldsymbol{Y}_1$ appearing in (4.38), $\boldsymbol{S}$ defined in (4.48) and input and transfer impedance matrices $\boldsymbol{Z}_i$ and $\boldsymbol{Z}_t$, input and transfer admittance matrices $\boldsymbol{Y}_i$ and $\boldsymbol{Y}_t$ ($\boldsymbol{Y}_t \neq \boldsymbol{Z}_t^{-1}$) as well as voltage transfer matrix $\boldsymbol{W}$, describing the relationships

$$\mathbf{U}_p = \boldsymbol{Z}_i \mathbf{I}_p; \qquad \mathbf{I}_p = \boldsymbol{Z}_i^{-1} \mathbf{U}_p = \boldsymbol{Y}_i \mathbf{U}_p, \tag{4.51}$$

$$\mathbf{U}_s = \boldsymbol{Z}_t \mathbf{I}_p, \tag{4.52}$$

$$\mathbf{I}_s = \boldsymbol{Y}_t \mathbf{U}_p, \tag{4.53}$$

$$\mathbf{U}_s = \boldsymbol{W} \mathbf{U}_p. \tag{4.54}$$

The calculation of $\boldsymbol{Z}_1$ and $\boldsymbol{Y}_1$ is first dealt with [53]. The lumped element two-terminal elements of the network considered are passive. (Source-current and source-voltage are both zero in the network drawn in Fig. 4.4.) Let the vertices of the network be classified into three groups. The connection points belong to the first. The vertices not in the first group with their terminations not short-circuits constitute the second group (i.e. the impedances connected to these vertices are non-zero). The vertices terminated by short-circuits are included in the third group. Order numbers are assigned to the vertices in accordance with the classification, i.e. the vertices in the first group are assigned order numbers (1), (2), . . ., (n_1), those in the second (n_1+1), (n_1+2), . . ., (n_1+n_2) and those in the third (n_1+n_2+2), (n_1+n_2+2), . . . (n).

Current-sources with their source-currents known are joined to connection-points. Since in the network thus terminated $\mathbf{U}_g = \mathbf{0}$, (4.27) takes the form

$$(\boldsymbol{Y}_c + \boldsymbol{Y}_g)\mathbf{U} = \mathbf{I}_g \tag{4.55}$$

in this case. Let the matrices appearing here be partitioned in accordance with the three groups of vertices

$$\boldsymbol{Y}_c = \begin{bmatrix} \boldsymbol{Y}_{11} & \boldsymbol{Y}_{12} & \boldsymbol{Y}_{13} \\ \boldsymbol{Y}_{21} & \boldsymbol{Y}_{22} & \boldsymbol{Y}_{23} \\ \boldsymbol{Y}_{31} & \boldsymbol{Y}_{32} & \boldsymbol{Y}_{33} \end{bmatrix}, \tag{4.56}$$

$$\boldsymbol{Y}_g = \begin{bmatrix} \boldsymbol{0} & \boldsymbol{0} & \boldsymbol{0} \\ \boldsymbol{0} & \boldsymbol{Y}_{g2} & \boldsymbol{0} \\ \boldsymbol{0} & \boldsymbol{0} & \boldsymbol{Y}_{g3} \end{bmatrix}, \tag{4.57}$$

$$\mathbf{U} = \begin{bmatrix} \mathbf{U}_1 \\ \mathbf{U}_2 \\ \mathbf{0} \end{bmatrix} \tag{4.58}$$

and

$$\mathbf{I}_g = \begin{bmatrix} \mathbf{I}_{g1} \\ \mathbf{0} \\ \mathbf{0} \end{bmatrix}, \tag{4.59}$$

where it has been taken into account that the terminating admittances at the connection points are zero, and the voltages of the vertices in the third group are zero, these being terminated by short-circuits. On substitution into (4.55)

$$\begin{bmatrix} \boldsymbol{Y}_{11} & \boldsymbol{Y}_{12} & \boldsymbol{Y}_{13} \\ \boldsymbol{Y}_{21} & \boldsymbol{Y}_{22}+\boldsymbol{Y}_{g2} & \boldsymbol{Y}_{23} \\ \boldsymbol{Y}_{31} & \boldsymbol{Y}_{32} & \boldsymbol{Y}_{33}+\boldsymbol{Y}_{g3} \end{bmatrix} \begin{bmatrix} \mathbf{U}_1 \\ \mathbf{U}_2 \\ \mathbf{0} \end{bmatrix} = \begin{bmatrix} \mathbf{I}_{g1} \\ \mathbf{0} \\ \mathbf{0} \end{bmatrix} \tag{4.60}$$

is obtained. $\boldsymbol{Y}_{g3}$ does not exist, but it is not needed for the calculation, since (4.60) yields

$$\begin{bmatrix} \boldsymbol{Y}_{11} & \boldsymbol{Y}_{12} \\ \boldsymbol{Y}_{21} & \boldsymbol{Y}_{22}+\boldsymbol{Y}_{g2} \end{bmatrix} \begin{bmatrix} \mathbf{U}_1 \\ \mathbf{U}_2 \end{bmatrix} = \begin{bmatrix} \mathbf{I}_{g1} \\ \mathbf{0} \end{bmatrix}. \tag{4.61}$$

Introducing the notation

$$\begin{bmatrix} \boldsymbol{Y}_{11} & \boldsymbol{Y}_{12} \\ \boldsymbol{Y}_{21} & \boldsymbol{Y}_{22}+\boldsymbol{Y}_{g2} \end{bmatrix}^{-1} = \begin{bmatrix} \boldsymbol{Z}_{11} & \boldsymbol{Z}_{12} \\ \boldsymbol{Z}_{21} & \boldsymbol{Z}_{22} \end{bmatrix}, \tag{4.62}$$

from (4.61)

$$\begin{bmatrix} \mathbf{U}_1 \\ \mathbf{U}_2 \end{bmatrix} = \begin{bmatrix} \boldsymbol{Z}_{11} & \boldsymbol{Z}_{12} \\ \boldsymbol{Z}_{21} & \boldsymbol{Z}_{22} \end{bmatrix} \begin{bmatrix} \mathbf{I}_{g1} \\ \mathbf{0} \end{bmatrix} \tag{4.63}$$

is obtained, i.e.

$$\mathbf{U}_1 = \boldsymbol{Z}_{11}\mathbf{I}_{g1}. \tag{4.64}$$

Since $\mathbf{I}_{g1} = \mathbf{I}_1$, the impedance matrix of the network is

$$\boldsymbol{Z}_1 = \boldsymbol{Z}_{11}, \tag{4.65}$$

and according to (4.39) its inverse is the admittance matrix:

$$\boldsymbol{Y}_1 = \boldsymbol{Z}_1^{-1} = \boldsymbol{Z}_{11}^{-1}. \tag{4.66}$$

The scattering matrix $\boldsymbol{S}$ is seen to be expressible from (4.48), knowing $\boldsymbol{Y}_1$ and $\boldsymbol{Y}_{c0}$.

To determine the other matrices characterizing the network the above classification of the vertices is employed. At the partitioning of the matrices however the forms (4.49) of $\mathbf{U}_1$ and $\mathbf{I}_1$ are observed. The terminations of the connection points in the following calculations are Norton generators, i.e. in the connection of Fig. 4.4 $u_{gi} = 0$. The primary connection points are terminated by current-sources ($Y_{gi} = 0$), and the secondary connection points by non-zero, finite admittances

$(i_{gi}=0)$. The latter yield the diagonal matrix $Z_s = Y_s^{-1}$ introduced in (4.50). Now, using the partitioning

$$Y_{11} = \begin{bmatrix} Y_{pp} & Y_{ps} \\ Y_{sp} & Y_{ss} \end{bmatrix}; \qquad \mathbf{U}_1 = \begin{bmatrix} \mathbf{U}_p \\ \mathbf{U}_s \end{bmatrix}; \qquad \mathbf{I}_{g1} = \begin{bmatrix} \mathbf{I}_p \\ \mathbf{0} \end{bmatrix} \tag{4.67}$$

(4.55) is written as follows:

$$\begin{bmatrix} Y_{pp} & Y_{ps} & Y_{12} & Y_{13} \\ Y_{sp} & Y_{ss}+Y_s & & \\ Y_{21} & & Y_{22}+Y_{g2} & Y_{23} \\ Y_{31} & & Y_{32} & Y_{33}+Y_{g3} \end{bmatrix} \begin{bmatrix} \mathbf{U}_p \\ \mathbf{U}_s \\ \mathbf{U}_2 \\ \mathbf{0} \end{bmatrix} = \begin{bmatrix} \mathbf{I}_p \\ \mathbf{0} \\ \mathbf{0} \\ \mathbf{0} \end{bmatrix}. \tag{4.68}$$

With the notation

$$\begin{bmatrix} Y_{pp} & Y_{ps} & Y_{12} \\ Y_{sp} & Y_{ss}+Y_s & \\ Y_{21} & & Y_{22}+Y_{g2} \end{bmatrix}^{-1} = \begin{bmatrix} Z_{11} & Z_{12} & Z_{13} \\ Z_{21} & Z_{22} & Z_{23} \\ Z_{31} & Z_{32} & Z_{33} \end{bmatrix}, \tag{4.69}$$

$$\begin{bmatrix} \mathbf{U}_p \\ \mathbf{U}_s \\ \mathbf{U}_2 \end{bmatrix} = \begin{bmatrix} Z_{11} & Z_{12} & Z_{13} \\ Z_{21} & Z_{22} & Z_{23} \\ Z_{31} & Z_{32} & Z_{33} \end{bmatrix} \begin{bmatrix} \mathbf{I}_p \\ \mathbf{0} \\ \mathbf{0} \end{bmatrix}. \tag{4.70}$$

Hence

$$\mathbf{U}_p = Z_{11}\mathbf{I}_p \tag{4.71}$$

i.e. on comparison with (4.51):

$$Z_i = Z_{11} \tag{4.72}$$

is the input impedance matrix and

$$Y_i = Z_{11}^{-1} \tag{4.73}$$

is the input admittance matrix. According to the relationship

$$\mathbf{U}_s = Z_{21}\mathbf{I}_1 \tag{4.74}$$

comparison with (4.52) yields

$$Z_t = Z_{21}, \tag{4.75}$$

the transfer impedance matrix. Using (4.71):

$$\mathbf{U}_s = Z_{21} Z_{11}^{-1} \mathbf{U}_p, \tag{4.76}$$

i.e. according to (4.54):

$$W = Z_{21} Z_{11}^{-1} \tag{4.77}$$

is the voltage transfer matrix. Observing (4.50), (4.76) yields

$$\mathbf{I}_s = Z_s^{-1} Z_{21} Z_{11}^{-1} \mathbf{U}_p, \tag{4.78}$$

i.e. on comparison with (4.53)

$$Y_t = Z_s^{-1} Z_{21} Z_{11}^{-1} \tag{4.79}$$

is the transfer admittance matrix.

Power distribution networks

In order to write the equations of multiphase power distribution networks, the theory of coupled transmission lines is reviewed, then using its results transmission line systems above earth are examined, finally the formulae relating to the whole network are presented.

Transmission line systems

Since the theory of coupled, parallel transmission lines is relatively less well-known, and its results will be used to write the equations of networks containing coupled transmission lines, this theory is first briefly presented.

Let us first consider a system of k pairs of conductors (Fig. 4.11), with their voltages and currents varying sinusoidally in time with angular frequency ω [18, 50, 52, 61]. The current of the j-th pair of conductors is denoted by i_j or $-i_j$, its voltage by u_j. Faraday's law is applied to the surface of length dz of the j-th pair of

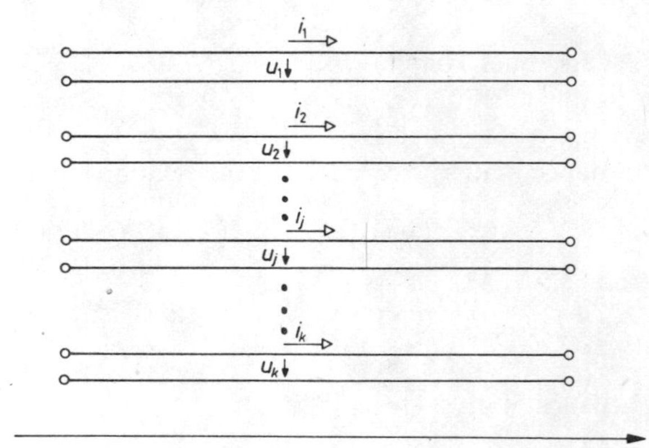

Fig. 4.11

conductors (Fig. 4.12). The integral of electric field intensity **E** along the circumference of the loop shown in Fig. 4.12 appears in the law of induction. At the side of the loop at a conductor, this integral equals the product of the current of the conductor and the internal impedance (skin impedance) of the conductor of length dz. If the internal impedance of unit length is Z_{bj}, the integral of **E** along length dz is $i_j Z_{bj} \mathrm{d}z$, thus along the two conductors $2 i_j Z_{bj} \mathrm{d}z$. The integral along the side of the

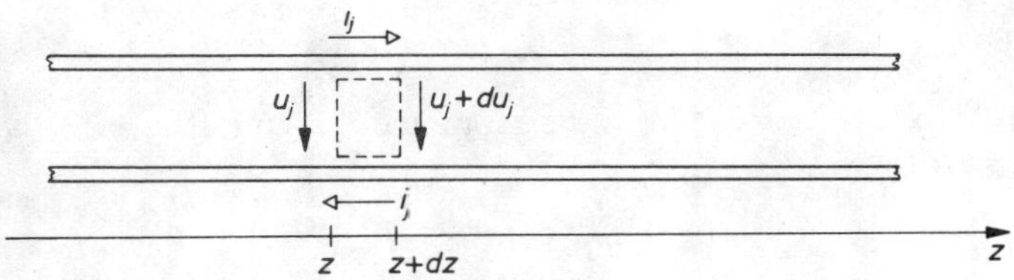

Fig. 4.12

loop at z is $-u_j$ with the reference directions chosen observed, while at $z+\mathrm{d}z$ it is $u_j +$ $+ \frac{\mathrm{d}u_j}{\mathrm{d}z}\mathrm{d}z$. This latter expression reflects the variation of voltage. The integral of field intensity is according to Faraday's law the negative value of the time-derivative of the flux Ψ_j of the surface surrounded by the loop, i.e. $-\frac{\mathrm{d}\Psi_j}{\mathrm{d}t} = -j\omega\Psi_j$:

$$\oint \mathbf{E}\mathrm{d}\mathbf{l} = -u_j + \left(u_j + \frac{\mathrm{d}u_j}{\mathrm{d}z}\mathrm{d}z\right) + 2i_j Z_{bj}\mathrm{d}z = -j\omega\Psi_j. \tag{4.80}$$

Ψ_j is expressible with the aid of the currents i_i of the conductor pairs, and the inductances L_{ij} of unit length:

$$\Psi_j = \sum_{i=1}^{k} L_{ij} i_i \mathrm{d}z. \tag{4.81}$$

On substitution into (4.80), and writing (4.80) for all conductor pairs, these are summarized in the following matrix equation:

$$-\frac{\mathrm{d}\mathbf{u}}{\mathrm{d}z} = \mathbf{Z}_s \mathbf{i}, \tag{4.82}$$

where

$$\mathbf{u} = \begin{bmatrix} u_1 \\ u_2 \\ \vdots \\ u_k \end{bmatrix}, \qquad \mathbf{i} = \begin{bmatrix} i_1 \\ i_2 \\ \vdots \\ i_k \end{bmatrix} \tag{4.83}$$

are the column matrices formed by the voltages and currents of the conductor pairs.

The series impedance matrix $\boldsymbol{Z}_s$ is a square matrix of order k:

$$\boldsymbol{Z}_s = \frac{j\omega\mu}{\pi}\boldsymbol{M} + \boldsymbol{Z}_b, \tag{4.84}$$

where μ is the permeability of the medium surrounding the conductors, $\boldsymbol{Z}_b$ is the diagonal matrix of the internal (skin) impedances of unit length of the conductor pairs:

$$\boldsymbol{Z}_b = 2\langle Z_{b1} \ Z_{b2} \ \dots \ Z_{bk}\rangle. \tag{4.85}$$

The square matrix $\boldsymbol{M}$ is derived from the geometrical layout (Fig. 4.13), with the j-th element of its i-th row:

$$m_{ij} = \ln\frac{r_{ij}}{a_{ij}}, \tag{4.86}$$

$$r_{ij} = \sqrt{d_1 d_2}, \qquad a_{ij} = \sqrt{b_1 b_2} \quad (i \neq j). \tag{4.87}$$

The interpretation of d_1, d_2, b_1, b_2, r_{ii}, a_{ii} is seen in Fig. 4.13.

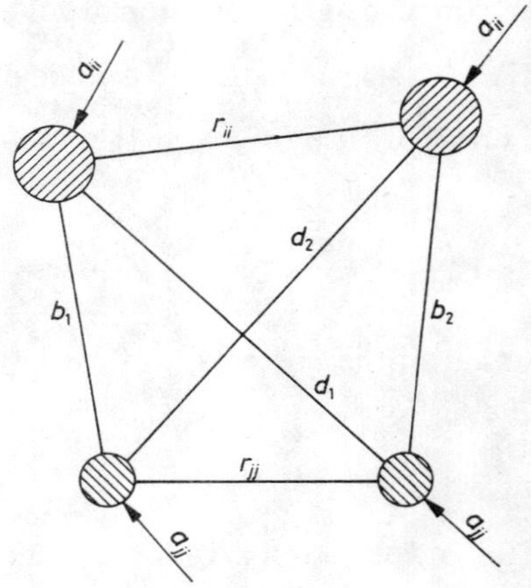

Fig. 4.13

The continuity equation yields a further relationship between **u** and **i**. Let the continuity equation be written for the volume of a cylinder of length dz surrounding one of the conductors of the j-th conductor pair (Fig. 4.14):

$$\oint \mathbf{J}_j \mathrm{d}\mathbf{A} + \frac{\partial Q_j}{\partial t} = 0, \tag{4.88}$$

where $\mathbf{J}_j$ is current density on the surface of the cylinder, and Q_j is the charge inside the cylinder. The integral of $\mathbf{J}_j$ on the surface of the cylinder at z is the current i_j of the conductor, while at $z+\mathrm{d}z$ it is $-\left(i_j+\frac{\mathrm{d}i_j}{\mathrm{d}z}\mathrm{d}z\right)$. If q_j is the charge on a unit length of the conductor, $Q_j=q_j\mathrm{d}z$, and thus from (4.88):

$$i_j-\left(i_j+\frac{\mathrm{d}i_j}{\mathrm{d}z}\mathrm{d}z\right)+j\omega q_j\mathrm{d}z=0\,. \tag{4.89}$$

The voltage u_j of the j-th conductor pair is expressible with the aid of the charges of the conductor pairs:

$$u_j=\sum_{i=1}^{k} p_{ij}q_i\,. \tag{4.90}$$

The coefficients p_{ij} appearing here are proportional to m_{ij} written in (4.86). Thus according to (4.90) the following matrix equation is written for the voltages of the

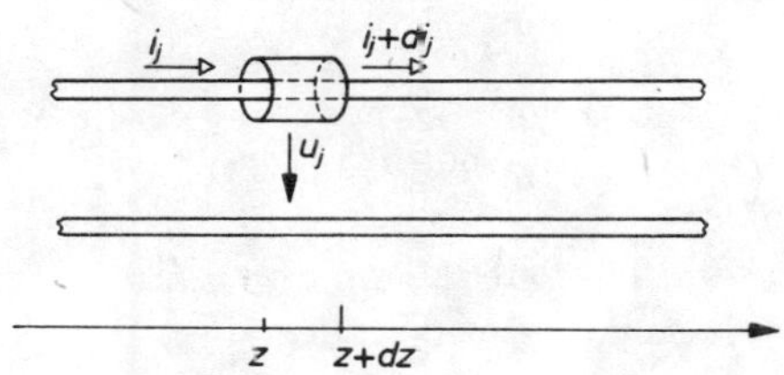

Fig. 4.14

conductor pairs:

$$\mathbf{u}=\frac{1}{\pi\varepsilon_k}\boldsymbol{M}\boldsymbol{q}\,, \tag{4.91}$$

where ε_k is the complex permittivity of the medium surrounding the conductors, and

$$\mathbf{q}=\begin{bmatrix} q_1 \\ q_2 \\ \vdots \\ q_k \end{bmatrix} \tag{4.92}$$

is the column matrix formed by the charges on unit length of the conductors. According to (4.91) and (4.89):

$$-\frac{\mathrm{d}\mathbf{i}}{\mathrm{d}z}=\boldsymbol{Y}_p\mathbf{u}\,, \tag{4.93}$$

where

$$\boldsymbol{Y}_p=j\omega\varepsilon_k\pi\boldsymbol{M}^{-1} \tag{4.94}$$

is the parallel admittance matrix.

(4.82) and (4.94) constitute a system of differential equations. If either **u** or **i** is eliminated from them, the telegraph equations generalized to transmission line systems are obtained:

$$\frac{d^2\mathbf{u}}{dz^2} = \boldsymbol{\Gamma}^2\mathbf{u}, \tag{4.95}$$

$$\frac{d^2\mathbf{i}}{dz^2} = \boldsymbol{\Gamma}^{+2}\mathbf{i}, \tag{4.96}$$

where

$$\boldsymbol{\Gamma}^2 = \boldsymbol{Z}_s\boldsymbol{Y}_p, \qquad \boldsymbol{\Gamma}^{+2} = \boldsymbol{Y}_p\boldsymbol{Z}_s, \tag{4.97}$$

$$\boldsymbol{\Gamma} = (\boldsymbol{Z}_s\boldsymbol{Y}_p)^{1/2}, \qquad \boldsymbol{\Gamma}^{+} = (\boldsymbol{Y}_p\boldsymbol{Z}_s)^{1/2}, \tag{4.98}$$

and $\boldsymbol{\Gamma}$ is the propagation coefficient matrix. The solutions of equations (4.95) and (4.96) are:

$$\mathbf{u}(z) = e^{-\boldsymbol{\Gamma}z}\mathbf{U}^{(+)} + e^{\boldsymbol{\Gamma}z}\mathbf{U}^{(-)}, \tag{4.99}$$

$$\mathbf{i}(z) = \boldsymbol{Y}_0(e^{-\boldsymbol{\Gamma}z}\mathbf{U}^{(+)} - e^{\boldsymbol{\Gamma}z}\mathbf{U}^{(-)}). \tag{4.100}$$

Here the constants $\mathbf{U}^{(+)}$ and $\mathbf{U}^{(-)}$ are the column matrices formed by the values at $z=0$ of voltage waves propagating in directions $+z$ and $-z$ respectively, and

$$\boldsymbol{Y}_0 = \boldsymbol{Z}_s^{-1}\boldsymbol{\Gamma} \tag{4.101}$$

is the characteristic admittance matrix.

The solutions of telegraph equations (4.95) and (4.96) are given by matrix functions (4.99) and (4.100). To interpret them the definition of a matrix function should be given. A function $\boldsymbol{f}(\boldsymbol{A})$ of a square matrix $\boldsymbol{A}$ can be obtained with the aid of the eigenvectors and eigenvalues of matrix $\boldsymbol{A}$. The eigenvector $\mathbf{X}_s$ of a square matrix $\boldsymbol{A}$ of order p satisfies the equation

$$\boldsymbol{A}\mathbf{X}_s = \lambda_s\mathbf{X}_s = \lambda_s\boldsymbol{1}\mathbf{X}_s, \tag{4.102}$$

i.e. the matrix transforms its eigenvector into a vector parallel with the eigenvector, and the transform is the eigenvector multiplied by λ_s. λ_s is the eigenvalue corresponding to the eigenvector. A square matrix may have several eigenvectors and eigenvalues. The eigenvalues may be determined from (4.102), since the equation

$$[\boldsymbol{A} - \lambda_s\boldsymbol{1}]\mathbf{X}_s = \mathbf{0} \tag{4.103}$$

may have nontrivial solutions ($\mathbf{X}_s \neq \mathbf{0}$) only if

$$\det|\boldsymbol{A} - \lambda_s\boldsymbol{1}| = 0, \tag{4.104}$$

i.e. denoting the j-th element of the i-th row of $\boldsymbol{A}$ by a_{ij}:

$$\begin{vmatrix} a_{11}-\lambda_s & a_{12} & \dots & a_{1p} \\ a_{21} & a_{22}-\lambda_s & \dots & a_{2p} \\ \dots & \dots & \dots & \dots \\ a_{p1} & a_{p2} & \dots & a_{pp}-\lambda_s \end{vmatrix} = 0\,. \tag{4.105}$$

The left-hand side is a polynomial of degree p in λ_s, the so-called characteristic equation, with at most p different solutions. Thus a square matrix of order p may have at most p different eigenvalues. One normed eigenvector corresponds to each eigenvalue. If the characteristic equation has multiple roots, an infinite number of eigenvectors correspond to these.

Knowing the distinct eigenvalues the matrix $\boldsymbol{A}$ may be written with the aid of matrix Lagrange-polynomials.* With Lagrange-polynomials, a function $y(x)$ can be written with values $y_1, y_2, \dots, y_k, \dots, y_p$ at given points $x_1, x_2, \dots, x_k, \dots, x_p$:

$$y(x) = \sum_{k=1}^{p} \prod_{\substack{j=1 \\ j\neq k}}^{p} \frac{x-x_j}{x_k-x_j}\, y_k = \frac{(x-x_2)(x-x_3)\dots(x-x_p)}{(x_1-x_2)(x_1-x_3)\dots(x_1-x_p)}\, y_1 +$$
$$+ \frac{(x-x_1)(x-x_3)\dots(x-x_p)}{(x_2-x_1)(x_2-x_3)\dots(x_2-x_p)}\, y_2 +$$
$$+ \dots + \frac{(x-x_1)(x-x_2)\dots(x-x_{p-1})}{(x_p-x_1)(x_p-x_2)\dots(x_p-x_{p-1})}\, y_p = \sum_{k=1}^{p} L_k y_k\,, \tag{4.106}$$

where L_k is the Lagrange-polynomial. The coefficient of y_k is seen to be 1 if $x=x_k$ and 0 if $x=x_j$ $(j\neq k)$. Thus this polynomial is indeed of value y_k at points x_k $(k=1, 2, \dots, p)$. Generalizing the interpolation formula (4.106), a matrix $\boldsymbol{L}_k$ may be written whose product with eigenvector $\mathbf{X}_j$ of the matrix $\boldsymbol{A}$ is $\mathbf{0}$ if $j\neq k$, and $\mathbf{X}_k$ if $j=k$, i.e.

$$\boldsymbol{L}_k \mathbf{X}_j = \begin{cases} \mathbf{X}_k, & \text{if} \quad j=k\,, \\ \mathbf{0}\,, & \text{if} \quad j\neq k\,. \end{cases} \tag{4.107}$$

The matrix polynomial sought is

$$\boldsymbol{L}_k = \prod_{\substack{j=1 \\ j\neq k}}^{p} \frac{\boldsymbol{A}-\lambda_j \boldsymbol{1}}{\lambda_k-\lambda_j} =$$
$$= \frac{(\boldsymbol{A}-\lambda_1\boldsymbol{1})(\boldsymbol{A}-\lambda_2\boldsymbol{1})\dots(\boldsymbol{A}-\lambda_{k-1}\boldsymbol{1})(\boldsymbol{A}-\lambda_{k+1}\boldsymbol{1})\dots(\boldsymbol{A}-\lambda_p\boldsymbol{1})}{(\lambda_k-\lambda_1)(\lambda_k-\lambda_2)\dots(\lambda_k-\lambda_{k-1})(\lambda_k-\lambda_{k+1})\dots(\lambda_k-\lambda_p)}\,. \tag{4.108}$$

* The method presented here, utilizing Lagrange-polynomials, is only applicable to cases with the characteristic equation having multiple roots, if the minimal equation has no multiple roots. However, only the case of distinct eigenvalues will be considered.

When determining the product $\boldsymbol{L}_k\mathbf{X}_j$, it should be noted that $\boldsymbol{A}\mathbf{X}_k=\lambda_k\mathbf{X}_k$, so

$$(\boldsymbol{A}-\boldsymbol{1}\lambda_j)\mathbf{X}_k=\boldsymbol{1}(\lambda_k-\lambda_j)\mathbf{X}_k\,. \tag{4.109}$$

Multiplying this by $(\boldsymbol{A}-\boldsymbol{1}\lambda_i)$, the product obtained is $\boldsymbol{1}(\lambda_k-\lambda_i)(\lambda_k-\lambda_j)\mathbf{X}_k$, i.e. the numerator of (4.108) multiplied by $\mathbf{X}_k$ is the product of $\boldsymbol{1}\mathbf{X}_k$ and the denominator. For $\boldsymbol{L}_k\mathbf{X}_j$ $(j\neq k)$, $\mathbf{0}$ is obtained after carrying out the multiplication because of the presence of the term $(\lambda_j-\lambda_j)$.

We note that the sum of all Lagrange-polynomials of a matrix $\boldsymbol{A}$ of order p is the unit matrix:

$$\sum_{k=1}^{p}\boldsymbol{L}_k=\boldsymbol{1}\,. \tag{4.110}$$

This fact provides a useful method for checking the calculations.

The product $\boldsymbol{A}\mathbf{X}$ may be expressed with the aid of matrix Lagrange-polynomials, namely

$$\boldsymbol{A}\mathbf{X}=\sum_{k=1}^{p}\boldsymbol{L}_k\lambda_k\mathbf{X}\,. \tag{4.111}$$

In the case of $\mathbf{X}=\mathbf{X}_j$ this equation holds according to the above considerations, since $\boldsymbol{L}_j\mathbf{X}_j=\boldsymbol{1}\mathbf{X}_j$, and the further terms of the sum are zero. Let the vector $\mathbf{X}$ be decomposed with the aid of the eigenvectors $\mathbf{X}_k$ $(k=1,2,\ldots,p)$ of the matrix $\boldsymbol{A}$, i.e. write it in the form

$$\mathbf{X}=\sum_{k=1}^{p}\alpha_k\mathbf{X}_k\,. \tag{4.112}$$

$\boldsymbol{L}_k\mathbf{X}$ is the component of vector $\mathbf{X}$ in the direction of eigenvector $\mathbf{X}_k$, i.e.

$$\boldsymbol{L}_k\mathbf{X}=\alpha_k\mathbf{X}_k\,. \tag{4.113}$$

Thus the terms of the sum is (4.111) are the components of $\boldsymbol{A}\mathbf{X}$ in the direction of the eigenvectors of $\boldsymbol{A}$, i.e. (4.111) yields the decomposition of $\boldsymbol{A}\mathbf{X}$ with the aid of the eigenvectors of $\boldsymbol{A}$. From (4.111):

$$\boldsymbol{A}=\sum_{k=1}^{p}\boldsymbol{L}_k\lambda_k\,. \tag{4.114}$$

A function of the matrix $\boldsymbol{A}$ is defined as follows:

$$\boldsymbol{f}(\boldsymbol{A})=\sum_{k=1}^{p}\boldsymbol{L}_k f(\lambda_k)\,. \tag{4.115}$$

This means that the function should be applied to the eigenvalues. Thus, for example,

$$\boldsymbol{A}^m=\sum_{k=1}^{p}\boldsymbol{L}_k\lambda_k^m\,,$$

$$\sin \boldsymbol{A} = \sum_{k=1}^{p} \boldsymbol{L}_k \sin \lambda_k ,$$

$$e^{\boldsymbol{A}z} = \sum_{k=1}^{p} \boldsymbol{L}_k e^{\lambda_k z} .$$

It is noted that the eigenvectors of $\boldsymbol{f}(\boldsymbol{A})$ coincide with those of $\boldsymbol{A}$.

In the present problem the eigenvalues γ_i^2 $(i = 1, 2, \ldots, k)$ of the matrix $\boldsymbol{\Gamma}^2$ are to be determined, i.e. the algebraic equation of order k

$$\det |\boldsymbol{\Gamma}^2 - \gamma_i^2 \boldsymbol{1}| = 0 \tag{4.116}$$

is to be solved. From a knowledge of γ_i^2 the matrix Lagrange-polynomials may be written according to (4.108):

$$\boldsymbol{L}_i(\boldsymbol{\Gamma}^2) = \prod_{\substack{j=1 \\ j \neq i}}^{k} \frac{\boldsymbol{\Gamma}^2 - \gamma_j^2 \boldsymbol{1}}{\gamma_i^2 - \gamma_j^2} \qquad (i = 1, 2, \ldots, k), \tag{4.117}$$

and hence from (4.115):

$$e^{\pm \boldsymbol{\Gamma} z} = \sum_{i=1}^{k} e^{\pm \gamma_i z} \boldsymbol{L}_i(\boldsymbol{\Gamma}^2) \tag{4.118}$$

and

$$\boldsymbol{\Gamma} = \sum_{i=1}^{k} \gamma_i \boldsymbol{L}_i(\boldsymbol{\Gamma}^2) \tag{4.119}$$

$\boldsymbol{Y}_0$ can be determined from (4.101). Substituting $\boldsymbol{Y}_0$ and (4.118) into (4.99) and (4.100) the following solution of the telegraph equations of the transmission line system is obtained:

$$\mathbf{u}(z) = \sum_{i=1}^{k} \boldsymbol{L}_i(\boldsymbol{\Gamma}^2) [e^{-\gamma_i z} \mathbf{U}^{(+)} + e^{\gamma_i z} \mathbf{U}^{(-)}] , \tag{4.120}$$

$$\mathbf{i}(z) = \boldsymbol{Y}_0 \sum_{i=1}^{k} \boldsymbol{L}_i(\boldsymbol{\Gamma}^2) [e^{-\gamma_i z} \mathbf{U}^{(+)} - e^{\gamma_i z} \mathbf{U}^{(-)}] . \tag{4.121}$$

The solution can be interpreted as follows. The formulae obtained for both voltage and current can be divided into two parts, describing in general damped waves propagating in directions $+z$ and $-z$, respectively. The quantity $-\gamma_i z$ appears in the exponents of the functions describing waves propagating in direction $+z$, while $+\gamma_i z$ appears for those propagating in direction $-z$. The wave travelling with propagation coefficient $\pm \gamma_i$ is called a mode. In general, two waves correspond to a mode, propagating in directions $+z$ and $-z$, respectively. The number of mode equals the number of eigenvalues γ_i^2 of the matrix $\boldsymbol{\Gamma}^2$. This coincides with the number of conductor pairs, except that in the case of multiple roots of the characteristic polynomial it may be less than that.

Equations (4.120) and (4.121) describe the decomposition according to modes of the waves propagating in the transmission line system. The voltage column matrix $\mathbf{u}_i(z)$ corresponding to the i-th mode is the i-th term of (4.120):

$$\mathbf{u}_i(z)=\boldsymbol{L}_i(\boldsymbol{\Gamma}^2)\left[e^{-\gamma_i z}\mathbf{U}^{(+)}+e^{\gamma_i z}\mathbf{U}^{(-)}\right]=e^{-\gamma_i z}\mathbf{U}_i^{(+)}+e^{\gamma_i z}\mathbf{U}_i^{(-)}. \qquad (4.122)$$

The voltage column matrices $\mathbf{U}_i^{(+)}=\boldsymbol{L}_i\mathbf{U}^{(+)}$ and $\mathbf{U}_i^{(-)}=\boldsymbol{L}_i\mathbf{U}^{(-)}$ are eigenvectors of matrix $\boldsymbol{\Gamma}^2$, i.e. (4.120) describes the decomposition of $\mathbf{u}(z)$ according to eigenvectors, whose physical interpretation is the decomposition according to modes. The components of $\mathbf{u}_i$ are the parts of the conductor pair voltages corresponding to the i-th mode. The decomposition of $\mathbf{i}$ according to the eigenvectors of $\boldsymbol{\Gamma}^{2+}$ can similarly be written.

The constants $\mathbf{U}^{(+)}$ and $\mathbf{U}^{(-)}$ in (4.120) can be determined from a knowledge of the boundary conditions prescribed by the terminations of the transmission lines.

Transmission line system above earth

In case of transmission line systems consisting of cylindrical conductors above earth [18, 50], the earth is considered to be homogeneous, with loss and bounded by a horizontal plane. The results presented here without showing the method of derivation relate to the case when displacement currents in the earth are negligible in comparison with conduction currents. In practice this is true for frequencies under 1 MHz.

In the case of earth-return transmission lines the current of the conductor parallel to the earth returns through the load impedance and earth. The current i of the conductor creates electromagnetic waves. This causes currents to flow in the earth, creating an additional field. In our further calculations the component of electric field intensity in the direction of wave propagation is needed. To evaluate it, the image of the conductor in the earth should be taken into account (Fig. 4.15). At the point with coordinates x, y, z the longitudinal component of electric field intensity is

$$E_z=I(Z-Z_f)e^{-\gamma z}=i(Z-Z_f), \qquad (4.123)$$

where i is the current of the conductor,

$$Z\approx\frac{g^2}{2\pi j\omega\varepsilon_0}\ln\frac{\rho}{r}, \qquad (4.124)$$

$$g^2=\gamma^2-j\omega\mu_0\, j\omega\varepsilon_0\,, \qquad (4.125)$$

r and ρ are the distances of the point with coordinates x, y, z from the conductor and its image respectively, γ is the propagation coefficient of the wave, ε_0 and μ_0 are the permittivity and permeability of vacuum, Z_f is earth impedance:

$$Z_f=\frac{\omega\mu_0}{\pi}\left\{\frac{\pi}{4}\left[\frac{\sqrt{j}}{\tau}e^{-j\Theta}\mathscr{H}_1(\sqrt{j}\tau e^{j\Theta})+e^{j\Theta}\mathscr{H}_1(\sqrt{j}\tau e^{-j\Theta})-\right.\right.$$

$$-e^{-j\Theta}N_1(\sqrt{j}\tau e^{j\Theta})-e^{j\Theta}N_1(\sqrt{j}\tau e^{-j\Theta})\Big]-\frac{\cos 2\Theta}{\tau^2}\Big\}=$$

$$=\frac{\omega\mu_0}{\pi}(P+jQ), \tag{4.126}$$

where $\mathscr{H}_1$ is the Struve-function and N_1 the Neumann-function of first order. Their series yield the following:

$$P=\frac{\pi}{8}-\frac{\tau}{\sqrt{2}\cdot 3}\cos\Theta-\frac{\tau^2}{16}\left(\ln m\tau-\frac{5}{4}\right)\cos 2\Theta+$$

$$+\frac{\tau^2}{16}\Theta\sin 2\Theta-\frac{\tau^4}{384}\Theta\sin 4\Theta+\ldots \tag{4.127}$$

$$Q=\frac{1}{4}-\frac{1}{2}\ln m\tau+\frac{\tau}{\sqrt{2}\cdot 3}\cos\Theta-\frac{\tau^2\pi}{64}\cos 2\Theta+$$

$$+\frac{\tau^3}{\sqrt{2}\cdot 45}\cos 3\Theta+\ldots \tag{4.128}$$

$$m=0{\cdot}890536\ldots \tag{4.129}$$

and

$$\tau=\omega\mu\sigma_f\rho^2\,. \tag{4.130}$$

σ_f is the conductivity of earth, and the meaning of Θ is shown in Fig. 4.15.

The first and second terms of (4.123) may be considered to originate from the currents of the conductor and the earth, respectively.

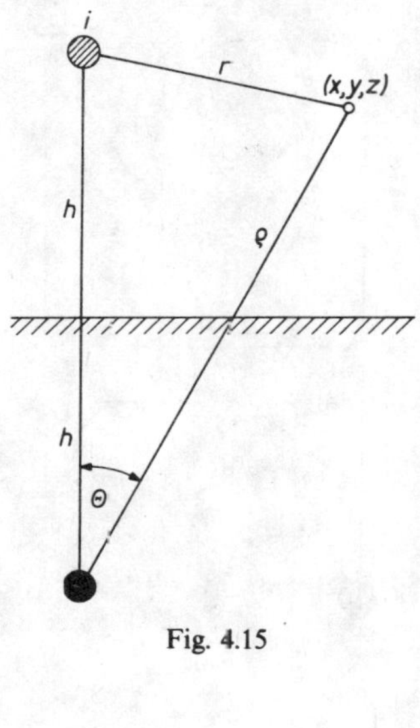

Fig. 4.15

On the basis of the formulae describing earth return transmission lines the calculation of a system of conductors horizontally running above earth (Fig. 4.16) can be reduced to the calculation of a system of conductor pairs.

The value of the longitudinal component of electric field intensity E_{zkj} on the surface of conductor k, as an effect of current i_j of conductor j as well as of earth currents induced by it can be calculated on the basis of (4.123) as

$$E_{zkj}=i_j(Z_{kj}-Z_{fkj}). \tag{4.131}$$

In case the distances between conductors are substantially greater than their radii, E_{zkj} may be considered to equal the electric field intensity arising in the axis of conductor k. The number of conductors being n, the tangential component of electric field intensity on the surface of conductor k according to (4.131) using the theorem of superposition is (Fig. 4.17):

$$E_{zk}=\sum_{j=1}^{n} i_j(Z_{kj}-Z_{fkj})=\sum_{j=1}^{n} i_j\left(\frac{g^2}{2\pi j\omega\varepsilon_0}\ln\frac{\rho_{kj}}{r_{kj}}-Z_{fkj}\right). \tag{4.132}$$

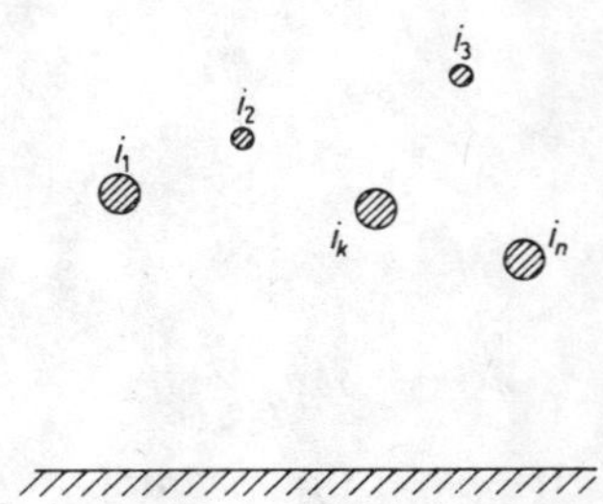

Fig. 4.16

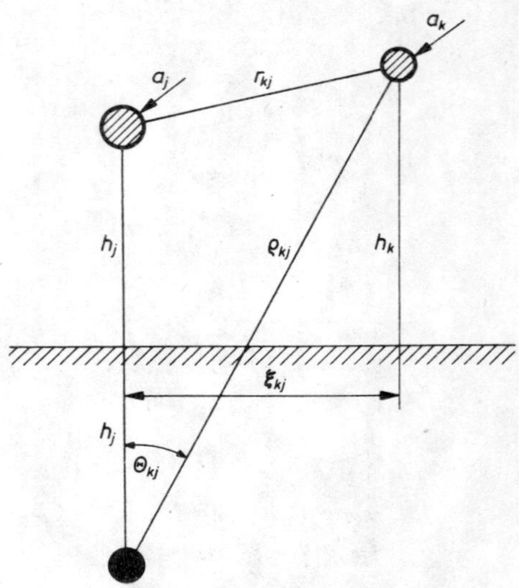

Fig. 4.17

The meaning of ρ_{kj} and r_{kj} is shown in Fig. 4.17. E_{zk} being continuous at the surface of the conductor, it can be written with the aid of the conductor's internal field as well:

$$E_{zk} = i_k Z_{bk}, \tag{4.133}$$

where Z_{bk} is the internal impedance of the conductor. Using the two latter equations the matrix equation

$$\left[\frac{g^2}{\pi j\,\omega\varepsilon_0}\boldsymbol{M} - (\boldsymbol{Z}_f + \boldsymbol{Z}_{bv})\right]\mathbf{i} = \mathbf{0} \tag{4.134}$$

is obtained. The matrix $\boldsymbol{M}$ appearing here equals the half of the matrix $\boldsymbol{M}$ in (4.84), if each conductor is considered to form a conductor pair with its image. The elements of diagonal matrix $\boldsymbol{Z}_{bv}$ are the skin impedances $Z_{bk}(k=1, 2, \ldots, n)$ of the conductors, while the elements of $\boldsymbol{Z}_f$ are the impedances Z_{fkj} appearing in (4.132), with their values calculated from (4.126). Let us introduce the matrix

$$\boldsymbol{Z}_b = \boldsymbol{Z}_{bv} - \boldsymbol{Z}_f . \tag{4.135}$$

On substitution into (4.134), and multiplication from the left by $-j\omega\varepsilon_0\pi\boldsymbol{M}^{-1}$:

$$(j\omega\varepsilon_0\pi\boldsymbol{M}^{-1}\boldsymbol{Z}_b - g^2\boldsymbol{1})\mathbf{i} = \mathbf{0}. \tag{4.136}$$

This is a homogeneous set of linear equations for the currents, with non-trivial solution existing only if the determinant formed by the coefficients equals zero. Thus the following equation is obtained:

$$\det|j\omega\varepsilon_0\pi\boldsymbol{M}^{-1}\boldsymbol{Z}_b - g^2\boldsymbol{1}| = 0. \tag{4.137}$$

Hence g^2 and thus γ^2 from (4.125) can be determined.

The voltages at the initial and final points of a section formed by coupled transmission lines of length l (Fig. 4.18) are written as follows in accordance with (4.99):

$$\mathbf{U}_1 = \mathbf{U}_1^{(+)} + e^{-\boldsymbol{\Gamma} l}\mathbf{U}_2^{(-)}, \tag{4.138}$$

$$\mathbf{U}_2 = e^{-\boldsymbol{\Gamma} l}\mathbf{U}_1^{(+)} + \mathbf{U}_2^{(-)}, \tag{4.139}$$

where $\mathbf{U}_1^{(+)} = \mathbf{U}^{(+)}$ and $\mathbf{U}_2^{(-)} = e^{\boldsymbol{\Gamma} l}\mathbf{U}^{(-)}$ are the column matrices of the amplitudes of waves propagating in directions $+z$ and $-z$ at locations $z=0$ and $z=l$, respectively.

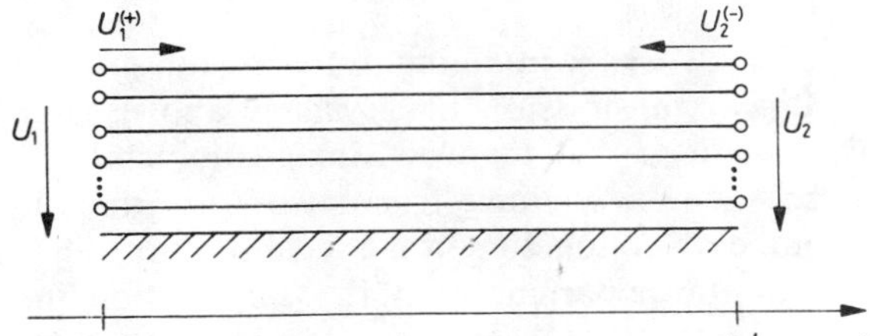

Fig. 4.18

(4.138) and (4.139) are summarized in matrix equation

$$\begin{bmatrix} \mathbf{U}_1 \\ \mathbf{U}_2 \end{bmatrix} = \begin{bmatrix} \mathbf{1} & e^{-\boldsymbol{\Gamma} l} \\ e^{-\boldsymbol{\Gamma} l} & \mathbf{1} \end{bmatrix} \begin{bmatrix} \mathbf{U}_1^{(+)} \\ \mathbf{U}_2^{(-)} \end{bmatrix}. \tag{4.140}$$

Similarly, the relationships

$$\mathbf{I}_1 = \boldsymbol{Y}_0(\mathbf{U}_1^{(+)} - e^{-\boldsymbol{\Gamma} l} U_2^{(-)}), \tag{4.141}$$

$$\mathbf{I}_2 = \boldsymbol{Y}_0(-e^{-\boldsymbol{\Gamma} l}\mathbf{U}_1^{(+)} + \mathbf{U}_2^{(-)}) \tag{4.142}$$

hold for currents according to (4.100), i.e.

$$\begin{bmatrix} \mathbf{I}_1 \\ \mathbf{I}_2 \end{bmatrix} = \boldsymbol{Y}_0 \begin{bmatrix} \mathbf{1} & -e^{-\boldsymbol{\Gamma} l} \\ -e^{-\boldsymbol{\Gamma} l} & \mathbf{1} \end{bmatrix} \begin{bmatrix} \mathbf{U}_1^{(+)} \\ \mathbf{U}_2^{(-)} \end{bmatrix}. \tag{4.143}$$

With the aid of (4.140):

$$\begin{bmatrix} \mathbf{I}_1 \\ \mathbf{I}_2 \end{bmatrix} = \boldsymbol{Y}_0 \begin{bmatrix} \mathbf{1} & -e^{-\boldsymbol{\Gamma} l} \\ -e^{-\boldsymbol{\Gamma} l} & \mathbf{1} \end{bmatrix} \begin{bmatrix} \mathbf{1} & e^{-\boldsymbol{\Gamma} l} \\ e^{-\boldsymbol{\Gamma} l} & \mathbf{1} \end{bmatrix}^{-1} \begin{bmatrix} \mathbf{U}_1 \\ \mathbf{U}_2 \end{bmatrix} = \boldsymbol{Y} \begin{bmatrix} \mathbf{U}_1 \\ \mathbf{U}_2 \end{bmatrix}, \tag{4.144}$$

where

$$\boldsymbol{Y} = \boldsymbol{Y}_0 \begin{bmatrix} \mathbf{1} & -e^{-\boldsymbol{\Gamma} l} \\ -e^{-\boldsymbol{\Gamma} l} & \mathbf{1} \end{bmatrix} \begin{bmatrix} \mathbf{1} & e^{-\boldsymbol{\Gamma} l} \\ e^{-\boldsymbol{\Gamma} l} & \mathbf{1} \end{bmatrix}^{-1} \tag{4.145}$$

is the admittance parameter matrix of the transmission line section. On carrying out the operations indicated,

$$\boldsymbol{Y} = \boldsymbol{Y}_0 \begin{bmatrix} \cosh \boldsymbol{\Gamma} l & -\mathbf{1} \\ -\mathbf{1} & \cosh \boldsymbol{\Gamma} l \end{bmatrix} \sinh^{-1} \boldsymbol{\Gamma} l = \begin{bmatrix} \boldsymbol{p} & \boldsymbol{r} \\ \boldsymbol{r} & \boldsymbol{p} \end{bmatrix} \tag{4.146}$$

is obtained, where

$$\boldsymbol{p} = \boldsymbol{Y}_0 \cosh \boldsymbol{\Gamma} l \sinh^{-1} \boldsymbol{\Gamma} l, \tag{4.147}$$

$$\boldsymbol{r} = -\boldsymbol{Y}_0 \sinh^{-1} \boldsymbol{\Gamma} l. \tag{4.148}$$

The section of the transmission line system will be characterized by matrices $\boldsymbol{p}$ and $\boldsymbol{r}$ in the following analysis.

Networks containing transmission line systems

In the analysis of networks containing transmission line systems [52] the conditions discussed for transmission line systems above earth relate to each section. It is further assumed that the electromagnetic fields of distinct sections create no coupling between the sections. The network is formed by b transmission line sections. The number of conductors in the sections may differ. The connection of the network, the admittance parameters of the transmission line sections as well as the characteristics of the $(k+1)$-terminal elements connected to connection points of transmission line sections, to vertices are considered to be known.

Let order numbers be associated with the conductors of the sections and the nodes of the vertices. The conductors above earth of the m-th section are assigned order numbers 1, 2, ..., k_m, while the nodes of vertex (i) are assigned order numbers 1, 2, ..., n_i. The connection of the conductors of the m-th section to the nodes of vertex (i) is characterized by matrix $\mathbf{a}_{im}$. The columns of $\mathbf{a}_{im}$ correspond to the conductors of section m, and its rows to the nodes of vertex (i). If the p-th

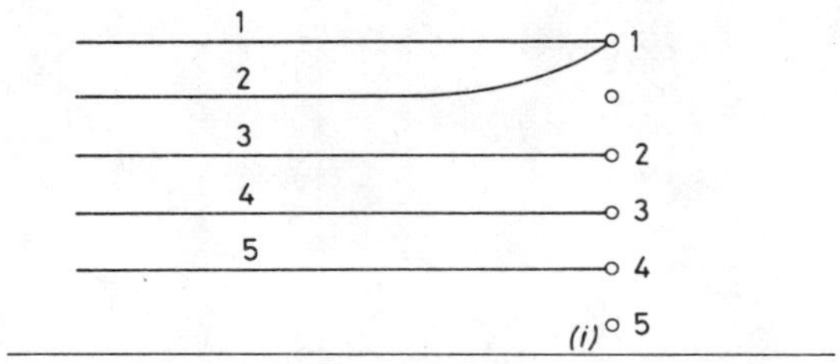

Fig. 4.19

conductor of section m is incident with the l-th node of vertex (i), the p-th element of the l-th row in $\mathbf{a}_{im}$ is 1, otherwise it is 0. E. g. in the case drawn in Fig. 4.19 $k_m = 5$, $n_i = 5$ and

$$\mathbf{a}_{im} = \begin{bmatrix} 1 & 1 & 0 & 0 & 0 \\ 0 & 0 & 1 & 0 & 0 \\ 0 & 0 & 0 & 1 & 0 \\ 0 & 0 & 0 & 0 & 1 \\ 0 & 0 & 0 & 0 & 0 \end{bmatrix}.$$

If section m does not connect to vertex (i) $\mathbf{a}_{im} = \mathbf{0}$.

Besides transmission line sections a lumped element $(k_i + 1)$-terminal network may connect to vertex (i) (Fig. 4.20). One of its terminals connects to earth and is not assigned any order numbers, while the further terminals are numbered $(1, 2, \ldots, k_i)$. The connection of these terminals to the nodes of the vertex are described by matrix $\mathbf{t}_i$. The r-th element of the l-th row of matrix $\mathbf{t}_i$ is 1 if the r-th terminal is incident with the l-th node of vertex (i), otherwise it is 0.

A graph may be associated with networks containing coupled transmission line sections as well. The edges of the graph correspond to transmission line sections, and its vertices to connection points. Edge m is equivalent to a $2 \times (k_m + 1)$-terminal element from the point of view of network theory, since the earth should also be considered a conductor.

The graph is once again characterized in our analysis by its (non-basis) incidence matrix. The incidence matrix is defined for our discussion to yield information about the interconnection of the conductors of transmission line sections and the nodes of vertices as well. This is given as a hypermatrix. A matrix row in the incidence matrix corresponds to a vertex, and a column to an edge. The m-th element of the i-th row is the matrix $\boldsymbol{a}_{im}$ defined above.

A network has been drawn in Fig. 4.21 with the earth represented by the thick line, the sections by thin lines, and the number of dashes indicates the number of

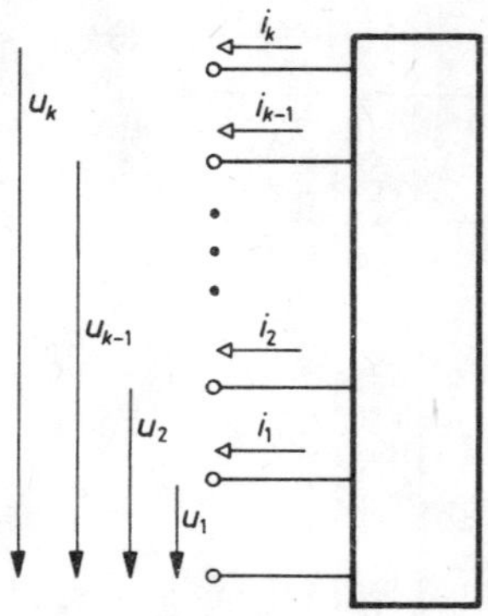

Fig. 4.20

conductors above earth in each section. This number is the number of columns of the corresponding block $\boldsymbol{a}_{im}$. The incidence matrix of the network is

$$\boldsymbol{A}_{t0} = \begin{bmatrix} \boldsymbol{a}_{11} & \boldsymbol{0} & \boldsymbol{0} & \boldsymbol{a}_{14} & \boldsymbol{0} & \boldsymbol{0} & \boldsymbol{0} & \boldsymbol{0} \\ \boldsymbol{a}_{21} & \boldsymbol{a}_{22} & \boldsymbol{a}_{23} & \boldsymbol{0} & \boldsymbol{0} & \boldsymbol{0} & \boldsymbol{0} & \boldsymbol{0} \\ \boldsymbol{0} & \boldsymbol{a}_{32} & \boldsymbol{0} & \boldsymbol{0} & \boldsymbol{a}_{35} & \boldsymbol{a}_{36} & \boldsymbol{0} & \boldsymbol{a}_{38} \\ \boldsymbol{0} & \boldsymbol{0} & \boldsymbol{a}_{43} & \boldsymbol{a}_{44} & \boldsymbol{a}_{45} & \boldsymbol{0} & \boldsymbol{a}_{47} & \boldsymbol{0} \\ \boldsymbol{0} & \boldsymbol{0} & \boldsymbol{0} & \boldsymbol{0} & \boldsymbol{0} & \boldsymbol{a}_{56} & \boldsymbol{0} & \boldsymbol{0} \\ \boldsymbol{0} & \boldsymbol{0} & \boldsymbol{0} & \boldsymbol{0} & \boldsymbol{0} & \boldsymbol{0} & \boldsymbol{0} & \boldsymbol{a}_{68} \\ \boldsymbol{0} & \boldsymbol{0} & \boldsymbol{0} & \boldsymbol{0} & \boldsymbol{0} & \boldsymbol{0} & \boldsymbol{a}_{77} & \boldsymbol{0} \end{bmatrix}.$$

In the case of $\boldsymbol{a}_{im} = \boldsymbol{0}$, $\boldsymbol{0}$ has been written in the matrix.

To write the equations of the network, orientations or reference directions should be associated with the sections of the network and the edges of the graph. The orientations are arbitrary. Thus, for example, the graph of the network drawn in Fig. 4.21 is shown in Fig. 4.22 with a possible choice of edge orientations indicated. For such a directed graph the directed incidence matrix $\boldsymbol{A}_t$ may be defined. Its construction is similar to that of incidence matrix $\boldsymbol{A}_{t0}$. The m-th element of its i-th row is $\boldsymbol{a}_{im}$ if section m is incident with vertex (i) with its orientation pointing away from the vertex. If section m is incident with vertex (i), directed towards vertex (i), the m-th element of the i-th row is $-\boldsymbol{a}_{im}$. The directed incidence matrix of the network characterized by the directed graph shown in Fig. 4.22 is

$$\boldsymbol{A}_t = \begin{bmatrix} \boldsymbol{a}_{11} & 0 & 0 & \boldsymbol{a}_{14} & 0 & 0 & 0 & 0 \\ -\boldsymbol{a}_{21} & -\boldsymbol{a}_{22} & -\boldsymbol{a}_{23} & 0 & 0 & 0 & 0 & 0 \\ 0 & \boldsymbol{a}_{32} & 0 & 0 & -\boldsymbol{a}_{35} & \boldsymbol{a}_{36} & 0 & -\boldsymbol{a}_{38} \\ 0 & 0 & \boldsymbol{a}_{43} & -\boldsymbol{a}_{44} & \boldsymbol{a}_{45} & 0 & \boldsymbol{a}_{47} & 0 \\ 0 & 0 & 0 & 0 & 0 & -\boldsymbol{a}_{56} & 0 & 0 \\ 0 & 0 & 0 & 0 & 0 & 0 & 0 & \boldsymbol{a}_{68} \\ 0 & 0 & 0 & 0 & 0 & 0 & -\boldsymbol{a}_{77} & 0 \end{bmatrix}.$$

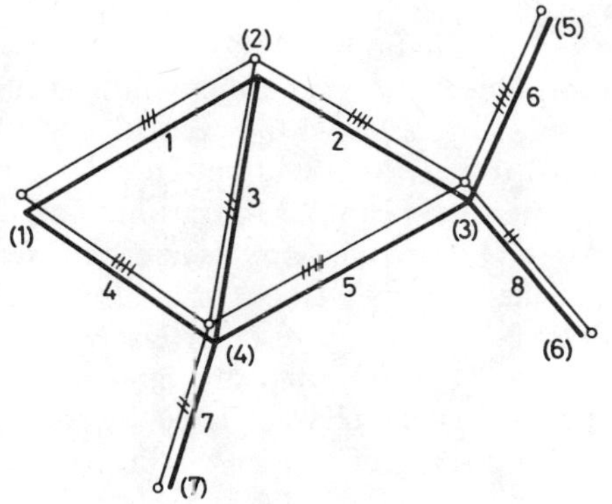

Fig. 4.21

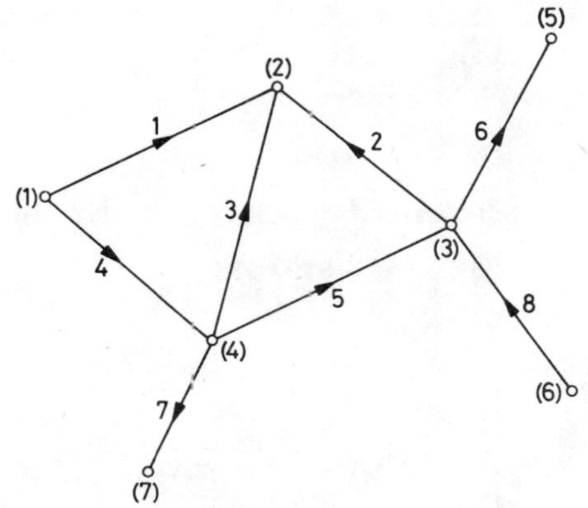

Fig. 4.22

Let the matrices $\boldsymbol{p}_m$ and $\boldsymbol{r}_m$ defined in (4.147) and (4.148) be written for each section of the network ($m=1, 2, \ldots, b$), and let these form hypermatrices.

$$\boldsymbol{P}=\begin{bmatrix} \boldsymbol{p}_1 & \boldsymbol{0} & \boldsymbol{0} & \ldots & \boldsymbol{0} \\ \boldsymbol{0} & \boldsymbol{p}_2 & \boldsymbol{0} & \ldots & \boldsymbol{0} \\ \ldots & \ldots & \ldots & \ldots & \ldots \\ \boldsymbol{0} & \boldsymbol{0} & \boldsymbol{0} & \ldots & \boldsymbol{p}_b \end{bmatrix}, \tag{4.149}$$

$$\boldsymbol{R}=\begin{bmatrix} \boldsymbol{r}_1 & \boldsymbol{0} & \boldsymbol{0} & \ldots & \boldsymbol{0} \\ \boldsymbol{0} & \boldsymbol{r}_2 & \boldsymbol{0} & \ldots & \boldsymbol{0} \\ \ldots & \ldots & \ldots & \ldots & \ldots \\ \boldsymbol{0} & \boldsymbol{0} & \boldsymbol{0} & \ldots & \boldsymbol{r}_b \end{bmatrix}, \tag{4.150}$$

which will be used in our calculations.

Coupled transmission lines are primarily employed in power distribution networks. Therefore, in accordance with practical considerations, our further calculations are restricted to networks without current-sources.

Generators and consumers may connect to transmission line sections at vertices, that can be considered $(k+1)$-terminal elements at each vertex (Fig. 4.20). At vertex (i), the voltages between the terminals and earth and the currents of the terminals constitute column matrices $\mathbf{u}_i$ and $\mathbf{i}_i$, respectively. The reference directions of voltages and currents have been indicated in Fig. 4.20. Let this (k_i+1)-terminal element connected to vertex (i) be characterized by its open-circuit impedance matrix $\boldsymbol{z}_{gi}$ and the column matrix $\mathbf{u}_{gi}$ of its open-circuit voltages (see chapter 3.). With these:

$$\mathbf{u}_i=\mathbf{u}_{gi}-\boldsymbol{z}_{gi}\mathbf{i}_i\,. \tag{4.151}$$

To write the equations of the network this should be transformed to obtain the voltages between the nodes and earth in the order of the numbering of the nodes at vertex (i). This is achieved by multiplication by $\boldsymbol{t}_i$:

$$\boldsymbol{t}_i\mathbf{u}_i=\boldsymbol{t}_i\mathbf{u}_{gi}-\boldsymbol{t}_i\boldsymbol{z}_{gi}\mathbf{i}_i\,. \tag{4.152}$$

This can be written for each vertex ($i=1, 2, \ldots, n$), and summarized in

$$\mathbf{U}=\mathbf{U}_g-\boldsymbol{Z}_g\mathbf{I}_c\,, \tag{4.153}$$

where

$$\mathbf{U}=\begin{bmatrix} \boldsymbol{t}_1\mathbf{u}_1 \\ \boldsymbol{t}_2\mathbf{u}_2 \\ \vdots \\ \boldsymbol{t}_n\mathbf{u}_n \end{bmatrix}; \qquad \mathbf{U}_g=\begin{bmatrix} \boldsymbol{t}_1\mathbf{u}_{g1} \\ \boldsymbol{t}_2\mathbf{u}_{g2} \\ \vdots \\ \boldsymbol{t}_n\mathbf{u}_{gn} \end{bmatrix}; \qquad \mathbf{I}_c=\begin{bmatrix} \boldsymbol{t}_1\mathbf{i}_1 \\ \boldsymbol{t}_2\mathbf{i}_2 \\ \vdots \\ \boldsymbol{t}_n\mathbf{i}_n \end{bmatrix}, \tag{4.154}$$

$$Z_g = \begin{bmatrix} t_1 z_{g1} t_1^+ & 0 & 0 & \dots & 0 \\ 0 & t_2 z_{g2} t_2^+ & 0 & \dots & 0 \\ \dots & \dots & \dots & \dots & \dots \\ 0 & 0 & 0 & \dots & t_n z_{gn} t_n^+ \end{bmatrix}. \tag{4.155}$$

Provided that Z_g is non-singular, (4.153) yields

$$\mathbf{I}_c = Y_g(\mathbf{U}_g - \mathbf{U}), \tag{4.156}$$

where

$$Y_g = Z_g^{-1}. \tag{4.157}$$

The elements of column matrix **U**, the voltages of vertices, are to be determined from the equations relating to the network. The voltages at the ends at vertex (i) of the transmission line sections incident with vertex (i), and the voltages of the (k_i+1)-terminal elements connected to vertex (i) equal the same column matrix $\mathbf{u}_i$. The network equations are written with the aid of these to satisfy Kirchhoff's loop equations at vertices automatically. At most k nodes belong to a vertex, with the node equations written for them being linearly independent. The node equations for the nodes at one vertex are written in one matrix equation.

To write the current flowing out of a particular node of a vertex, the currents of the branches incident with the vertex are classified into three groups, in the same way as for the transmission line networks discussed earlier. The first group is formed by the currents of the sections incident with the vertex whose reference directions coincide with the orientation of the section. The currents with reference directions opposite to the orientation of the relevant section constitute the second group. Finally the currents of the $(k+1)$-terminal elements incident with the vertex appear in the third group.

If section m is incident with vertices (i) and (j) directed from (i) to (j) the current in the first group of the section is:

$$\mathbf{i}_{mi} = p_m \mathbf{u}_i + r_m \mathbf{u}_j. \tag{4.158}$$

Similar equations may be written for each section. The set of equations thus derived can be summarized as follows:

$$\mathbf{I}' = \frac{1}{2} P(A_{t0} + A_t)^+ \mathbf{U} + \frac{1}{2} R(A_{t0} - A_t)^+ \mathbf{U}. \tag{4.159}$$

The current in the second group of section m is:

$$\mathbf{i}_{mj} = r_m \mathbf{u}_i + p_m \mathbf{u}_j. \tag{4.160}$$

Writing the same equation for each section the formula

$$\mathbf{I}'' = \frac{1}{2} R(A_{t0} + A_t)^+ \mathbf{U} + \frac{1}{2} P(A_{t0} - A_t)^+ \mathbf{U} \tag{4.161}$$

is obtained. The currents of the generators and impedances at the vertices are as in (4.156).

The currents at each node should satisfy the node equation. The currents written in $\mathbf{I}'$ and $\mathbf{I}''$ flow out of the nodes of vertices. Let us construct the sum of these at each vertex, and let $\mathbf{I}'_c$ and $\mathbf{I}''_c$ denote the column vectors formed by these:

$$\mathbf{I}'_c = \frac{1}{2}(\boldsymbol{A}_{t0} + \boldsymbol{A}_t)\mathbf{I}' ; \qquad \mathbf{I}''_c = \frac{1}{2}(\boldsymbol{A}_{t0} - \boldsymbol{A}_t)\mathbf{I}'' . \tag{4.162}$$

The currents appearing in $\mathbf{I}_c$ flow towards the node of the vertex, having the currents corresponding to the elements of $\mathbf{I}'_c$ and $\mathbf{I}''_c$ flow outwards. Thus

$$\mathbf{I}'_c + \mathbf{I}''_c - \mathbf{I}_c = \mathbf{0} . \tag{4.163}$$

Substituting the above formulae, with the notation

$$\boldsymbol{Y}_c = \frac{1}{2}\boldsymbol{A}_{t0}(\boldsymbol{P} + \boldsymbol{R})\boldsymbol{A}_{t0}^{+} + \frac{1}{2}\boldsymbol{A}_t(\boldsymbol{P} - \boldsymbol{R})\boldsymbol{A}_t^{+} \tag{4.164}$$

the following equation is obtained on rearrangement:

$$(\boldsymbol{Y}_c + \boldsymbol{Y}_g)\mathbf{U} = \boldsymbol{Y}_g \boldsymbol{U}_g . \tag{4.165}$$

$\boldsymbol{Y}_c$ is a symmetric hypermatrix. Its main diagonal is formed by the sums of matrices $\boldsymbol{p}$ of the sections incident with the vertex corresponding to the relevant row (column). The further matrix elements are the matrices $\boldsymbol{r}$ of the sections connecting the vertices corresponding to the row and column. If no section connects the two vertices, the corresponding block in the hypermatrix is $\boldsymbol{0}$. $\boldsymbol{Y}_c$ is the vertex admittance matrix of the network.

If $\boldsymbol{Y}_g$ does not exist due to the properties of the $(k+1)$-terminal elements connected to the vertices, but $\boldsymbol{Z}_g$ does exist, the voltages of the vertices are determined from

$$(\boldsymbol{Z}_g \boldsymbol{Y}_c + \boldsymbol{1})\mathbf{U} = \mathbf{U}_g \tag{4.166}$$

obtained similarly to (4.30) from (4.165) on multiplication by $\boldsymbol{Z}_g$.

If on deactivation of the $(k+1)$-terminal elements connected to the vertices both open-circuits and short-circuits appear between the terminals and the earth, the nodes are classified into two groups. The first group is formed by the nodes connected to the earth by non-zero impedance in the deactivated network. The nodes in the second group and the earth are connected by short-circuits in the deactivated network (the impedance between the node and the earth is zero). Two order numbers are assigned to the nodes. At the first numbering the order numbers are given to the nodes in the order of the vertices, i.e. the nodes of vertex (1) are assigned order numbers $1, 2, \ldots, k_1$, and those of vertex (2) the order numbers $k_1+1, k_1+2, \ldots, k_1+k_2$, etc. The matrices in (4.154) and (4.155) are arranged in accordance with this numbering. At the second numbering the order numbers $1, 2, \ldots$ are given to nodes in the first group defined above, while the nodes in the second group are assigned the remaining order numbers. The connection between the two numberings is described by a matrix $\boldsymbol{H}$ with its rows corresponding to the first and its columns to the second numbering. The j-th element of the i-th row in matrix $\boldsymbol{H}$ is 1 if the i-th node in the first numbering obtains order number j in the second numbering, otherwise it is 0.

Let the matrices $\boldsymbol{Y}_c$ and $\boldsymbol{Y}_g$ be partitioned in accordance with the two groups of nodes. To this end the matrix $\boldsymbol{H}$ is employed:

$$\boldsymbol{H}\boldsymbol{Y}_c\boldsymbol{H}^+ = \begin{bmatrix} \boldsymbol{Y}_{c1} & \boldsymbol{Y}_{c2} \\ \boldsymbol{Y}_{c3} & \boldsymbol{Y}_{c4} \end{bmatrix}, \tag{4.167}$$

$$\boldsymbol{H}\boldsymbol{Y}_g\boldsymbol{H}^+ = \begin{bmatrix} \boldsymbol{Y}_{g1} & \boldsymbol{0} \\ \boldsymbol{0} & \boldsymbol{Y}_{g2} \end{bmatrix}. \tag{4.168}$$

Introducing the notations

$$\mathbf{U}_h = \boldsymbol{H}\mathbf{U} = \begin{bmatrix} \mathbf{U}_{h1} \\ \mathbf{U}_{h2} \end{bmatrix} \tag{4.169}$$

and

$$\mathbf{U}_{gh} = \boldsymbol{H}\mathbf{U}_g = \begin{bmatrix} \mathbf{U}_{gh1} \\ \mathbf{U}_{gh2} \end{bmatrix} \tag{4.170}$$

(4.165) is written in the following transformed form:

$$\begin{bmatrix} \boldsymbol{Y}_{c1} + \boldsymbol{Y}_{g1} & \boldsymbol{Y}_{c2} \\ \boldsymbol{Y}_{c3} & \boldsymbol{Y}_{c4} + \boldsymbol{Y}_{g2} \end{bmatrix} \begin{bmatrix} \mathbf{U}_{h1} \\ \mathbf{U}_{h2} \end{bmatrix} = \begin{bmatrix} \boldsymbol{Y}_{g1} & \boldsymbol{0} \\ \boldsymbol{0} & \boldsymbol{Y}_{g2} \end{bmatrix} \begin{bmatrix} U_{gh1} \\ U_{gh2} \end{bmatrix}. \tag{4.171}$$

Here $\mathbf{U}_{h2} = \mathbf{U}_{gh2}$, and thus

$$(\boldsymbol{Y}_{c1} + \boldsymbol{Y}_{g1})\mathbf{U}_{h1} = \boldsymbol{Y}_{g1}\mathbf{U}_{gh1} - \boldsymbol{Y}_{c2}\mathbf{U}_{gh2}. \tag{4.172}$$

Hence

$$\mathbf{U}_{h1} = (\boldsymbol{Y}_{c1} + \boldsymbol{Y}_{g1})^{-1}(\boldsymbol{Y}_{g1}\mathbf{U}_{gh1} - \boldsymbol{Y}_{c2}\mathbf{U}_{gh2}) \tag{4.173}$$

is the voltage of the nodes connected to the earth by non-zero impedance. The voltages of the nodes to earth in accordance with the first numbering are obtained from here as

$$\mathbf{U} = \boldsymbol{H}^+\mathbf{U}_h = \boldsymbol{H}^+ \begin{bmatrix} \mathbf{U}_{h1} \\ \mathbf{U}_{gh2} \end{bmatrix}. \tag{4.174}$$

Knowing the voltages at the vertices the current and voltage at any point of the system examined can be calculated. These results make it possible to determine the powers of the voltage-sources and the consumers as well.

Examples

1. In the network consisting of seven transmission line sections drawn in Fig. 4.23 the characteristic admittances and propagation coefficients at the operating frequency and the lengths of each section, as well as the admittance of the capacitor connected to vertex (4) are known. The admittance matrix of the transmission line network is to be determined.

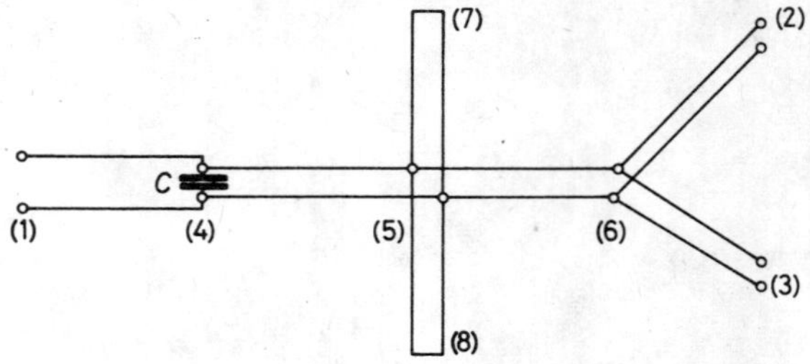

Fig. 4.23

The vertices of the passive network are classified into three groups for the application of the method presented for the determination of matrix $\boldsymbol{Y}_1$ characterizing the network. Vertices (1), (2), (3) being connection points belong to the first group. The second group is formed by vertices (4), (5), (6) with their terminations not short-circuits, and finally vertices (7) and (8) terminated by short-circuits are included in the third group.

The graph of the network with the orientations and order numbers chosen for the edges indicated is shown in Fig. 4.24. Thus the directed (non-basis) incidence matrix of the network is

$$\boldsymbol{A}_t = \begin{bmatrix} 1 & 0 & 0 & 0 & 0 & 0 & 0 \\ 0 & 0 & 0 & 0 & 0 & -1 & 0 \\ 0 & 0 & 0 & 0 & 0 & 0 & -1 \\ -1 & 1 & 0 & 0 & 0 & 0 & 0 \\ 0 & -1 & 1 & 1 & 1 & 0 & 0 \\ 0 & 0 & 0 & 0 & -1 & 1 & 1 \\ 0 & 0 & -1 & 0 & 0 & 0 & 0 \\ 0 & 0 & 0 & -1 & 0 & 0 & 0 \end{bmatrix}.$$

To write the vertex admittance matrix the values p_i and r_i $(i=1, 2, \ldots, 7)$ are calculated according to (4.16), and the matrices $\boldsymbol{P}$ and $\boldsymbol{R}$ in (4.17) are formed from them. Thus in accordance with (4.28):

$$Y_c=\left[\begin{array}{ccc|ccc|cc} p_1 & 0 & 0 & r_1 & 0 & 0 & 0 & 0 \\ 0 & p_6 & 0 & 0 & 0 & r_6 & 0 & 0 \\ 0 & 0 & p_7 & 0 & 0 & r_7 & 0 & 0 \\ \hline r_1 & 0 & 0 & p_1+p_2 & r_2 & 0 & 0 & 0 \\ 0 & 0 & 0 & r_2 & p_2+p_3+p_4+p_5 & r_5 & r_3 & r_4 \\ 0 & r_6 & r_7 & 0 & r_5 & p_5+p_6+p_7 & 0 & 0 \\ \hline 0 & 0 & 0 & 0 & r_3 & 0 & p_3 & 0 \\ 0 & 0 & 0 & 0 & r_4 & 0 & 0 & p_4 \end{array}\right],$$

where the partitioning of (4.56) has been indicated. The submatrix Y_{g2} appearing in (4.57) in our example is

$$Y_{g2}=\langle j\omega C\ \ 0\ \ 0\rangle .$$

Hence, according to the notations of (4.62) Z_{11} can be determined from the equation

$$\left[\begin{array}{ccc|ccc} p_1 & 0 & 0 & r_1 & 0 & 0 \\ 0 & p_6 & 0 & 0 & 0 & r_6 \\ 0 & 0 & p_7 & 0 & 0 & r_7 \\ \hline r_1 & 0 & 0 & p_1+p_2+j\omega C & r_2 & 0 \\ 0 & 0 & 0 & r_2 & p_2+p_3+p_4+p_5 & r_5 \\ 0 & r_6 & r_7 & 0 & r_5 & p_5+p_6+p_7 \end{array}\right]\begin{bmatrix} Z_{11} & Z_{12} \\ Z_{21} & Z_{22} \end{bmatrix}=$$

$$=\begin{bmatrix} \mathit{1} & \mathit{0} \\ \mathit{0} & \mathit{1} \end{bmatrix}.$$

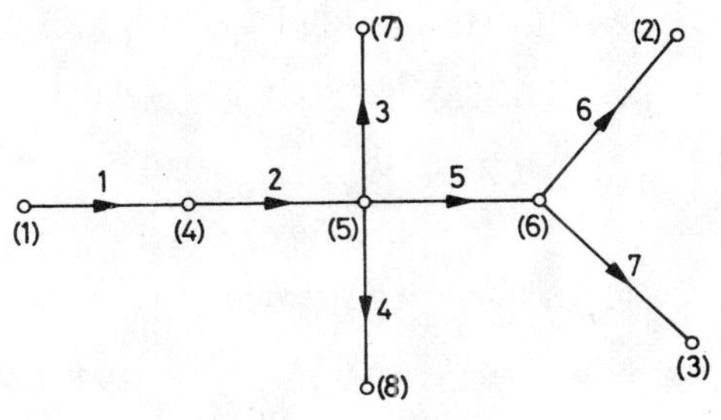

Fig. 4.24

Introducing the notations

$$s_1 = p_1 + p_2 + j\omega C\,,$$
$$s_2 = p_2 + p_3 + p_4 + p_5\,,$$
$$s_3 = p_5 + p_6 + p_7\,,$$
$$D = s_1 s_2 s_3 - s_1 r_5^2 - s_3 r_2^2$$

the admittance matrix sought is

$$Y_1 = \frac{1}{D}\begin{bmatrix} Dp_1 - r_1^2(s_2 s_3 - r_5^2) & -r_1 r_2 r_5 r_6 & -r_1 r_2 r_5 r_7 \\ -r_1 r_2 r_5 r_6 & Dp_6 - (s_1 s_2 - r_2^2) r_6^2 & -(s_1 s_2 - r_2^2) r_6 r_7 \\ -r_1 r_2 r_5 r_7 & -(s_1 s_2 - r_2^2) r_6 r_7 & Dp_7 - (s_1 s_2 - r_2^2) r_7^2 \end{bmatrix}.$$

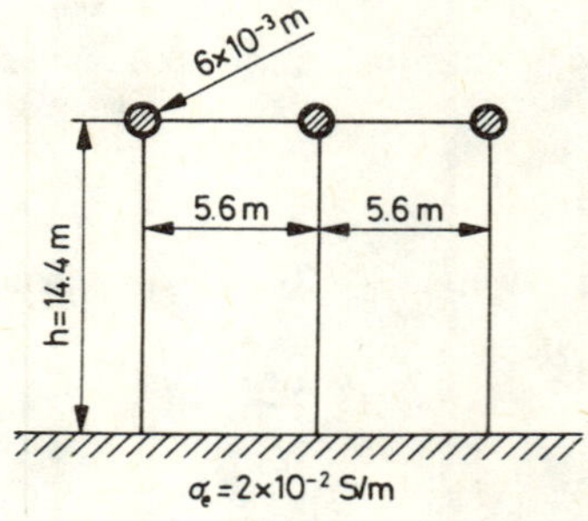

Fig. 4.25

2. The admittance parameter matrix of the transmission line system shown in Fig. 4.25 consisting of three copper ($\sigma = 57 \cdot 10^6$ S/m) conductors of length $l = 300$ km arranged parallel to each other as well as to the surface of the earth is to be determined. The conductivity of earth is $\sigma_e = 2 \cdot 10^{-2}$ S/m.

From the geometry of the layout:

$$M = \begin{bmatrix} 4{\cdot}238 & 0{\cdot}8281 & 0{\cdot}5074 \\ 0{\cdot}8281 & 4{\cdot}238 & 0{\cdot}8281 \\ 0{\cdot}5074 & 0{\cdot}8281 & 4{\cdot}238 \end{bmatrix},$$

$$M^{-1} = \begin{bmatrix} 0{\cdot}2471 & -0{\cdot}04419 & -0{\cdot}02095 \\ -0{\cdot}04419 & 0{\cdot}2532 & -0{\cdot}04419 \\ -0{\cdot}02095 & -0{\cdot}04419 & 0{\cdot}2471 \end{bmatrix}.$$

The matrix appearing in (4.135) formed by internal impedances of unit length and earth impedances is

$$Z_b = \begin{bmatrix} 0{\cdot}2027 \ +j0{\cdot}2151 & 0{\cdot}04714+j0{\cdot}1978 & 0{\cdot}04711+j0{\cdot}1946 \\ 0{\cdot}04714+j0{\cdot}1978 & 0{\cdot}2027 \ +j0{\cdot}2151 & 0{\cdot}04714+j0{\cdot}1978 \\ 0{\cdot}04711+j0{\cdot}1946 & 0{\cdot}04714+j0{\cdot}1978 & 0{\cdot}2027 \ +j0{\cdot}2151 \end{bmatrix} 10^{-3}\ \Omega/\text{m}.$$

From the matrices written, taking into account that in our example $\mu=\mu_0=$ $=4\pi\cdot 10^{-7}$ Vs/Am and $\varepsilon_k=\varepsilon_0=8{\cdot}86\cdot 10^{-12}$ As/Vm, Z_s and Y_p may be determined from (4.84) and (4.94), and hence Γ^2 obtained from (4.97):

$$\Gamma^2 = \begin{bmatrix} -1{\cdot}448 \ +j0{\cdot}4094 & -0{\cdot}2794+j0{\cdot}008013 & -0{\cdot}3041+j0{\cdot}04647 \\ -0{\cdot}3077+j0{\cdot}01505 & -1{\cdot}419 \ +j0{\cdot}4106 & -0{\cdot}3077+j0{\cdot}01505 \\ -0{\cdot}3041+j0{\cdot}04647 & -0{\cdot}2794+j0{\cdot}008013 & -1{\cdot}448 \ +j0{\cdot}4094 \end{bmatrix} 10^{-12}\,\text{m}^{-2}$$

The eigenvalues $\gamma_i^2 (i=1, 2, 3)$ of matrix Γ^2 are the roots of the equation

$$\det|\Gamma^2-\gamma_i^2 \mathbf{1}|=0,$$

i.e.

$$\gamma_1^2=(-1{\cdot}139+j0{\cdot}4099)\cdot 10^{-12}\,\text{m}^{-2},$$

$$\gamma_2^2=(-1{\cdot}144+j0{\cdot}3629)\cdot 10^{-12}\,\text{m}^{-2},$$

$$\gamma_3^2=(-2{\cdot}032+j0{\cdot}4566)\cdot 10^{-12}\,\text{m}^{-2}$$

are the squares of the propagation coefficients of the transmission line system.

The appropriate Lagrange-polynomials may be written with the aid of the eigenvalues of Γ^2 according to (4.117):

$$L_1(\Gamma^2) = \begin{bmatrix} 0{\cdot}1563+j0{\cdot}007752 & -0{\cdot}3124-j0{\cdot}007372 & 0{\cdot}1563+j0{\cdot}007752 \\ -0{\cdot}3444-j0{\cdot}001173 & 0{\cdot}6874-j0{\cdot}01550 & -0{\cdot}3444-j0{\cdot}001173 \\ 0{\cdot}1563+j0{\cdot}007752 & -0{\cdot}3124-j0{\cdot}007372 & 0{\cdot}1563+j0{\cdot}007752 \end{bmatrix},$$

$$L_2(\Gamma^2) = \begin{bmatrix} 0{\cdot}5 & 0 & -0{\cdot}5 \\ 0 & 0 & 0 \\ -0{\cdot}5 & 0 & 0{\cdot}5 \end{bmatrix},$$

$$L_3(\boldsymbol{\Gamma}^2)=\begin{bmatrix} 0{\cdot}3437-j0{\cdot}007752 & 0{\cdot}3124+j0{\cdot}007372 & 0{\cdot}3437-j0{\cdot}007752 \\ 0{\cdot}3444+j0{\cdot}001173 & 0{\cdot}3126+j0{\cdot}01550 & 0{\cdot}3444+j0{\cdot}001173 \\ 0{\cdot}3437-j0{\cdot}007752 & 0{\cdot}3124+j0{\cdot}007372 & 0{\cdot}3437-j0{\cdot}007752 \end{bmatrix}.$$

Thus in accordance with (4.115):

$$\boldsymbol{\Gamma}=\sum_{i=1}^{3} \boldsymbol{L}_i\gamma_i=$$

$$=\begin{bmatrix} 0{\cdot}1708\ +j1{\cdot}204 & -0{\cdot}01194+j0{\cdot}1093 & 0{\cdot}00317+j0{\cdot}1213 \\ -0{\cdot}01072+j0{\cdot}1207 & 0{\cdot}1743\ +j1{\cdot}193 & -0{\cdot}01072+j0{\cdot}1207 \\ 0{\cdot}00317+j0{\cdot}1213 & -0{\cdot}01194+j0{\cdot}1093 & 0{\cdot}1708\ +j1{\cdot}204 \end{bmatrix} 10^{-6}\,\mathrm{m}^{-1},$$

and the characteristic admittance matrix from (4.101)is

$$\boldsymbol{Y}_0=\begin{bmatrix} 1{\cdot}825\ +j0{\cdot}2773 & -0{\cdot}4472-j0{\cdot}1016 & -0{\cdot}2856-j0{\cdot}04948 \\ -0{\cdot}4472-j0{\cdot}1016 & 1{\cdot}888\ +j0{\cdot}3035 & -0{\cdot}4472-j0{\cdot}1016 \\ -0{\cdot}2856-j0{\cdot}04948 & -0{\cdot}4472-j0{\cdot}1016 & 1{\cdot}825\ +j0{\cdot}2773 \end{bmatrix} 10^{-3}\,\mathrm{S}.$$

Hence, from $\boldsymbol{\Gamma}l$ and $\boldsymbol{Y}_0$, (4.147) and (4.148) yield the submatrices of the admittance parameter matrix in (4.146):

$$\boldsymbol{p}=\begin{bmatrix} 1{\cdot}602\ -j4{\cdot}866 & -0{\cdot}6573+j1{\cdot}467 & -0{\cdot}3633+j1{\cdot}093 \\ -0{\cdot}6573+j1{\cdot}467 & 1{\cdot}771\ -j5{\cdot}058 & -0{\cdot}6573+j1{\cdot}467 \\ -0{\cdot}3633+j1{\cdot}093 & -0{\cdot}6573+j1{\cdot}467 & 1{\cdot}602\ -j4{\cdot}866 \end{bmatrix} 10^{-3}\,\mathrm{S},$$

$$\boldsymbol{r}=\begin{bmatrix} -1{\cdot}601\ +j5{\cdot}193 & 0{\cdot}6572-j1{\cdot}525 & 0{\cdot}3633-j1{\cdot}120 \\ 0{\cdot}6572-j1{\cdot}525 & -1{\cdot}770\ +j5{\cdot}392 & 0{\cdot}6572-j1{\cdot}525 \\ 0{\cdot}3633-j1{\cdot}120 & 0{\cdot}6572-j1{\cdot}525 & -1{\cdot}601\ +j5{\cdot}193 \end{bmatrix} 10^{-3}\,\mathrm{S}.$$

3. In the course of the discussion of networks containing transmission line systems the incidence matrix of the network drawn in Fig. 4.21 has been determined. In the following example the currents of the transmission line sections are to be calculated. To this end the admittance parameters of the sections written in (4.147) and (4.148) are employed.

To write the currents of the sections the orientations of the sections are chosen as in the graph in Fig. 4.22. The currents with reference directions coinciding with the orientations of the relevant sections are written according to (4.159):

$$\mathbf{I}'=\begin{bmatrix} \mathbf{i}_{11} \\ \mathbf{i}_{23} \\ \mathbf{i}_{34} \\ \mathbf{i}_{41} \\ \mathbf{i}_{54} \\ \mathbf{i}_{63} \\ \mathbf{i}_{74} \\ \mathbf{i}_{86} \end{bmatrix} = \begin{bmatrix} \boldsymbol{p}_1\mathbf{u}_1+\boldsymbol{r}_1\mathbf{u}_2 \\ \boldsymbol{p}_2\mathbf{u}_3+\boldsymbol{r}_2\mathbf{u}_2 \\ \boldsymbol{p}_3\mathbf{u}_4+\boldsymbol{r}_3\mathbf{u}_2 \\ \boldsymbol{p}_4\mathbf{u}_1+\boldsymbol{r}_4\mathbf{u}_4 \\ \boldsymbol{p}_5\mathbf{u}_4+\boldsymbol{r}_5\mathbf{u}_3 \\ \boldsymbol{p}_6\mathbf{u}_3+\boldsymbol{r}_6\mathbf{u}_5 \\ \boldsymbol{p}_7\mathbf{u}_4+\boldsymbol{r}_7\mathbf{u}_7 \\ \boldsymbol{p}_8\mathbf{u}_6+\boldsymbol{r}_8\mathbf{u}_3 \end{bmatrix}.$$

The indices of $\boldsymbol{p}$ and $\boldsymbol{r}$ are the order numbers of the sections, those of $\mathbf{u}$ refer to the relevant vertex, while the first index of $\mathbf{i}$ relates to the section and the second to the vertex.

The currents in the second group, i.e. those with reference directions opposite to the orientations of the relevant sections are, in accordance with (4.161):

$$\mathbf{I}''=\begin{bmatrix} \mathbf{i}_{12} \\ \mathbf{i}_{22} \\ \mathbf{i}_{32} \\ \mathbf{i}_{44} \\ \mathbf{i}_{53} \\ \mathbf{i}_{65} \\ \mathbf{i}_{77} \\ \mathbf{i}_{83} \end{bmatrix} = \begin{bmatrix} \boldsymbol{r}_1\mathbf{u}_1+\boldsymbol{p}_1\mathbf{u}_2 \\ \boldsymbol{r}_2\mathbf{u}_3+\boldsymbol{p}_2\mathbf{u}_2 \\ \boldsymbol{r}_3\mathbf{u}_4+\boldsymbol{p}_3\mathbf{u}_2 \\ \boldsymbol{r}_4\mathbf{u}_1+\boldsymbol{p}_4\mathbf{u}_4 \\ \boldsymbol{r}_5\mathbf{u}_4+\boldsymbol{p}_5\mathbf{u}_3 \\ \boldsymbol{r}_6\mathbf{u}_3+\boldsymbol{p}_6\mathbf{u}_5 \\ \boldsymbol{r}_7\mathbf{u}_4+\boldsymbol{p}_7\mathbf{u}_7 \\ \boldsymbol{r}_8\mathbf{u}_6+\boldsymbol{p}_8\mathbf{u}_3 \end{bmatrix}.$$

The matrices in (4.162) are

$$\mathbf{I}'_c=\begin{bmatrix} \boldsymbol{p}_1\mathbf{u}_1+\boldsymbol{r}_1\mathbf{u}_2+\boldsymbol{p}_4\mathbf{u}_1+\boldsymbol{r}_4\mathbf{u}_4 \\ \mathbf{0} \\ \boldsymbol{p}_2\mathbf{u}_3+\boldsymbol{r}_2\mathbf{u}_2+\boldsymbol{p}_6\mathbf{u}_3+\boldsymbol{r}_6\mathbf{u}_5 \\ \boldsymbol{p}_3\mathbf{u}_4+\boldsymbol{r}_3\mathbf{u}_2+\boldsymbol{p}_5\mathbf{u}_4+\boldsymbol{r}_5\mathbf{u}_3+\boldsymbol{p}_7\mathbf{u}_4+\boldsymbol{r}_7\mathbf{u}_7 \\ \mathbf{0} \\ \boldsymbol{p}_8\mathbf{u}_6+\boldsymbol{r}_8\mathbf{u}_3 \\ \mathbf{0} \end{bmatrix},$$

$$\mathbf{I}_c'' = \begin{bmatrix} 0 \\ r_1\mathbf{u}_1 + p_1\mathbf{u}_2 + r_2\mathbf{u}_3 + p_2\mathbf{u}_2 + r_3\mathbf{u}_4 + p_3\mathbf{u}_2 \\ r_5\mathbf{u}_4 + p_5\mathbf{u}_3 + r_8\mathbf{u}_6 + p_8\mathbf{u}_3 \\ r_4\mathbf{u}_1 + p_4\mathbf{u}_4 \\ r_6\mathbf{u}_3 + p_6\mathbf{u}_5 \\ 0 \\ r_7\mathbf{u}_4 + p_7\mathbf{u}_7 \end{bmatrix}.$$

According to (4.164), the vertex admittance matrix of the network is

$$Y_c = \begin{bmatrix} p_1+p_4 & r_1 & 0 & r_4 & 0 & 0 & 0 \\ r_1 & p_1+p_2+p_3 & r_2 & r_3 & 0 & 0 & 0 \\ 0 & r_2 & p_2+p_5+p_6+p_8 & r_5 & r_6 & r_8 & 0 \\ r_4 & r_3 & r_5 & p_3+p_4+p_5+p_7 & 0 & 0 & r_7 \\ 0 & 0 & r_6 & 0 & p_6 & 0 & 0 \\ 0 & 0 & r_8 & 0 & 0 & p_8 & 0 \\ 0 & 0 & 0 & r_7 & 0 & 0 & p_7 \end{bmatrix}.$$

CHAPTER 5

NETWORKS CONTAINING TWO-PORTS WITH EXTREME PARAMETERS, ELECTRONIC CIRCUITS

The analyses presented in this chapter concern networks which include linear two-ports in addition to two-terminal elements and coupled impedances and admittances.

Linear two-ports are known to be characterized by parameter matrices: impedance, admittance, hybrid, inverse hybrid, chain parameter matrices. For a given two-port, however, the existence of each parameter matrix listed is not ensured, i.e. a parameter matrix which does exist may be singular. In the following sections, two-ports described by such singular matrices will be called two-ports with extreme parameters.

The network may consist of impedances, admittances, sources as well as electronic components (transistors, electron tubes, etc.). Networks containing electronic components are called electronic circuits. Electronic components are usually active nonlinear elements, most of them being three-terminal elements treatable as nonreciprocal two-ports from the point of view of network theory.

The time-variation of the signals (currents, voltages) appearing in electronic circuits is periodic in most practical cases, and may be decomposed into a part constant in time (direct current component) and into one periodically changing with zero mean value. The working points of nonlinear network elements and thus of electronic components are determined by the direct current voltages and currents, hence the parameters needed for calculations concerning the periodic signals may be obtained from a knowledge of the characteristics of the nonlinear elements.

In the networks examined the task of the electronic components is the processing of the periodic signals superposed on the direct current signals. In some cases the amplitude of the periodic signal with zero mean value is small enough to allow the approximation of the characteristic describing the component by a straight line in the appropriate neighbourhood of the working point. This means that the output signal caused by a sinusoidal input signal is approximately sinusoidal, and the electronic component can be approximated by a linear two-port, possibly one with extreme parameters. The following discussion is restricted to networks containing elements that may be treated as linear. At first, attention is drawn to the question of setting the working points of electronic components at prescribed values. Subsequently, some methods for the analysis of electronic circuits are discussed for case of sinusoidal signals.

Adjustment of working points in electronic circuits

In case of direct current signals electronic circuits consist of voltage-sources (possibly current-sources), resistances (conductances), nonlinear elements forming two-terminal networks or two-ports, provided that the non-ideal generators are substituted by Thevenin or Norton generators. In such networks the necessity may arise to choose the value of some resistances to adjust the working points of the electronic components employed. Although the network is nonlinear, this problem leads to a linear set of equations.

The electronic component with a prescribed working point is modelled for the solution of the problem by two-terminal elements (in case of a diode by one element) with their currents and voltages prescribed. A two-terminal element with a prescribed current and voltage is called in the following *fixator*,* and is denoted in connection diagrams as shown in Fig. 5.1. Since both the current and voltage of a fixator are specified, its power is also determined, being positive or negative depending upon whether the element is consuming or producing power.

Accordingly, both the primary and secondary sides of an electronic component with prescribed working point may be substituted by fixators to establish their currents and voltages. Thus, for example, in the case of the transistor connected in the common-emitter configuration shown in Fig. 5.2, a, the voltage and current of the fixator representing the primary side are U_{BE} and I_B, while those of the fixator modelling the secondary side are U_{CE} and I_C (Fig. 5.2, b). Similarly two fixators may

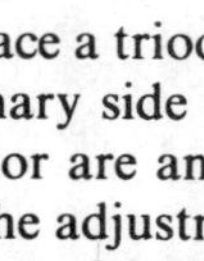

Fig. 5.1

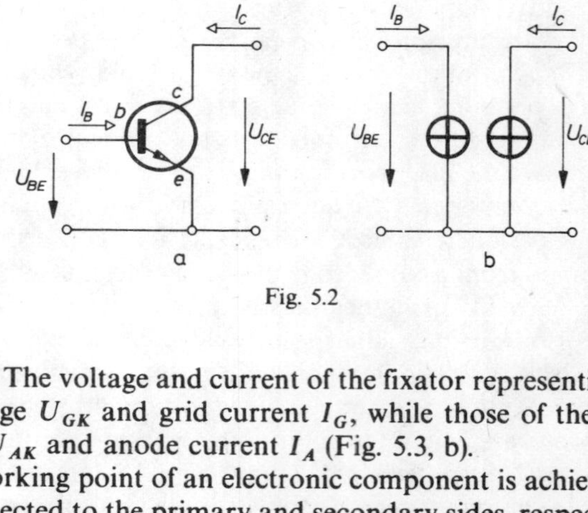

Fig. 5.2

replace a triode (Fig. 5.3, a). The voltage and current of the fixator representing the primary side are grid voltage U_{GK} and grid current I_G, while those of the other fixator are anode voltage U_{AK} and anode current I_A (Fig. 5.3, b).

The adjustment of the working point of an electronic component is achieved by means of two resistors connected to the primary and secondary sides, respectively. The values of these resistors which produce the prescribed working point are to be determined. For our calculations these resistors are considered as norators [3, 21].

* The name fixator is due to Prof. Dr György Fodor.

A norator is a two-terminal element with both its current and voltage of arbitrary value. In connection diagrams norators are shown as in Fig. 5.4.

By following this procedure, a network model is obtained, consisting of voltage-sources (current-sources), resistances, conductances, fixators and norators. The Kirchhoff equations written for this model may have a unique solution or they may be redundant or indefinite. To determine whether the problem is uniquely solvable it is expedient to substitute the following parallel, and series connections by single two-terminal elements in the network model.

(a) One fixator may replace several fixators with identical currents connected in series or fixators with identical voltages connected in parallel, a fixator and a resistance or conductance connected parallel or in series (Figs 5.5, a, b) a voltage-source or Thevenin-generator connected in series with a fixator, a current-source or Norton-generator connected in parallel with a fixator, a parallel connection of a fixator and a Thevenin generator and a series connection of a fixator and a Norton generator.

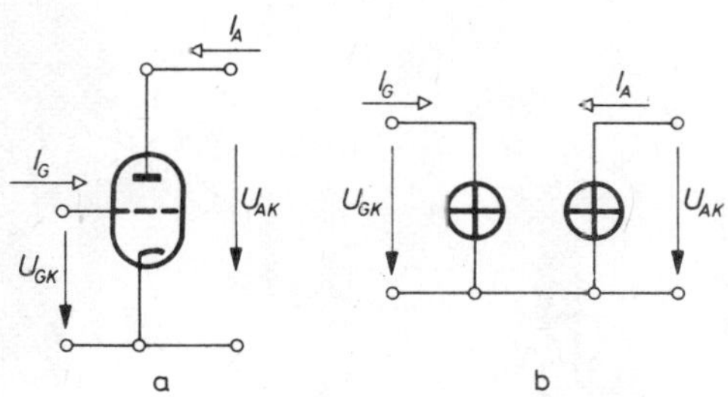

Fig. 5.3

Fig. 5.4

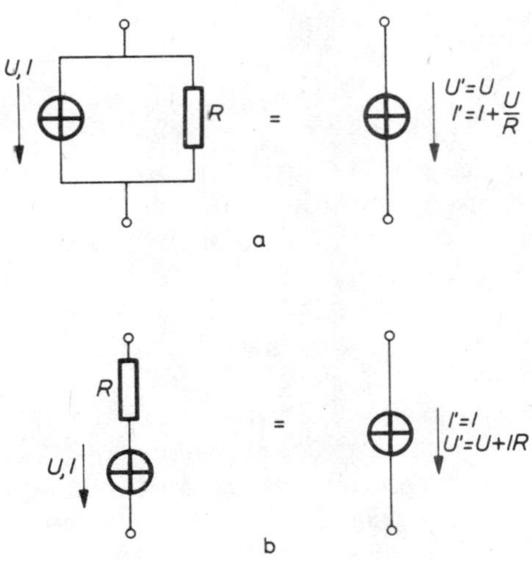

Fig. 5.5

(b) One norator is equivalent to parallel or series connections of norators, to any two-terminal network constructed of norators only, to a parallel or series connection of a norator and a resistance or conductance (Fig. 5.6), a series connection of a norator and a voltage-source, to a parallel connection of a norator and a current-source and to parallel or series connections of a norator and a Thevenin or Norton generator.

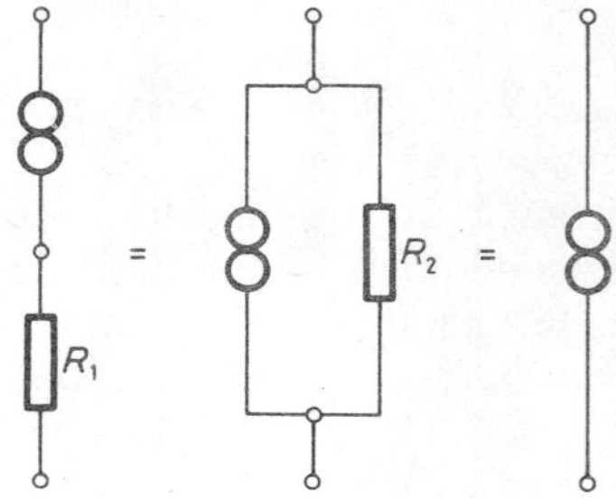

Fig. 5.6

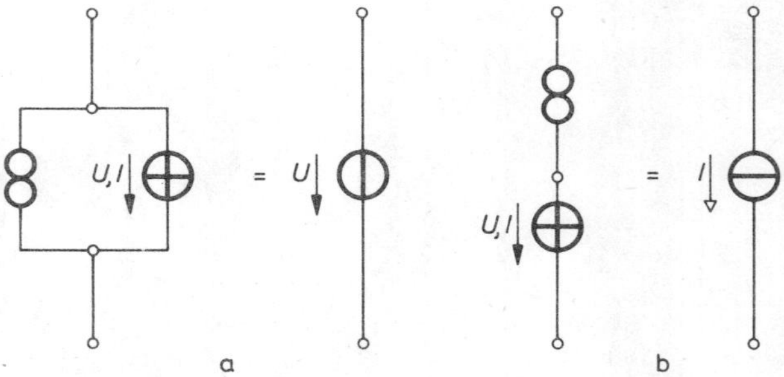

Fig. 5.7

(c) One voltage-source may replace a norator connected parallel with a voltage-source and a parallel connection of a norator and a fixator (Fig. 5.7, a).

(d) One current-source may replace a norator connected in series with a current-source and a series connection of a norator and a fixator (Fig. 5.7, b).

The investigation into the solvability of the network calculation problem is carried out using the reduced network model obtained by making as many as possible of the substitutions listed above.

There is one constraint for the current and voltage of a resistance or conductance, and the number of independent equations written for a network consisting of resistances, conductances, sources equals the number of unknowns. In this respect a fixator implies two constraints, while the norator implies none, i.e. the inclusion of a fixator causes the set of equations regarding the currents and voltages of the network to be redundant and they are indefinite on the inclusion of a norator.

Therefore, a necessary condition for the unique solvability of the problem is that the numbers of fixators and norators in the reduced network model must be equal.

For a loop consisting of fixators and voltage-sources only, and for cutsets formed exclusively of fixators and current-sources the relevant Kirchhoff equation is not necessarily satisfied. The equations of a network containing such a loop or cutset may be contradictory or redundant.

If a resistance (conductance) belongs to a loop formed by voltage-sources and fixators only except for the resistance (conductance), the voltage of this resistance (conductance) is determined by the sources and fixators of the loop. If this resistance (conductance) belongs at the same time to a cutset with all of its other branches current-sources and fixators, these latter determine the current of the resistance (conductance). The voltage and current are thus independently specified and in such cases Ohm's law is not necessarily satisfied, i.e. the relationships written may be contradictory.

If the network includes loops formed by norators and voltage-sources only or cutsets formed exclusively by norators and current-sources, the equations written for the network are indefinite.

Thus, it may be ascertained by such considerations whether the network examined presents one of the redundant, contradictory or indefinite problems listed above. Such problems without unique solution will be excluded from our discussion.

In the reduced network model Thevenin and Norton generators are considered to form two branches: a source and a resistance, and short-circuits and open-circuits are taken into account by regarding them as voltage-sources with zero source-voltage, and current-sources with zero source-current, respectively. After drawing the graph of the network, a tree of the graph is first chosen, with each fixator and voltage-source represented by tree-branches and each norator and current-source by chords. The remaining branches of the network may either be tree-branches or chords. The branches of the network are first classified into six groups as follows:

1. chords containing current-sources;
2. chords containing norators;
3. chords containing finite conductances;
4. tree-branches containing finite resistances;
5. tree-branches containing fixators;
6. tree-branches containing voltage-sources.

If such a classification of the branches is not possible, either loops are formed by voltage-sources and fixators or cutsets by current-sources and norators in the reduced network model. If such a classification of the branches is possible a tree of the network graph is chosen for further investigation, with each voltage-source and norator represented by tree-branches and each current-source and fixator by chords. The branches of the network are next classified into the following six groups:

1. chords containing current-sources;
2. chords containing fixators;
3. chords containing finite conductances;
4. tree-branches containing resistances;
5. tree-branches containing norators;
6. tree-branches containing voltage-sources.

Such a classification is impossible if current-sources and fixators form cutsets or norators and voltage-sources form loops.

If both classifications are possible, the branches in groups 3. and 4. should be chosen with the same branches forming group 3. in both cases. Thus naturally group 4. is also formed by identical branches in both classifications. If this is possible the network equations are solvable.

The analysis may be carried out with the aid of either of the above classifications. The following example uses the first classification. The number of branches in the groups is denoted by $b_1, b_2, b_3, b_4, b_5, b_6$, respectively. The number of branches in groups 2. and 5. is identical, i.e. $b_2 = b_5$.

The branches are numbered in the order of the classification. The order numbers of the loops and cutsets in the sets of loops and cutsets generated by the tree chosen are determined by the order numbers of the relevant chords and tree-branches. The sets of loops and cutsets are characterized by loop matrix $\boldsymbol{B}$ and cutset matrix $\boldsymbol{Q}$ respectively.

The linearly independent loop equations valid for the network are summarized in

$$\boldsymbol{B}\mathbf{U} = \mathbf{0} \tag{5.1}$$

where $\mathbf{U}$ is the column matrix of branch-voltages. Let $\boldsymbol{B}$ and $\mathbf{U}$ be partitioned in accordance with the classification of the branches. Thus (5.1) is written as

$$\begin{array}{c} \\ b_1 \\ b_2 \\ b_3 \end{array} \overset{\begin{array}{cccccc} b_1 & b_2 & b_3 & b_4 & b_5 & b_6 \end{array}}{\begin{bmatrix} \mathbf{1} & \mathbf{0} & \mathbf{0} & \boldsymbol{F}_{11} & \boldsymbol{F}_{12} & \boldsymbol{F}_{13} \\ \mathbf{0} & \mathbf{1} & \mathbf{0} & \boldsymbol{F}_{21} & \boldsymbol{F}_{22} & \boldsymbol{F}_{23} \\ \mathbf{0} & \mathbf{0} & \mathbf{1} & \boldsymbol{F}_{31} & \boldsymbol{F}_{32} & \boldsymbol{F}_{33} \end{bmatrix}} \begin{bmatrix} \mathbf{U}_1 \\ \mathbf{U}_2 \\ \mathbf{U}_3 \\ \mathbf{U}_4 \\ \mathbf{U}_5 \\ \mathbf{U}_{g6} \end{bmatrix} = \mathbf{0}, \tag{5.2}$$

where $\mathbf{U}_6 = \mathbf{U}_{g6}$ is the column matrix formed by the source-voltages of voltage-sources. The numbers of columns and rows in the sub-matrices of $\boldsymbol{B}$ are indicated above, and on the left-hand side of the matrix, respectively. (5.2) is decomposed into the following three matrix equations:

$$\mathbf{U}_1 + \boldsymbol{F}_{11}\mathbf{U}_4 + \boldsymbol{F}_{12}\mathbf{U}_5 + \boldsymbol{F}_{13}\mathbf{U}_{g6} = \mathbf{0}, \tag{5.3}$$

$$\mathbf{U}_2 + \boldsymbol{F}_{21}\mathbf{U}_4 + \boldsymbol{F}_{22}\mathbf{U}_5 + \boldsymbol{F}_{23}\mathbf{U}_{g6} = \mathbf{0}, \tag{5.4}$$

$$\mathbf{U}_3 + \boldsymbol{F}_{31}\mathbf{U}_4 + \boldsymbol{F}_{32}\mathbf{U}_5 + \boldsymbol{F}_{33}\mathbf{U}_{g6} = \mathbf{0}. \tag{5.5}$$

The linearly independent cutset equations of the network are summarized in

$$\boldsymbol{Q}\mathbf{I} = \mathbf{0}, \tag{5.6}$$

where $\mathbf{I}$ is the column matrix of the branch-currents of the network. As a consequence of the above numbering of branches, loops and cutsets, $\boldsymbol{B}$ and $\boldsymbol{Q}$ are of the forms

$$\boldsymbol{B} = [\boldsymbol{1}\ \boldsymbol{F}], \qquad \boldsymbol{Q} = [-\boldsymbol{F}^{+}\ \boldsymbol{1}], \tag{5.7}$$

due to the orthogonal relationship between them. According to (5.2), (5.6) and (5.7):

$$\begin{bmatrix} -\boldsymbol{F}_{11}^{+} & -\boldsymbol{F}_{21}^{+} & -\boldsymbol{F}_{31}^{+} & \boldsymbol{1} & \boldsymbol{0} & \boldsymbol{0} \\ -\boldsymbol{F}_{12}^{+} & -\boldsymbol{F}_{22}^{+} & -\boldsymbol{F}_{32}^{+} & \boldsymbol{0} & \boldsymbol{1} & \boldsymbol{0} \\ -\boldsymbol{F}_{13}^{+} & -\boldsymbol{F}_{23}^{+} & -\boldsymbol{F}_{33}^{+} & \boldsymbol{0} & \boldsymbol{0} & \boldsymbol{1} \end{bmatrix} \begin{bmatrix} \mathbf{I}_{g1} \\ \mathbf{I}_2 \\ \mathbf{I}_3 \\ \mathbf{I}_4 \\ \mathbf{I}_5 \\ \mathbf{I}_6 \end{bmatrix} = \mathbf{0}, \tag{5.8}$$

where $\mathbf{I}_1 = \mathbf{I}_{g1}$ is the column matrix formed by the source-currents of current-sources. Hence:

$$-\boldsymbol{F}_{11}^{+}\mathbf{I}_{g1} - \boldsymbol{F}_{21}^{-}\mathbf{I}_2 - \boldsymbol{F}_{31}^{-}\mathbf{I}_3 + \mathbf{I}_4 = \mathbf{0}, \tag{5.9}$$

$$-\boldsymbol{F}_{12}^{+}\mathbf{I}_{g1} - \boldsymbol{F}_{22}^{+}\mathbf{I}_2 - \boldsymbol{F}_{32}^{-}\mathbf{I}_3 + \mathbf{I}_5 = \mathbf{0}, \tag{5.10}$$

$$-\boldsymbol{F}_{13}^{+}\mathbf{I}_{g1} - \boldsymbol{F}_{23}^{+}\mathbf{I}_2 - \boldsymbol{F}_{33}^{+}\mathbf{I}_3 + \mathbf{I}_6 = \mathbf{0}. \tag{5.11}$$

The known quantities in Eqs (5.3) . . . (5.5) and (5.9) . . . (5.11) are the source-voltages $\mathbf{U}_{g6}$ of voltage-sources, the source-currents $\mathbf{I}_{g1}$ of current-sources, the voltages $\mathbf{U}_5$ and currents $\mathbf{I}_5$ of fixators, as well as the coefficient matrices $\boldsymbol{F}_{ik}$ ($i, k = 1, 2, 3$). The relationships between the currents and voltages of the branches in groups 3. and 4.

$$\mathbf{I}_3 = \boldsymbol{G}_3\mathbf{U}_3, \tag{5.12}$$

$$\mathbf{U}_4 = \boldsymbol{R}_4\mathbf{I}_4 \tag{5.13}$$

are further known, with $\boldsymbol{G}_3$ and $\boldsymbol{R}_4$ being the diagonal matrices of conductances in the branches of group 3. and of resistances in the branches of group 4. respectively. The voltages $\mathbf{U}_2$ and currents $\mathbf{I}_2$ of the norators are to be determined.

Since $b_2 = b_5$, the matrix F_{22}, and thus F_{22}^{+}, is square. If F_{22} is non-singular, the network analysis problem is solvable, and $\mathbf{I}_2$ may be expressed from (5.10):

$$\mathbf{I}_2 = -{F_{22}^{+}}^{-1} F_{12}^{+}\mathbf{I}_{g1} - {F_{22}^{+}}^{-1} F_{32}^{+}\mathbf{I}_3 + {F_{22}^{+}}^{-1}\mathbf{I}_5 . \tag{5.14}$$

On substitution into (5.9):

$$\begin{aligned}(F_{21}^{+}{F_{22}^{+}}^{-1}F_{32} - F_{31}^{+})\mathbf{I}_3 + \mathbf{I}_4 &= \\ = (F_{11}^{+} - F_{21}^{+}{F_{22}^{+}}^{-1}F_{12}^{+})\mathbf{I}_{g1} &+ F_{21}^{+}{F_{22}^{+}}^{-1}\mathbf{I}_5 .\end{aligned} \tag{5.15}$$

Hence, using (5.13):

$$\begin{aligned}\mathbf{U}_4 = -R_4(F_{21}^{+}{F_{22}^{+}}^{-1}F_{32}^{+} - F_{31}^{+})\mathbf{I}_3 &+ \\ + R_4(F_{11}^{+} - F_{21}^{+}{F_{22}^{+}}^{-1}F_{12})\mathbf{I}_{g1} &+ R_4F_{21}^{+}{F_{22}^{+}}^{-1}\mathbf{I}_5 .\end{aligned} \tag{5.16}$$

$\mathbf{I}_3$ is expressed from (5.5), observing (5.12):

$$\mathbf{I}_3 = -G_3F_{31}\mathbf{U}_4 - G_3F_{32}\mathbf{U}_5 - G_3F_{33}\mathbf{U}_{g6} . \tag{5.17}$$

Substitution into (5.16) yields

$$\begin{aligned}&[1 - R_4(F_{21}^{+}{F_{22}^{+}}^{-1}F_{32}^{+} - F_{31}^{+})G_3F_{31}]\mathbf{U}_4 = R_4(F_{21}^{+}{F_{22}^{+}}^{-1}F_{32}^{+} - F_{31}^{+})G_3F_{32}\mathbf{U}_5 + \\ &\quad + R_4(F_{21}^{+}{F_{22}^{+}}^{-1}F_{32}^{+} - F_{31}^{+})G_3F_{33}\mathbf{U}_{g6} + R_4(F_{11}^{+} - F_{21}^{+}{F_{22}^{+}}^{-1}F_{12}^{+})\mathbf{I}_{g1} + \\ &\quad + R_4F_{21}^{+}{F_{22}^{+}}^{-1}\mathbf{I}_5 .\end{aligned} \tag{5.18}$$

Hence:

$$\begin{aligned}&\mathbf{U}_4 = [1 - R_4(F_{21}^{+}{F_{22}^{+}}^{-1}F_{32}^{+} - F_{31}^{+})G_3F_{31}]^{-1}[R_4\{(F_{21}^{+}{F_{22}^{+}}^{-1}F_{32}^{+} - \\ &- F_{31}^{+})G_3(F_{32}\mathbf{U}_5 + F_{33}\mathbf{U}_{g6}) + (F_{11}^{+} - F_{21}^{+}{F_{22}^{+}}^{-1}F_{12}^{+})\mathbf{I}_{g1} + F_{21}^{+}{F_{22}^{+}}^{-1}\mathbf{I}_5\}] .\end{aligned} \tag{5.19}$$

Writing this into (5.4) a form of $\mathbf{U}_2$ is obtained, expressed with the aid of known quantities:

$$\begin{aligned}\mathbf{U}_2 = &-F_{21}[1 - R_4(F_{21}^{+}{F_{22}^{+}}^{-1}F_{32}^{+} - F_{31}^{+})G_3F_{31}]^{-1}R_4(F_{11}^{+} - F_{21}^{+}{F_{22}^{+}}^{-1}F_{12}^{+})\mathbf{I}_{g1} - \\ &-\{F_{21}[1 - R_4(F_{21}^{+}{F_{22}^{+}}^{-1}F_{32}^{+} - F_{31}^{+})G_3F_{31}]^{-1}R_4(F_{21}^{+}{F_{22}^{+}}^{-1}F_{32}^{+} - \\ &-F_{31}^{+})G_3F_{33} + F_{23}\}\mathbf{U}_{g6} - \\ &-\{F_{21}[1 - R_4(F_{21}^{+}{F_{22}^{+}}^{-1}F_{32}^{+} - F_{31}^{+})G_3F_{31}]^{-1}R_4(F_{21}^{+}{F_{22}^{+}}^{-1}F_{32}^{+} - \\ &-F_{31}^{+})G_3F_{32} + F_{22}\}\mathbf{U}_5 - \\ &-F_{21}[1 - R_4(F_{21}^{+}{F_{22}^{+}}^{-1}F_{32}^{+} - F_{31}^{+})G_3F_{31}]^{-1}R_4F_{21}^{+}{F_{22}^{+}}^{-1}\mathbf{I}_5 ,\end{aligned} \tag{5.20}$$

i.e. the voltages of the norators have been determined.

To calculate $\mathbf{I}_2$ (5.19) is at first substituted into (5.17):

$$\begin{aligned}\mathbf{I}_3 = &-G_3F_{31}[1-R_4(F_{21}^{+}{F_{22}^{+}}^{-1}F_{32}^{+}-F_{31}^{+})G_3F_{31}]^{-1}R_4(F_{11}^{+}-F_{21}^{+}{F_{22}^{+}}^{-1}F_{12}^{+})\mathbf{I}_{g1}-\\ &-G_3\{F_{31}[1-R_4(F_{21}^{+}{F_{22}^{+}}^{-1}F_{32}^{+}-F_{31}^{+})G_3F_{31}]^{-1}R_4(F_{21}^{+}{F_{22}^{+}}^{-1}F_{32}^{+}-\\ &-F_{31}^{+})G_3F_{33}+F_{33}\}\mathbf{U}_{g6}-\\ &-G_3\{F_{31}[1-R_4(F_{21}^{+}{F_{22}^{+}}^{-1}F_{32}^{+}-F_{31}^{+})G_3F_{31}]^{-1}R_4(F_{21}^{+}{F_{22}^{+}}^{-1}F_{32}^{+}-\\ &-F_{31}^{+})G_3F_{32}+F_{32}\}\mathbf{U}_5-\\ &-G_3F_{31}[1-R_4(F_{21}^{+}{F_{22}^{+}}^{-1}F_{32}^{+}-F_{31}^{+})G_3F_{31}]^{-1}R_4F_{21}^{+}{F_{22}^{+}}^{-1}\mathbf{I}_5 .\end{aligned} \tag{5.21}$$

Writing this into (5.14)

$$\begin{aligned}\mathbf{I}_2 = &{F_{22}^{+}}^{-1}\{F_{32}^{+}G_3F_{31}[1-R_4(F_{21}^{+}{F_{22}^{+}}^{-1}F_{32}^{+}-\\ &-F_{31}^{+})G_3F_{31}]^{-1}R_4(F_{11}^{+}-F_{21}^{+}{F_{22}^{+}}^{-1}F_{12}^{+})-F_{12}^{+}\}\mathbf{I}_{g1}+\\ &+{F_{22}^{+}}^{-1}F_{32}^{+}G_3\{F_{31}[1-R_4(F_{21}^{+}{F_{22}^{+}}^{-1}F_{32}^{+}-\\ &-F_{31}^{+})G_3F_{31}]^{-1}R_4(F_{21}^{+}{F_{22}^{+}}^{-1}F_{32}^{+}-F_{31}^{+})G_3F_{33}+F_{33}\}\mathbf{U}_{g6}+\\ &+{F_{22}^{+}}^{-1}F_{32}^{+}G_3\{F_{31}[1-R_4(F_{21}^{+}{F_{22}^{+}}^{-1}F_{32}^{+}-\\ &-F_{31}^{+})G_3F_{31}]^{-1}R_4(F_{21}^{+}{F_{22}^{+}}^{-1}F_{32}^{+}-F_{31}^{+})G_3F_{32}+F_{32}\}\mathbf{U}_5+\\ &+{F_{22}^{+}}^{-1}\{F_{32}^{+}G_3F_{31}[1-R_4(F_{21}^{+}{F_{22}^{+}}^{-1}F_{32}-\\ &-F_{31}^{+})G_3F_{31}]^{-1}R_4F_{21}^{+}{F_{22}^{+}}^{-1}+1\}\mathbf{I}_5\end{aligned} \tag{5.22}$$

is the expression of the currents of the norators.

From a knowledge of $\mathbf{U}_2$ and $\mathbf{I}_2$, the values of the adjusting resistors, the appropriate elements of diagonal matrix R_2 in Ohm's law

$$\mathbf{U}_2 = R_2\mathbf{I}_2 \tag{5.23}$$

are easily obtained.

The small-signal analysis of electronic circuits

In the analysis of electronic circuits the unknown branch-currents and branch-voltages of the network are to be determined from a knowledge of the interconnections of the network, and of the characteristics of the network's voltage- and current-sources, generators, self and mutual impedances, and electronic components. Our discussion is restricted to the case when the amplitudes of the sinusoidally varying signals are small enough to allow the approximation of the characteristics of the electronic components by straight lines in the appropriate neighbourhood of the working points. Thus, the input and output voltages of an electronic component may be approximated by the first two terms of their Taylor

Table 5.1

Number	Characterizing equation	Equivalent circuit
1	$\begin{bmatrix} U_1 \\ U_2 \end{bmatrix} = \begin{bmatrix} z_{11} & z_{12} \\ z_{21} & z_{22} \end{bmatrix} \begin{bmatrix} I_1 \\ I_2 \end{bmatrix}$	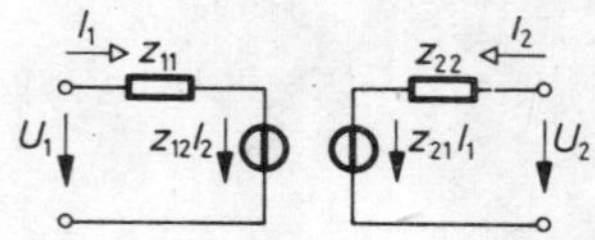
2	$\begin{bmatrix} I_1 \\ I_2 \end{bmatrix} = \begin{bmatrix} y_{11} & y_{12} \\ y_{21} & y_{22} \end{bmatrix} \begin{bmatrix} U_1 \\ U_2 \end{bmatrix}$	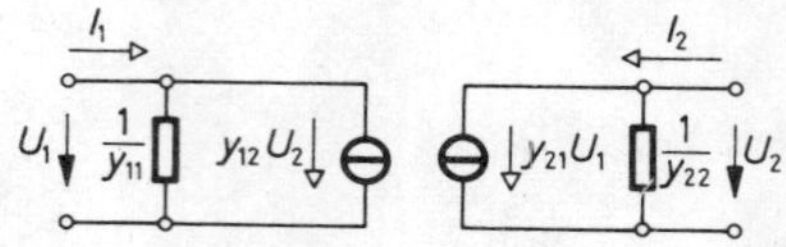
3	$\begin{bmatrix} U_1 \\ I_2 \end{bmatrix} = \begin{bmatrix} h_{11} & h_{12} \\ h_{21} & h_{22} \end{bmatrix} \begin{bmatrix} I_1 \\ U_2 \end{bmatrix}$	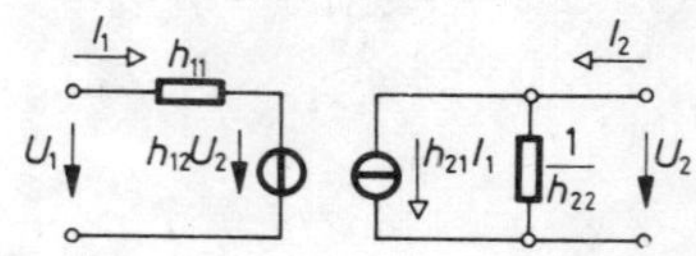
4	$\begin{bmatrix} I_1 \\ U_2 \end{bmatrix} = \begin{bmatrix} d_{11} & d_{12} \\ d_{21} & d_{22} \end{bmatrix} \begin{bmatrix} U_1 \\ I_2 \end{bmatrix}$	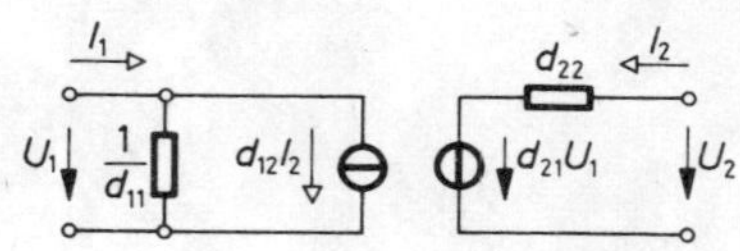

series as functions of the relevant currents. The relationships between the complex effective or peak values U_1, I_1, U_2, I_2 of the sinusoidally varying signals are

$$U_1 = z_{11} I_1 + z_{12} I_2 \tag{5.24}$$

$$U_2 = z_{21} I_1 + z_{22} I_2 \,. \tag{5.25}$$

z_{11}, z_{12}, z_{21}, z_{22} are seen to denote the impedance parameters of the electronic component as a two-port. (5.24) and (5.25) may be written in the form

$$\mathbf{U} = \boldsymbol{Z}\mathbf{I} \,, \tag{5.26}$$

where

$$\mathbf{U} = \begin{bmatrix} U_1 \\ U_2 \end{bmatrix}, \quad \mathbf{I} = \begin{bmatrix} I_1 \\ I_2 \end{bmatrix}, \quad \boldsymbol{Z} = \begin{bmatrix} z_{11} & z_{12} \\ z_{21} & z_{22} \end{bmatrix}. \tag{5.27}$$

On the basis of (5.26) the electronic component may be modelled by the connection shown in the first row of Table 5.1. The connection consists of two Thevenin-generators, with their source-voltages proportional to the currents I_2 and I_1, respectively. A generator with its source-voltage proportional to a branch-current of the network is called current-controlled voltage-generator, and is denoted as shown in Table 5.1.

The current I_1 and I_2 may be expressed with the aid of the voltages U_1 and U_2:

$$\mathbf{I} = \boldsymbol{YU}, \tag{5.28}$$

or in a more detailed form:

$$\begin{bmatrix} I_1 \\ I_2 \end{bmatrix} = \begin{bmatrix} y_{11} & y_{12} \\ y_{21} & y_{22} \end{bmatrix} \begin{bmatrix} U_1 \\ U_2 \end{bmatrix}, \tag{5.29}$$

where $y_{11}, y_{12}, y_{21}, y_{22}$ are the admittance parameters. (5.29) yields an alternative model of the electronic component (Table 5.1, second row), consisting of two Norton generators, namely two voltage-controlled current-generators.

The relationships between the input and output currents and voltages of the two-port may also be written with the aid of hybrid and inverse-hybrid parameters:

$$\begin{bmatrix} U_1 \\ I_2 \end{bmatrix} = \begin{bmatrix} h_{11} & h_{12} \\ h_{21} & h_{22} \end{bmatrix} \begin{bmatrix} I_1 \\ U_2 \end{bmatrix} = \boldsymbol{H} \begin{bmatrix} I_1 \\ U_2 \end{bmatrix} \tag{5.30}$$

and

$$\begin{bmatrix} I_1 \\ U_2 \end{bmatrix} = \begin{bmatrix} d_{11} & d_{12} \\ d_{21} & d_{22} \end{bmatrix} \begin{bmatrix} U_1 \\ I_2 \end{bmatrix} = \boldsymbol{D} \begin{bmatrix} U_1 \\ I_2 \end{bmatrix}. \tag{5.31}$$

These allow the establishment of further equivalent models (Table 5.1, third and fourth rows), containing current-controlled current-generator and voltage-controlled voltage-generator in addition to the controlled generators mentioned already.

In addition to impedance, admittance, hybrid and inverse hybrid parameters, the chain parameter matrix $\boldsymbol{A}$ may also serve for the characterization of the two-port, namely*

$$\begin{bmatrix} U_1 \\ I_1 \end{bmatrix} = \begin{bmatrix} a_{11} & a_{12} \\ a_{21} & a_{22} \end{bmatrix} \begin{bmatrix} U_2 \\ I_2 \end{bmatrix} = \boldsymbol{A} \begin{bmatrix} U_2 \\ I_2 \end{bmatrix}. \tag{5.32}$$

In the case of certain two-ports with extreme parameters not all of the above parameter matrices may exist. Thus, for example, the impedance, admittance and hybrid parameter matrices of the connection shown in Fig. 5.8 all exist, while its chain parameter matrix does not exist. Such two-ports with extreme parameters are discussed below.

* $\boldsymbol{A}$ is also referred to as the transmission matrix or the cascade matrix. Elements $a_{11}, a_{12}, a_{21}, a_{22}$ are often denoted A, B, C, D, and the orientation of I_2 is often reversed in the defining equations.

If $z_{12}=0$ in the impedance parameter matrix, a current-controlled voltage-generator shown in Table 5.2 is obtained. Similarly, the other equivalents drawn show a voltage-controlled current-generator for $y_{12}=0$, a current-controlled current-generator for $h_{12}=0$ and a voltage-controlled voltage-generator for $d_{12}=0$. From the diagrams of the table it can be seen that identical controlled generators are obtained if $z_{21}=0$, $y_{21}=0$, $h_{21}=0$, $d_{21}=0$, provided that the primary sides with voltage U_1 and current I_1 and the secondary sides with voltage U_2 and current I_2 have their roles reversed.

If the elements on the main diagonals of the parameter matrices are also zero in addition to the elements z_{12}, y_{12}, h_{12}, d_{12}, the four kinds of ideal, controlled

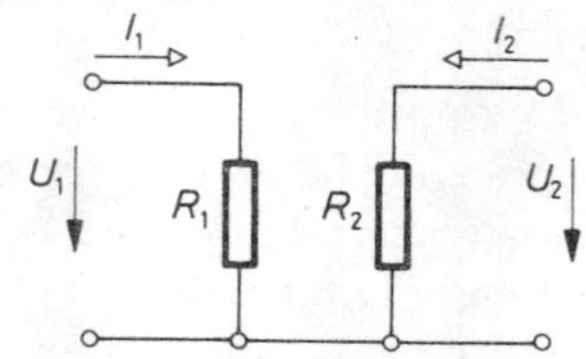

Fig. 5.8

Table 5.2

Name	Characterizing equation	Equivalent circuit
Current-controlled voltage-generator	$\begin{bmatrix} U_1 \\ U_2 \end{bmatrix} = \begin{bmatrix} z_{11} & 0 \\ z_{21} & z_{22} \end{bmatrix} \begin{bmatrix} I_1 \\ I_2 \end{bmatrix}$	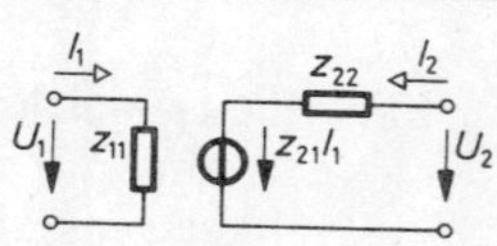
Voltage-controlled current-generator	$\begin{bmatrix} I_1 \\ I_2 \end{bmatrix} = \begin{bmatrix} y_{11} & 0 \\ y_{21} & y_{22} \end{bmatrix} \begin{bmatrix} U_1 \\ U_2 \end{bmatrix}$	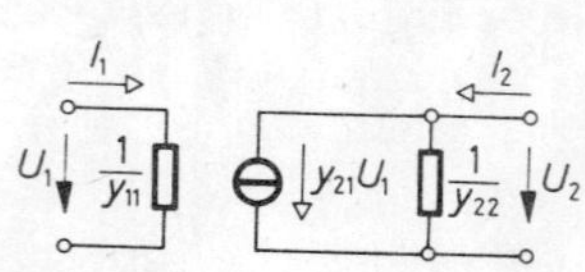
Current-controlled current-generator	$\begin{bmatrix} U_1 \\ I_2 \end{bmatrix} = \begin{bmatrix} h_{11} & 0 \\ h_{21} & h_{22} \end{bmatrix} \begin{bmatrix} I_1 \\ U_2 \end{bmatrix}$	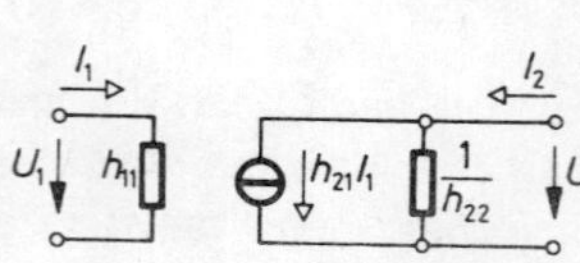
Voltage-controlled voltage-generator	$\begin{bmatrix} I_1 \\ U_2 \end{bmatrix} = \begin{bmatrix} d_{11} & 0 \\ d_{21} & d_{22} \end{bmatrix} \begin{bmatrix} U_1 \\ I_2 \end{bmatrix}$	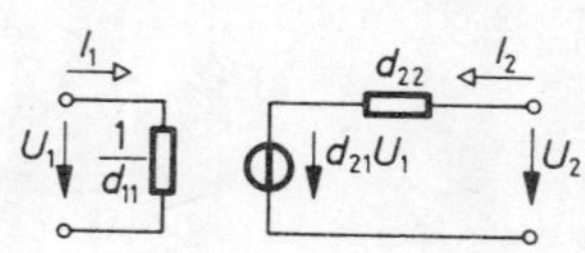

Table 5.3

Name	Characterizing equation		Equivalent circuit
Current-controlled voltage-source	$\begin{bmatrix} U_1 \\ U_2 \end{bmatrix} = \begin{bmatrix} 0 & 0 \\ z_{21} & 0 \end{bmatrix} \begin{bmatrix} I_1 \\ I_2 \end{bmatrix}$	$\begin{bmatrix} U_1 \\ I_1 \end{bmatrix} = \begin{bmatrix} 0 & 0 \\ 1/z_{21} & 0 \end{bmatrix} \begin{bmatrix} U_2 \\ I_2 \end{bmatrix}$	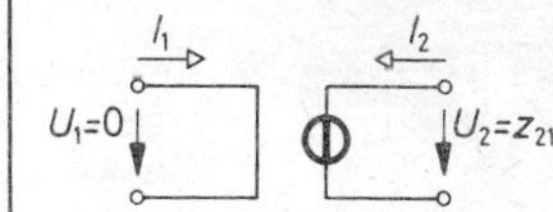
Voltage-controlled current-source	$\begin{bmatrix} I_1 \\ I_2 \end{bmatrix} = \begin{bmatrix} 0 & 0 \\ y_{21} & 0 \end{bmatrix} \begin{bmatrix} U_1 \\ U_2 \end{bmatrix}$	$\begin{bmatrix} U_1 \\ I_1 \end{bmatrix} = \begin{bmatrix} 0 & 1/y_{21} \\ 0 & 0 \end{bmatrix} \begin{bmatrix} U_2 \\ I_2 \end{bmatrix}$	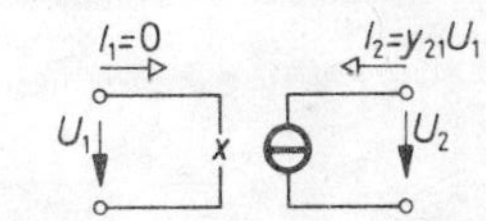
Current-controlled current-source	$\begin{bmatrix} U_1 \\ I_2 \end{bmatrix} = \begin{bmatrix} 0 & 0 \\ h_{21} & 0 \end{bmatrix} \begin{bmatrix} I_1 \\ U_2 \end{bmatrix}$	$\begin{bmatrix} U_1 \\ I_1 \end{bmatrix} = \begin{bmatrix} 0 & 0 \\ 0 & 1/h_{21} \end{bmatrix} \begin{bmatrix} U_2 \\ I_2 \end{bmatrix}$	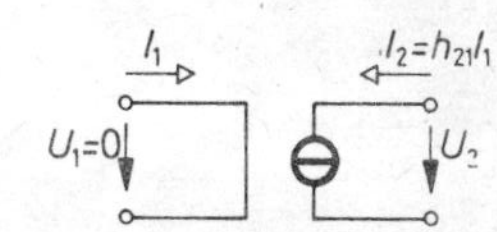
Voltage-controlled voltage-source	$\begin{bmatrix} I_1 \\ U_2 \end{bmatrix} = \begin{bmatrix} 0 & 0 \\ d_{21} & 0 \end{bmatrix} \begin{bmatrix} U_1 \\ I_2 \end{bmatrix}$	$\begin{bmatrix} U_1 \\ I_1 \end{bmatrix} = \begin{bmatrix} 1/d_{21} & 0 \\ 0 & 0 \end{bmatrix} \begin{bmatrix} U_2 \\ I_2 \end{bmatrix}$	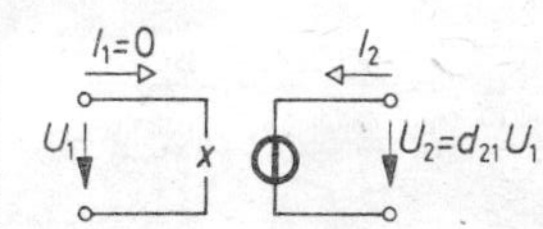

Table 5.4

Name	Characterizing equation		Common notation
Gyrator	$\begin{bmatrix} U_1 \\ U_2 \end{bmatrix} = \begin{bmatrix} 0 & -R_1 \\ R_2 & 0 \end{bmatrix} \begin{bmatrix} I_1 \\ I_2 \end{bmatrix}$	$\begin{bmatrix} U_1 \\ I_1 \end{bmatrix} = \begin{bmatrix} 0 & -R_1 \\ 1/R_2 & 0 \end{bmatrix} \begin{bmatrix} U_2 \\ I_2 \end{bmatrix}$	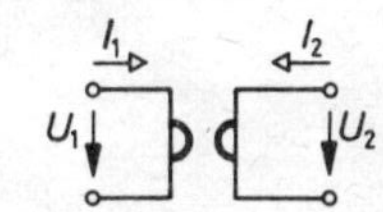
Negative impedance converter	$\begin{bmatrix} U_1 \\ I_2 \end{bmatrix} = \begin{bmatrix} 0 & k \\ k & 0 \end{bmatrix} \begin{bmatrix} I_1 \\ U_2 \end{bmatrix}$	$\begin{bmatrix} U_1 \\ I_1 \end{bmatrix} = \begin{bmatrix} k & 0 \\ 0 & 1/k \end{bmatrix} \begin{bmatrix} U_2 \\ I_2 \end{bmatrix}$	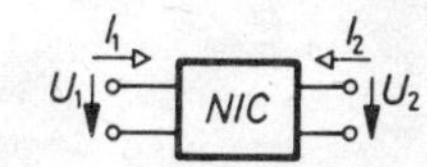
Ideal transformer	$\begin{bmatrix} U_1 \\ I_2 \end{bmatrix} = \begin{bmatrix} 0 & n \\ -n & 0 \end{bmatrix} \begin{bmatrix} I_1 \\ U_2 \end{bmatrix}$	$\begin{bmatrix} U_1 \\ I_1 \end{bmatrix} = \begin{bmatrix} n & 0 \\ 0 & -1/n \end{bmatrix} \begin{bmatrix} U_2 \\ I_2 \end{bmatrix}$	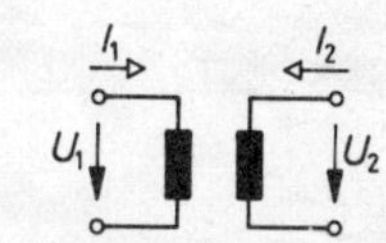
Nullor		$\begin{bmatrix} U_1 \\ I_1 \end{bmatrix} = \begin{bmatrix} 0 & 0 \\ 0 & 0 \end{bmatrix} \begin{bmatrix} U_2 \\ I_2 \end{bmatrix}$	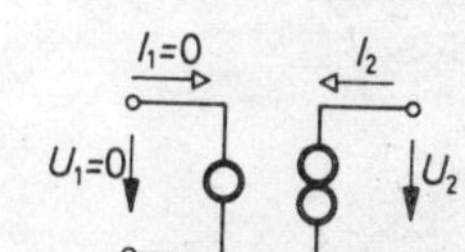

sources are described by the parameter matrices. These have been summarized in Table 5.3 in which the equations in terms of chain parameters have also been included for each case, in addition to impedance, admittance, hybrid or inverse hybrid parameters.

If all the elements off the main diagonal of an impedance, admittance, hybrid or inverse hybrid parameter matrix are zero, the equivalent-circuit is reduced to two immitances (impedances or admittances).

In the case of certain special two-ports, the elements off the main diagonals of the parameter matrices written in Table 5.1 are nonzero, real quantities, while the two elements on the main diagonals are zero. The gyrator is described by such impedance and admittance parameter matrices (Table 5.4), with one parameter positive, the other negative. If the nonzero elements in the hybrid or inverse hybrid parameter matrix are equal, a negative impedance converter is obtained, while if their absolute value is identical with different signs, an ideal transformer is obtained. The chain parameters of the gyrator, negative impedance converter and ideal transformer are also given in Table 5.4.

The nullor is also shown in Table 5.4. It is a two-port, that can be characterized only by its chain parameters, all of which equal zero. The nullor can be regarded as a model of an ideal operational amplifier (Fig. 5.9).

With the aid of the equivalents given in Tables 5.1, 5.2, 5.3 electronic circuits are modelled by networks containing controlled sources or controlled generators. A controlled voltage-generator consists of a series connection of a controlled voltage-source and an impedance, while a controlled current-generator is formed by a parallel connection of a controlled current-source and an admittance. Controlled

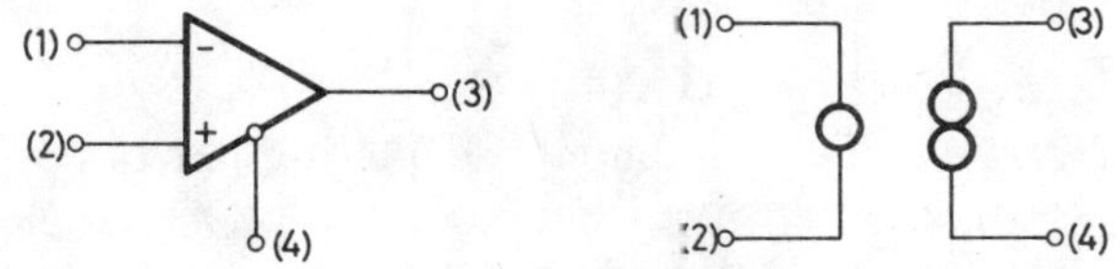

Fig. 5.9

sources have been seen to have four types: current-controlled voltage-source, voltage-controlled current-source, current-controlled current-source and voltage-controlled voltage-source. The network elements formerly called generators and sources will from now on be referred to as independent generators and independent sources, to distinguish them from controlled generators and controlled sources.

In the course of our further analyses the two-ports will be described by one of their parameter matrices.

Calculation of currents and voltages with the aid of chain parameter matrices

Initially a method is discussed which uses the chain parameters of two-ports to write the equations [20]. This method is therefore suitable for the analysis of networks containing two-ports with extreme parameters, for which the chain parameters exist.

The graph of the network is chosen for this procedure so as to associate a distinct edge of the graph with each two-terminal element as well as with the primary and secondary sides of each two-port. The two-terminal elements are classified into two groups. Each zero impedance branch as well as the branches containing Thevenin generators are z-branches. The relationship between their voltage U_z and current I_z may be written in the form

$$U_z = ZI_z + U_g. \tag{5.33}$$

The group formed by y-branches consists of zero admittance branches as well as of Norton generators. Their current I_y and voltage U_y are in the relation

$$I_y = YU_y + I_g. \tag{5.34}$$

Passive branches with nonzero, finite impedances and admittances are to be optionally assigned to the group of z- or y-branches.

The relations between the primary voltages and currents U_p and I_p and secondary voltages and currents U_s and I_s of two-ports with either extreme or not extreme parameters are given by chain parameters:

$$\begin{bmatrix} U_p \\ I_p \end{bmatrix} = \begin{bmatrix} a_{11} & a_{12} \\ a_{21} & a_{22} \end{bmatrix} \begin{bmatrix} U_s \\ I_s \end{bmatrix}. \tag{5.35}$$

Order numbers are so assigned to the edges of the graph as to give order numbers $1, 2, \ldots, b_1$ to z-branches, order numbers $b_1+1, b_1+2, \ldots, b_1+b_2$ to branches associated with the primary sides of two-ports, order numbers $b_1+b_2+1, b_1+b_2+2, \ldots, b_1+2b_2$ to secondary sides and, finally, the order numbers of y-branches are $b_1+2b_2+1, b_1+2b_2+2, \ldots, b_1+2b_2+b_3$. The column matrices of branch-voltages and branch-currents in the network are partitioned in accordance with the classification:

$$\mathbf{U} = \begin{bmatrix} \mathbf{ZI}_z + \mathbf{U}_g \\ \mathbf{U}_p \\ \mathbf{U}_s \\ \mathbf{U}_y \end{bmatrix}, \qquad \mathbf{I} = \begin{bmatrix} \mathbf{I}_z \\ \mathbf{I}_p \\ \mathbf{I}_s \\ \mathbf{YU}_y + \mathbf{I}_g \end{bmatrix}, \tag{5.36}$$

where $\boldsymbol{Z}$ and $\boldsymbol{Y}$ are diagonal matrices formed by the impedances of z-branches and admittances of y-branches respectively, and the column matrices consist of the voltages or currents of the branches in the groups indicated by their indices.

Coupled impedances and admittances have been presumed to have been taken into account as two-ports represented by their chain parameters in order to allow the above description of $\boldsymbol{Z}$ and $\boldsymbol{Y}$. The loop matrix $\boldsymbol{B}$ and cutset matrix $\boldsymbol{Q}$ of the graph are also partitioned according to the classification of the edges, i.e.

$$\boldsymbol{B}=[\boldsymbol{B}_z \ \boldsymbol{B}_p \ \boldsymbol{B}_s \ \boldsymbol{B}_y], \tag{5.37}$$

$$\boldsymbol{Q}=[\boldsymbol{Q}_z \ \boldsymbol{Q}_p \ \boldsymbol{Q}_s \ \boldsymbol{Q}_y]. \tag{5.38}$$

Accordingly, the Kirchhoff equations of the network are

$$[\boldsymbol{B}_z \ \boldsymbol{B}_p \ \boldsymbol{B}_s \ \boldsymbol{B}_y]\begin{bmatrix} \boldsymbol{Z}\mathbf{I}_z+\mathbf{U}_g \\ \mathbf{U}_p \\ \mathbf{U}_s \\ \mathbf{U}_y \end{bmatrix}=\mathbf{0}. \tag{5.39}$$

$$[\boldsymbol{Q}_z \ \boldsymbol{Q}_p \ \boldsymbol{Q}_s \ \boldsymbol{Q}_y]\begin{bmatrix} \mathbf{I}_z \\ \mathbf{I}_p \\ \mathbf{I}_s \\ \boldsymbol{Y}\mathbf{U}_y+\mathbf{I}_g \end{bmatrix}=0, \tag{5.40}$$

i.e.

$$\boldsymbol{B}_z\boldsymbol{Z}\mathbf{I}_z+\boldsymbol{B}_p\mathbf{U}_p+\boldsymbol{B}_s\mathbf{U}_s+\boldsymbol{B}_y\mathbf{U}_y=-\boldsymbol{B}_z\mathbf{U}_g, \tag{5.41}$$

$$\boldsymbol{Q}_z\mathbf{I}_z+\boldsymbol{Q}_p\mathbf{I}_p+\boldsymbol{Q}_s\mathbf{I}_s+\boldsymbol{Q}_y\boldsymbol{Y}\mathbf{U}_y=-\boldsymbol{Q}_y\mathbf{I}_g. \tag{5.42}$$

According to (5.35):

$$\mathbf{U}_p=\boldsymbol{A}_{11}\mathbf{U}_s+\boldsymbol{A}_{12}\mathbf{I}_s, \tag{5.43}$$

$$\mathbf{I}_p=\boldsymbol{A}_{21}\mathbf{U}_s+\boldsymbol{A}_{22}\mathbf{I}_s, \tag{5.44}$$

where $\boldsymbol{A}_{11}, \boldsymbol{A}_{12}, \boldsymbol{A}_{21}, \boldsymbol{A}_{22}$ are diagonal matrices formed by the chain parameters $a_{11}, a_{12}, a_{21}, a_{22}$ of the two-ports respectively. Substituting these into (5.41) and (5.42):

$$\boldsymbol{B}_z\boldsymbol{Z}\mathbf{I}_z+(\boldsymbol{B}_p\boldsymbol{A}_{11}+\boldsymbol{B}_s)\mathbf{U}_s+\boldsymbol{B}_p\boldsymbol{A}_{12}\mathbf{I}_s+\boldsymbol{B}_y\mathbf{U}_y=-\boldsymbol{B}_z\mathbf{U}_g \tag{5.45}$$

$$\boldsymbol{Q}_z\mathbf{I}_z+\boldsymbol{Q}_p\boldsymbol{A}_{21}\mathbf{U}_s+(\boldsymbol{Q}_p\boldsymbol{A}_{22}+\boldsymbol{Q}_s)\mathbf{I}_s+\boldsymbol{Q}_y\boldsymbol{Y}\mathbf{U}_y=-\boldsymbol{Q}_y\mathbf{I}_g. \tag{5.46}$$

These two equations may be written as follows:

$$\begin{bmatrix} \boldsymbol{B}_z\boldsymbol{Z} & \boldsymbol{B}_p\boldsymbol{A}_{11}+\boldsymbol{B}_s & \boldsymbol{B}_p\boldsymbol{A}_{12} & \boldsymbol{B}_y \\ \boldsymbol{Q}_z & \boldsymbol{Q}_p\boldsymbol{A}_{21} & \boldsymbol{Q}_p\boldsymbol{A}_{22}+\boldsymbol{Q}_s & \boldsymbol{Q}_y\boldsymbol{Y} \end{bmatrix}\begin{bmatrix} \mathbf{I}_z \\ \mathbf{U}_s \\ \mathbf{I}_s \\ \mathbf{U}_y \end{bmatrix}=-\begin{bmatrix} \boldsymbol{B}_z & \boldsymbol{0} \\ \boldsymbol{0} & \boldsymbol{Q}_y \end{bmatrix}\begin{bmatrix} \mathbf{U}_g \\ \mathbf{I}_g \end{bmatrix}. \tag{5.47}$$

The currents of z- and secondary branches and the voltages of y- and secondary branches may be expressed from (5.47):

$$\begin{bmatrix} \mathbf{I}_z \\ \mathbf{U}_s \\ \mathbf{I}_s \\ \mathbf{U}_y \end{bmatrix} = \begin{bmatrix} \boldsymbol{B}_z\boldsymbol{Z} & \boldsymbol{B}_p\boldsymbol{A}_{11}+\boldsymbol{B}_s & \boldsymbol{B}_p\boldsymbol{A}_{12} & \boldsymbol{B}_y \\ \boldsymbol{Q}_z & \boldsymbol{Q}_p\boldsymbol{A}_{21} & \boldsymbol{Q}_p\boldsymbol{A}_{22}+\boldsymbol{Q}_s & \boldsymbol{Q}_y\boldsymbol{Y} \end{bmatrix}^{-1} \begin{bmatrix} \boldsymbol{B}_z & \boldsymbol{0} \\ \boldsymbol{0} & \boldsymbol{Q}_y \end{bmatrix} \begin{bmatrix} \mathbf{U}_g \\ \mathbf{I}_g \end{bmatrix}. \tag{5.48}$$

Hence,

$$\begin{bmatrix} \mathbf{U}_z \\ \mathbf{U}_p \\ \mathbf{I}_p \\ \mathbf{I}_y \end{bmatrix} = \begin{bmatrix} \boldsymbol{Z}\mathbf{I}_z+\mathbf{U}_g \\ \boldsymbol{A}_{11}\mathbf{U}_s+\boldsymbol{A}_{12}\mathbf{I}_s \\ \boldsymbol{A}_{21}\mathbf{U}_s+\boldsymbol{A}_{22}\mathbf{I}_s \\ \boldsymbol{Y}\mathbf{U}_y+\mathbf{I}_g \end{bmatrix}, \tag{5.49}$$

i.e. the voltages of z- and primary branches and the currents of y- and primary branches have also been determined.

The solution necessitated the inversion of a matrix with order equalling the number of edges in the graph associated with the network (see (5.48)), i.e. a system of equations with $b_1+2b_2+b_3$ unknowns had to be solved, where b_1+b_3 was the number of two-terminal elements, and b_2 that of the two-ports in the network.

Analysis with the aid of a model containing controlled generators

For the application of the method to be presented the two-ports are characterized by their impedance, admittance, hybrid or inverse hybrid parameter matrices, and these are employed to determine currents and voltages. Since none of these parameters of a nullor exist, the present method of calculation, also called the method of mixed parameters, is only suitable for the analysis of networks not containing nullors [7, 43, 44, 57, 58].

Two-ports not including independent sources have been seen to be replaceable by controlled generators by using one of the parameter matrices listed above, with the primary and secondary sides substituted by distinct controlled generators. Two-ports are taken into account by these equivalent connections, and two-terminal elements by their Thevenin or Norton equivalents. Thevenin and Norton generators are considered to consist of two branches.

The branches of the network are classified into four groups (Fig. 5.10). The first group is formed by the branches containing independent current-sources. Controlled current-sources (controlled current-generators), zero admittance

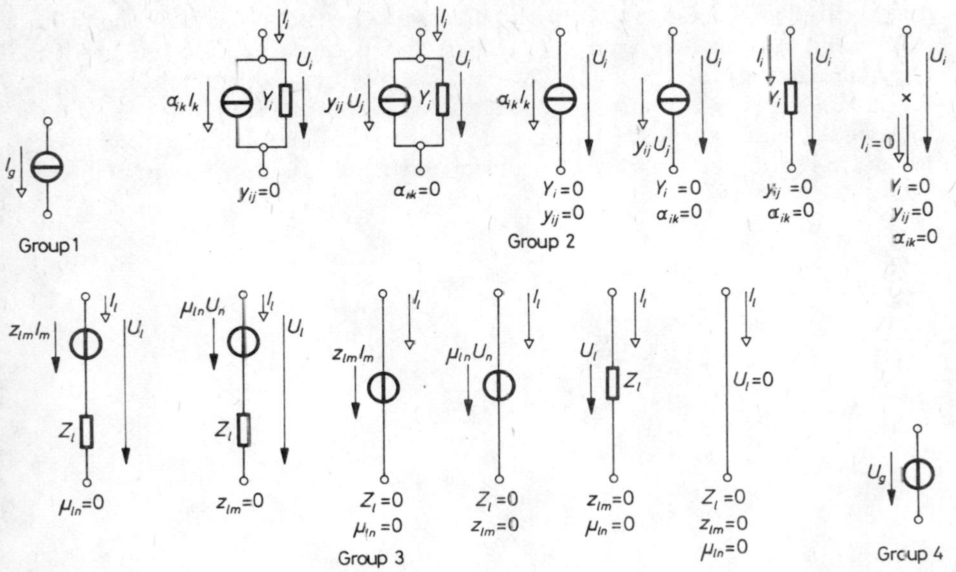

Fig. 5.10

branches, and branches with their voltages being controlling voltage are assigned to the second group. Controlled voltage-sources (controlled voltage-generators), zero impedance branches, and branches with their currents being controlling current are included in the third group. The fourth group consists of branches containing independent voltage-sources. Passive two-terminal elements with finite admittance and impedance belong to the second and third groups, taking into account that the branches of the first and second groups are chords, while those of the third and fourth groups are tree-branches. The passive branches in the second group are described by their admittances, and those in the third group by their impedances in the course of the analysis.

Accordingly, the primary and secondary sides of two-ports characterized by their admittance or impedance parameters belong to the second and third groups, respectively (Table 5.1, second and first row). If a two-port is described by its hybrid parameters, its primary side is included in the third and its secondary side in the second group (Table 5.1, third row), while in case of the application of inverse hybrid parameters the primary side belongs to the second and the secondary side to the third group (Table 5.1, fourth row).

The above classification of the branches of the network may prove to be impossible, because the independent voltage-sources and the controlled voltage-generators may form at least one loop or the independent current-sources and the controlled current generators may form at least one cutset. (Thus it is not feasible to associate tree-branches only with the former and chords only with the latter.) In this case controlled generators are decomposed into two branches, into a controlled source and an impedance, and the tree is chosen to include the branch of the impedance in the other group. If the desired classification still remains impossible, the independent and controlled voltage-sources form at least one loop or the independent and controlled current-sources form at least one cutset in the network, usually indicating a contradictory problem and being outside the scope of interest.

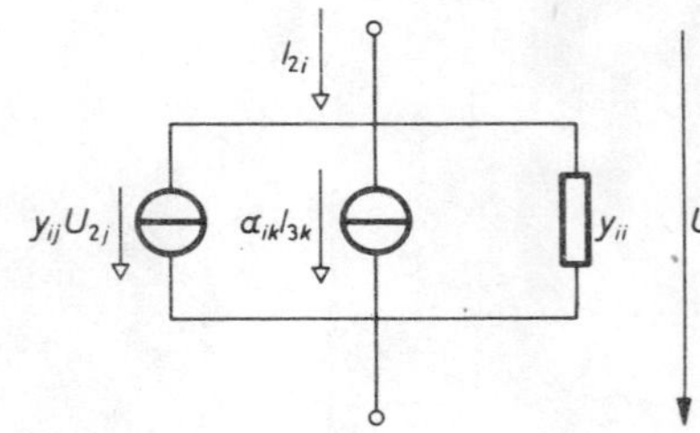

Fig. 5.11

The branches of the network are assigned order numbers in accordance with the classification, i.e. the branches in the first group are given order numbers $1, 2, \ldots, b_1$, those in the second group $b_1+1, b_1+2, \ldots, b_1+b_2$, those in the third group $b_1+b_2+1,\ b_1+b_2+2,\ \ldots,\ b_1+b_2+b_3$, and finally the branches of the fourth group are assigned $b_1+b_2+b_3+1,\ b_1+b_2+b_3+2,\ \ldots,\ b_1+b_2+b_3+b_4$. The number of branches in the groups has been assumed to be b_1, b_2, b_3, b_4, respectively.

The relationships concerning the currents and voltages of the branches in the second and third groups are written at first. The current of branch i in the second group (Fig. 5.11) is written in the form

$$I_{2i}=y_{ii}U_{2i}+y_{ij}U_{2j}+\alpha_{ik}I_{3k}, \tag{5.50}$$

where $y_{ii}=Y_i$ is the admittance of the branch, U_{2i} is its voltage, $y_{ij}U_{2j}$ is the source-current of the voltage-controlled current-source in the branch with controlling voltage U_{2j}, $\alpha_{ik}I_{3k}$ is the source-current of the current-controlled current-source in the branch with controlling current I_{3k}. The controlling voltage U_{2j} is the voltage of branch j in the second group, and the controlling current I_{3k} is the current of branch k in the third group. If the controlling voltage is the source-voltage of a voltage-source, a zero admittance branch is connected in parallel with the voltage-source, and its voltage is considered to be the controlling voltage U_{2j}. Similarly, if the controlling current is the source-current of a current-source, the controlling current I_{3k} is the current of a zero impedance branch connected in series with the current-source. If branch i is passive, i.e. it consists of an admittance only, $y_{ij}=0$ and $\alpha_{ik}=0$. If only a voltage-controlled current-generator is included in branch i, $\alpha_{ik}=0$, and finally if it is a current-controlled current-generator, $y_{ij}=0$. The column matrix $\mathbf{I}_2$

formed by the currents of the branches in the second group is written according to (5.50) as follows:

$$\mathbf{I}_2 = \boldsymbol{Y}_2\mathbf{U}_2 + \boldsymbol{K}\mathbf{I}_3\,, \tag{5.51}$$

where $\mathbf{U}_2$ and $\mathbf{I}_3$ are column matrices formed by the voltages of the branches in the second group and the currents of the branches in the third group respectively, $\boldsymbol{Y}_2$ is a square matrix of order b_2 with the j-th element of its i-th row equalling y_{ij}, while the number of rows in matrix $\boldsymbol{K}$ is b_2, the number of columns is b_3, and the k-th element of its i-th row is α_{ik}.

The column matrix $\mathbf{U}_3$ formed by the voltages of the branches in the third group may similarly be written. Namely, the voltage of branch l (Fig. 5.12) is

$$U_{3l} = z_{ll} I_{3l} + z_{lm} I_{3m} + \mu_{ln} U_{2n}\,, \tag{5.52}$$

where $z_{ll} = Z_l$ is the impedance of the branch, I_{3l} is its current, $z_{lm} I_{3m}$ is the source-voltage of the current-controlled voltage-source in the branch with controlling current I_{3m}, $\mu_{ln} U_{2n}$ is the source-voltage of the voltage-controlled voltage-source in the branch with controlling voltage U_{2n}. I_{3m} is the current of branch m in the third group (in the case of a controlling current-source, it is the current of a zero impedance branch connected in series with the current-source), U_{2n} is the voltage of branch n in the second group (in the case of a controlling voltage-source, it is the

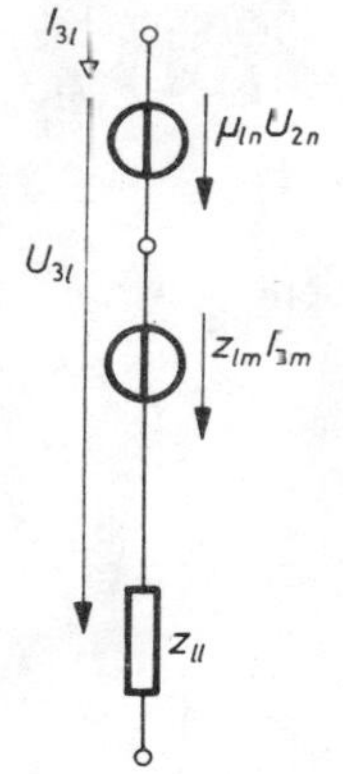

Fig. 5.12

voltage of a zero admittance branch connected in parallel with the voltage-source). If branch l consists of an impedance only, $z_{lm} = 0$ and $\mu_{ln} = 0$. If there is no current-controlled voltage-source in the branch, $z_{lm} = 0$, and in a branch without a voltage-controlled voltage-source $\mu_{ln} = 0$. According to (5.52) the column matrix $\mathbf{U}_3$ of the voltages of the branches in the third group is

$$\mathbf{U}_3 = \boldsymbol{M}\mathbf{U}_2 + \boldsymbol{Z}_3\mathbf{I}_3\,, \tag{5.53}$$

where $\boldsymbol{Z}_3$ is a square matrix of order b_3, with the m-th element of its l-th row equalling z_{lm}, while the number of rows in matrix $\boldsymbol{M}$ is b_3, the number of columns is b_2, and the n-th element of its l-th row is μ_{ln}. The relationships (5.51) and (5.53) will be used for writing the Kirchhoff equations of the network.

As mentioned above the tree of the network graph has been chosen for the analysis so as to associate chords with the branches in the first and second groups, and tree-edges with those in the third and fourth groups. With the aid of loop matrix $\boldsymbol{B}$ of the fundamental set of loops generated by the chosen tree, the linearly independent loop equations of the network can be summarized in the matrix equation

$$\boldsymbol{B}\mathbf{U}=\begin{bmatrix} \boldsymbol{1} & \boldsymbol{0} & \boldsymbol{F}_{11} & \boldsymbol{F}_{12} \\ \boldsymbol{0} & \boldsymbol{1} & \boldsymbol{F}_{21} & \boldsymbol{F}_{22} \end{bmatrix} \begin{bmatrix} \mathbf{U}_1 \\ \mathbf{U}_2 \\ \mathbf{U}_3 \\ \mathbf{U}_4 \end{bmatrix} = \mathbf{0}, \tag{5.54}$$

where loop matrix $\boldsymbol{B}$ and column matrix $\mathbf{U}$ of branch voltages have been partitioned in accordance with the four groups of branches, and the indices of the submatrices chosen correspondingly. Taking into account that $\mathbf{U}_4=\mathbf{U}_{g4}$ is the column matrix of the source-voltages of the independent voltage-sources, (5.54) can be written in the form of the two following equations:

$$\mathbf{U}_1+\boldsymbol{F}_{11}\mathbf{U}_3+\boldsymbol{F}_{12}\mathbf{U}_{g4}=\mathbf{0}, \tag{5.55}$$

$$\mathbf{U}_2+\boldsymbol{F}_{21}\mathbf{U}_3+\boldsymbol{F}_{22}\mathbf{U}_{g4}=\mathbf{0}. \tag{5.56}$$

Similarly, the linearly independent cutset equations of the network are

$$\boldsymbol{Q}\mathbf{I}=\begin{bmatrix} -\boldsymbol{F}_{11}^{+} & -\boldsymbol{F}_{21}^{+} & \boldsymbol{1} & \boldsymbol{0} \\ -\boldsymbol{F}_{12}^{+} & -\boldsymbol{F}_{22}^{+} & \boldsymbol{0} & \boldsymbol{1} \end{bmatrix} \begin{bmatrix} \mathbf{I}_1 \\ \mathbf{I}_2 \\ \mathbf{I}_3 \\ \mathbf{I}_4 \end{bmatrix} = \mathbf{0}. \tag{5.57}$$

With two equations, introducing $\mathbf{I}_1=\mathbf{I}_{g1}$:

$$-\boldsymbol{F}_{11}^{+}\mathbf{I}_{g1}-\boldsymbol{F}_{21}^{+}\mathbf{I}_2+\mathbf{I}_3=\mathbf{0}, \tag{5.58}$$

$$-\boldsymbol{F}_{12}^{+}\mathbf{I}_{g1}-\boldsymbol{F}_{22}^{+}\mathbf{I}_2+\mathbf{I}_4=\mathbf{0}. \tag{5.59}$$

Let us substitute (5.51) into (5.58) and (5.53) into (5.56):

$$-\boldsymbol{F}_{21}^{+}\boldsymbol{Y}_3\mathbf{U}_2+(\boldsymbol{1}-\boldsymbol{F}_{21}^{+}\boldsymbol{K})\mathbf{I}_3=\boldsymbol{F}_{11}^{+}\mathbf{I}_{g1}, \tag{5.60}$$

$$(\boldsymbol{1}+\boldsymbol{F}_{21}\boldsymbol{M})\mathbf{U}_2+\boldsymbol{F}_{21}\boldsymbol{Z}_3\mathbf{I}_3=-\boldsymbol{F}_{22}\mathbf{U}_{g4}. \tag{5.61}$$

Summarizing these in one equation, $\mathbf{U}_2$ and $\mathbf{I}_3$ are expressible:

$$\begin{bmatrix} \mathbf{U}_2 \\ \mathbf{I}_3 \end{bmatrix} = \begin{bmatrix} -\boldsymbol{F}_{21}^{+}\mathbf{Y}_2 & \boldsymbol{1}-\boldsymbol{F}_{21}^{+}\boldsymbol{K} \\ \boldsymbol{1}+\boldsymbol{F}_{21}\boldsymbol{M} & \boldsymbol{F}_{21}\mathbf{Z}_3 \end{bmatrix}^{-1} \begin{bmatrix} \boldsymbol{F}_{11}^{+} & \boldsymbol{0} \\ \boldsymbol{0} & -\boldsymbol{F}_{22} \end{bmatrix} \begin{bmatrix} \mathbf{I}_{g1} \\ \mathbf{U}_{g4} \end{bmatrix}. \tag{5.62}$$

Thus the voltages of the branches in the second group, and the currents of the branches in the third group have been expressed with the aid of known quantities. Hence the currents of the branches in the second group may be obtained from (5.51), and the voltages of the branches in the third group from (5.53). Knowing these the voltages of the independent current-sources may be calculated from (5.55), and the currents of the independent voltage-sources from (5.59).

It is worth noting that the independent sources in networks are often voltage-sources only. In this case $\boldsymbol{F}_{11}$ does not exist (the number of its rows is zero), however, in the second matrix on the right side of (5.62) a $\boldsymbol{0}$ submatrix appears with the number of its columns equalling the number of columns in $\boldsymbol{F}_{22}$, and the number of its rows equals the number of branches in the third group. Similarly, if there is no voltage-source in the network, $\boldsymbol{F}_{22}$ does not exist (the number of its rows is zero), and there is a $\boldsymbol{0}$ submatrix in the second matrix on the right side of (5.62), with the number of its columns equalling the number of rows in $\boldsymbol{F}_{11}$, and the number of its rows equals the number of branches in the second group. Accordingly, if there is no current-source in the network, the form of (5.62) is

$$\begin{bmatrix} \mathbf{U}_2 \\ \mathbf{I}_3 \end{bmatrix} = \begin{bmatrix} -\boldsymbol{F}_{21}^{+}\mathbf{Y}_2 & \boldsymbol{1}-\boldsymbol{F}_{21}^{+}\boldsymbol{K} \\ \boldsymbol{1}+\boldsymbol{F}_{21}\boldsymbol{M} & \boldsymbol{F}_{21}\mathbf{Z}_3 \end{bmatrix}^{-1} \begin{bmatrix} \boldsymbol{0} \\ -\boldsymbol{F}_{22} \end{bmatrix} \mathbf{U}_{g4}, \tag{5.63}$$

and for a network without voltage-sources:

$$\begin{bmatrix} \mathbf{U}_2 \\ \mathbf{I}_3 \end{bmatrix} = \begin{bmatrix} -\boldsymbol{F}_{21}^{+}\mathbf{Y}_2 & \boldsymbol{1}-\boldsymbol{F}_{21}^{+}\boldsymbol{K} \\ \boldsymbol{1}+\boldsymbol{F}_{21}\boldsymbol{M} & \boldsymbol{F}_{21}\mathbf{Z}_3 \end{bmatrix}^{-1} \begin{bmatrix} \boldsymbol{F}_{11}^{+} \\ \boldsymbol{0} \end{bmatrix} \mathbf{I}_{g1}. \tag{5.64}$$

To sum up, a method has been presented for the determination of all branch-currents and branch-voltages sought in the network, with the aid of a model containing controlled generators. The solution has been obtained by the inversion of a matrix of order b_2+b_3.

The equations written by the application of mixed parameters may be arranged to allow the determination of the unknowns by the inversion of a matrix of lower order. To this end branches are classified into six groups:

1. independent current-sources (chords);
2. controlled current-generators, current-sources and the branches with their voltages controlling other sources (chords);
3. finite admittances (chords);
4. finite impedances (tree-branches);

5. controlled voltage-generators, voltage-sources and the branches with their currents controlling other sources (tree-branches);
6. independent voltage-sources (tree-branches).

(The place of the two groups of controlled generators and passive branches in the former method has here been taken by four groups.) The tree is chosen for the analysis, i.e. branches are assigned to groups 3 and 4 to associate chords with the branches of the first three groups, and tree-branches with those in groups 4–6. This classification of the branches is always possible if independent and controlled current-generators form no cutset, and independent and controlled voltage-generators form no loop.

In the case of the above classification, the column matrices $\mathbf{I}_2$ and $\mathbf{U}_5$ of the branch-currents in the second group and of the branch-voltages in the fifth group may be written with the aid of the column matrices $\mathbf{U}_2$ and $\mathbf{I}_5$ of the branch-voltages in group 2 and of the branch-currents in group 5, similarly to (5.51) and (5.53):

$$\mathbf{I}_2 = \boldsymbol{Y}_2 \mathbf{U}_2 + \boldsymbol{K} \mathbf{I}_5 \,, \tag{5.65}$$

and

$$\mathbf{U}_5 = \boldsymbol{M} \mathbf{U}_2 + \boldsymbol{Z}_5 \mathbf{I}_5 \,. \tag{5.66}$$

The order of $\boldsymbol{Y}_2$ as well as the number of rows in $\boldsymbol{K}$ and that of columns in $\boldsymbol{M}$ equals the number of branches in group 2, while the order of $\boldsymbol{Z}_5$ as well as the number of columns in $\boldsymbol{K}$ and that of rows in $\boldsymbol{M}$ equals the number of branches in the fifth group. Ohm's law yields the relationships between the branch-currents and branch-voltages of groups 3 and 4:

$$\mathbf{I}_3 = \boldsymbol{Y}_3 \mathbf{U}_3 \,, \tag{5.67}$$

$$\mathbf{U}_4 = \boldsymbol{Z}_4 \mathbf{I}_4 \,, \tag{5.68}$$

i.e. $\boldsymbol{Y}_3$ is the admittance matrix of the branches in the third group, while $\boldsymbol{Z}_4$ is the impedance matrix of those in the fourth group.

Kirchhoff's equations are written for the fundamental sets of loops and cutsets generated by the tree chosen with the aid of matrices partitioned in accordance with the six groups of branches. With the notation $\mathbf{U}_6 = \mathbf{U}_{g6}$ and $\mathbf{I}_1 = \mathbf{I}_{g1}$:

$$\begin{bmatrix} \boldsymbol{1} & \boldsymbol{0} & \boldsymbol{0} & \boldsymbol{F}_{11} & \boldsymbol{F}_{12} & \boldsymbol{F}_{13} \\ \boldsymbol{0} & \boldsymbol{1} & \boldsymbol{0} & \boldsymbol{F}_{21} & \boldsymbol{F}_{22} & \boldsymbol{F}_{23} \\ \boldsymbol{0} & \boldsymbol{0} & \boldsymbol{1} & \boldsymbol{F}_{31} & \boldsymbol{F}_{32} & \boldsymbol{F}_{33} \end{bmatrix} \begin{bmatrix} \mathbf{U}_1 \\ \mathbf{U}_2 \\ \mathbf{U}_3 \\ \mathbf{U}_4 \\ \mathbf{U}_5 \\ \mathbf{U}_{g6} \end{bmatrix} = \mathbf{0} \,, \tag{5.69}$$

$$\begin{bmatrix} -F_{11}^{+} & -F_{21}^{+} & -F_{31}^{+} & 1 & 0 & 0 \\ -F_{12}^{+} & -F_{22}^{+} & -F_{32}^{+} & 0 & 1 & 0 \\ -F_{13}^{+} & -F_{23}^{+} & -F_{33}^{+} & 0 & 0 & 1 \end{bmatrix} \begin{bmatrix} \mathbf{I}_{g1} \\ \mathbf{I}_2 \\ \mathbf{I}_3 \\ \mathbf{I}_4 \\ \mathbf{I}_5 \\ \mathbf{I}_6 \end{bmatrix} = \mathbf{0}. \tag{5.70}$$

Hence

$$\mathbf{U}_1 + F_{11}\mathbf{U}_4 + F_{12}\mathbf{U}_5 + F_{13}\mathbf{U}_{g6} = \mathbf{0}, \tag{5.71}$$

$$\mathbf{U}_2 + F_{21}\mathbf{U}_4 + F_{22}\mathbf{U}_5 + F_{23}\mathbf{U}_{g6} = \mathbf{0}, \tag{5.72}$$

$$\mathbf{U}_3 + F_{31}\mathbf{U}_4 + F_{32}\mathbf{U}_5 + F_{33}\mathbf{U}_{g6} = \mathbf{0}, \tag{5.73}$$

$$-F_{11}^{+}\mathbf{I}_{g1} - F_{21}^{+}\mathbf{I}_2 - F_{31}^{+}\mathbf{I}_3 + \mathbf{I}_4 = \mathbf{0}, \tag{5.74}$$

$$-F_{12}^{+}\mathbf{I}_{g1} - F_{22}^{+}\mathbf{I}_2 - F_{32}^{+}\mathbf{I}_3 + I_5 = \mathbf{0}, \tag{5.75}$$

$$-F_{13}^{+}\mathbf{I}_{g1} - F_{23}^{+}\mathbf{I}_2 - F_{33}^{+}\mathbf{I}_3 + \mathbf{I}_6 = \mathbf{0}, \tag{5.76}$$

$\mathbf{U}_3$, $\mathbf{I}_3$ or $\mathbf{U}_4$, $\mathbf{I}_4$ may be eliminated from the equations.

The solution of the above set of equations is at first obtained after the elimination of $\mathbf{U}_3$ and $\mathbf{I}_3$. Applying (5.67), (5.73) yields:

$$\mathbf{I}_3 = -Y_3 F_{31}\mathbf{U}_4 - Y_3 F_{32}\mathbf{U}_5 - Y_3 F_{32}\mathbf{U}_5 - Y_3 F_{33}\mathbf{U}_{g6}. \tag{5.77}$$

Substituting this into (5.74) and (5.75):

$$\begin{aligned} -F_{21}^{+}\mathbf{I}_2 + F_{31}^{+} Y_3 F_{31}\mathbf{U}_4 + F_{31}^{+} Y_3 F_{32}\mathbf{U}_5 + \mathbf{I}_4 = \\ = F_{11}^{+}\mathbf{I}_{g1} - F_{31}^{+} Y_3 F_{33}\mathbf{U}_{g6}, \end{aligned} \tag{5.78}$$

$$\begin{aligned} -F_{22}^{+}\mathbf{I}_2 + F_{32}^{+} Y_3 F_{31}\mathbf{U}_4 + F_{32}^{+} Y_3 F_{32}\mathbf{U}_5 + \mathbf{I}_5 = \\ = F_{12}^{+}\mathbf{I}_{g1} - F_{32}^{+} Y_3 F_{33}\mathbf{U}_{g6}. \end{aligned} \tag{5.79}$$

With the application of (5.65), (5.66) and (5.68), (5.72), (5.78) and (5.79) yield:

$$(\mathbf{1} + F_{22}\mathbf{M})\mathbf{U}_2 + F_{21} Z_4 \mathbf{I}_4 + F_{22} Z_5 \mathbf{I}_5 = -F_{23}\mathbf{U}_{g6}, \tag{5.80}$$

$$\begin{aligned} (F_{31}^{+} Y_3 F_{32} M - F_{21}^{+} Y_2)\mathbf{U}_2 + (F_{31}^{+} Y_3 F_{31} Z_4 + \mathbf{1})\mathbf{I}_4 + \\ + (F_{31}^{+} Y_3 F_{32} Z_5 - F_{21}^{+} \mathbf{K})\mathbf{I}_5 = F_{11}^{+}\mathbf{I}_{g1} - F_{31}^{+} Y_3 F_{33}\mathbf{U}_{g6}, \end{aligned} \tag{5.81}$$

$$\begin{aligned} (F_{32}^{+} Y_3 F_{32} M - F_{22}^{+} Y_2)\mathbf{U}_2 + F_{32}^{+} Y_3 F_{31} Z_4 \mathbf{I}_4 + \\ + (F_{32}^{+} Y_3 F_{32} Z_5 - F_{22}^{+} \mathbf{K} + \mathbf{1})\mathbf{I}_5 = F_{12}^{+}\mathbf{I}_{g1} - F_{32}^{+} Y_3 F_{33}\mathbf{U}_{g6}. \end{aligned} \tag{5.82}$$

These can be summarized in

$$\boldsymbol{N}_1 \begin{bmatrix} \mathbf{U}_2 \\ \mathbf{I}_4 \\ \mathbf{I}_5 \end{bmatrix} = \boldsymbol{N}_2 \begin{bmatrix} \mathbf{I}_{g1} \\ \mathbf{U}_{g6} \end{bmatrix}, \tag{5.83}$$

where

$$\boldsymbol{N}_1 = \begin{bmatrix} \boldsymbol{1} + \boldsymbol{F}_{22}\boldsymbol{M} & \boldsymbol{F}_{21}\boldsymbol{Z}_4 & \boldsymbol{F}_{22}\boldsymbol{Z}_5 \\ \boldsymbol{F}_{31}^{+}\boldsymbol{Y}_3\boldsymbol{F}_{32}\boldsymbol{M} - \boldsymbol{F}_{21}^{+}\boldsymbol{Y}_2 & \boldsymbol{F}_{31}^{+}\boldsymbol{Y}_3\boldsymbol{F}_{31}\boldsymbol{Z}_4 + \boldsymbol{1} & \boldsymbol{F}_{31}^{+}\boldsymbol{Y}_3\boldsymbol{F}_{32}\boldsymbol{Z}_5 - \boldsymbol{F}_{21}^{+}\boldsymbol{K} \\ \boldsymbol{F}_{32}^{+}\boldsymbol{Y}_3\boldsymbol{F}_{32}\boldsymbol{M} - \boldsymbol{F}_{22}^{+}\boldsymbol{Y}_2 & \boldsymbol{F}_{32}^{+}\boldsymbol{Y}_3\boldsymbol{F}_{31}\boldsymbol{Z}_4 & \boldsymbol{F}_{32}^{+}\boldsymbol{Y}_3\boldsymbol{F}_{32}\boldsymbol{Z}_5 - \boldsymbol{F}_{22}^{+}\boldsymbol{K} + \boldsymbol{1} \end{bmatrix} \tag{5.84}$$

$$\boldsymbol{N}_2 = \begin{bmatrix} \boldsymbol{0} & -\boldsymbol{F}_{23} \\ \boldsymbol{F}_{11}^{+} & -\boldsymbol{F}_{31}^{+}\boldsymbol{Y}_3\boldsymbol{F}_{33} \\ \boldsymbol{F}_{12}^{+} & -\boldsymbol{F}_{32}^{+}\boldsymbol{Y}_3\boldsymbol{F}_{33} \end{bmatrix}. \tag{5.85}$$

By the inversion of $\boldsymbol{N}_1$ the column matrices $\mathbf{U}_2$, $\mathbf{I}_4$ and $\mathbf{I}_5$ may hence be determined. (5.65), (5.66) and (5.68) then yield $\mathbf{I}_2$, $\mathbf{U}_5$ and $\mathbf{U}_4$, allowing the calculation of $\mathbf{I}_3$ and $\mathbf{U}_3$ from (5.77) and (5.67). Hence, $\mathbf{U}_1$ and $\mathbf{I}_6$ are obtained from (5.71) and (5.76), respectively.

The procedure is similar, if $\mathbf{U}_4$ and $\mathbf{I}_4$ are eliminated. Now, (5.74) yields with the application of (5.68):

$$\mathbf{U}_4 = \boldsymbol{Z}_4\boldsymbol{F}_{11}^{+}\mathbf{I}_{g1} + \boldsymbol{Z}_4\boldsymbol{F}_{21}^{+}\mathbf{I}_2 + \boldsymbol{Z}_4\boldsymbol{F}_{31}^{+}\mathbf{I}_3. \tag{5.86}$$

On substitution into (5.72) and (5.73):

$$\begin{aligned} &\mathbf{U}_2 + \boldsymbol{F}_{21}\boldsymbol{Z}_4\boldsymbol{F}_{21}^{+}\mathbf{I}_2 + \boldsymbol{F}_{21}\boldsymbol{Z}_4\boldsymbol{F}_{31}^{+}\mathbf{I}_3 + \boldsymbol{F}_{22}\mathbf{U}_5 = \\ &= -\boldsymbol{F}_{21}\boldsymbol{Z}_4\boldsymbol{F}_{11}^{+}\mathbf{I}_{g1} - \boldsymbol{F}_{23}\mathbf{U}_{g6}, \end{aligned} \tag{5.87}$$

$$\begin{aligned} &\mathbf{U}_3 + \boldsymbol{F}_{31}\boldsymbol{Z}_4\boldsymbol{F}_{21}^{+}\mathbf{I}_2 + \boldsymbol{F}_{31}\boldsymbol{Z}_4\boldsymbol{F}_{31}^{+}\mathbf{I}_3 + \boldsymbol{F}_{32}\mathbf{U}_5 = \\ &= -\boldsymbol{F}_{31}\boldsymbol{Z}_4\boldsymbol{F}_{11}^{+}\mathbf{I}_{g1} - \boldsymbol{F}_{33}\mathbf{U}_{g6}. \end{aligned} \tag{5.88}$$

Applying (5.65), (5.66) and (5.67), (5.75), (5.87) and (5.88) yield:

$$\boldsymbol{F}_{22}^{+}\boldsymbol{Y}_2\mathbf{U}_2 + \boldsymbol{F}_{32}^{+}\boldsymbol{Y}_3\mathbf{U}_3 + (\boldsymbol{F}_{22}^{+}\boldsymbol{K} - \boldsymbol{1})\mathbf{I}_5 = -\boldsymbol{F}_{12}^{+}\mathbf{I}_{g1}, \tag{5.89}$$

$$\begin{aligned} &(\boldsymbol{1} + \boldsymbol{F}_{21}\boldsymbol{Z}_4\boldsymbol{F}_{21}^{+}\boldsymbol{Y}_2 + \boldsymbol{F}_{22}\boldsymbol{M})\mathbf{U}_2 + \boldsymbol{F}_{21}\boldsymbol{Z}_4\boldsymbol{F}_{31}^{+}\boldsymbol{Y}_3\mathbf{U}_3 + \\ &+ (\boldsymbol{F}_{21}\boldsymbol{Z}_4\boldsymbol{F}_{21}^{+}\boldsymbol{K} + \boldsymbol{F}_{22}\boldsymbol{Z}_5)\mathbf{I}_5 = -\boldsymbol{F}_{21}\boldsymbol{Z}_4\boldsymbol{F}_{11}^{+}\mathbf{I}_{g1} - \boldsymbol{F}_{23}\mathbf{U}_{g6}, \end{aligned} \tag{5.90}$$

$$\begin{aligned} &(\boldsymbol{F}_{31}\boldsymbol{Z}_4\boldsymbol{F}_{21}^{+}\boldsymbol{Y}_2 + \boldsymbol{F}_{32}\boldsymbol{M})\mathbf{U}_2 + (\boldsymbol{1} + \boldsymbol{F}_{31}\boldsymbol{Z}_4\boldsymbol{F}_{31}^{+}\boldsymbol{Y}_3)\mathbf{U}_3 + \\ &+ (\boldsymbol{F}_{31}\boldsymbol{Z}_4\boldsymbol{F}_{21}^{+}\boldsymbol{K} + \boldsymbol{F}_{32}\boldsymbol{Z}_5)\mathbf{I}_5 = -\boldsymbol{F}_{31}\boldsymbol{Z}_4\boldsymbol{F}_{11}^{+}\mathbf{I}_{g1} - \boldsymbol{F}_{33}\mathbf{U}_{g6}. \end{aligned} \tag{5.91}$$

In summary:

$$\boldsymbol{P}_1 \begin{bmatrix} \mathbf{U}_2 \\ \mathbf{U}_3 \\ \mathbf{I}_5 \end{bmatrix} = \boldsymbol{P}_2 \begin{bmatrix} \mathbf{I}_{g1} \\ \mathbf{U}_{g6} \end{bmatrix}, \tag{5.92}$$

where

$$\boldsymbol{P}_1 = \begin{bmatrix} \boldsymbol{F}_{22}^{+} \boldsymbol{Y}_2 & \boldsymbol{F}_{32}^{+} \boldsymbol{Y}_3 & \boldsymbol{F}_{22}^{+} \boldsymbol{K} - \boldsymbol{1} \\ \boldsymbol{1} + \boldsymbol{F}_{21} \boldsymbol{Z}_4 \boldsymbol{F}_{21}^{+} \boldsymbol{Y}_2 + \boldsymbol{F}_{22} \boldsymbol{M} & \boldsymbol{F}_{21} \boldsymbol{Z}_4 \boldsymbol{F}_{31}^{+} \boldsymbol{Y}_3 & \boldsymbol{F}_{21} \boldsymbol{Z}_4 \boldsymbol{F}_{21}^{+} \boldsymbol{K} + \boldsymbol{F}_{22} \boldsymbol{Z}_5 \\ \boldsymbol{F}_{31} \boldsymbol{Z}_4 \boldsymbol{F}_{21}^{+} \boldsymbol{Y}_2 + \boldsymbol{F}_{32} \boldsymbol{M} & \boldsymbol{1} + \boldsymbol{F}_{31} \boldsymbol{Z}_4 \boldsymbol{F}_{31}^{+} \boldsymbol{Y}_3 & \boldsymbol{F}_{31} \boldsymbol{Z}_4 \boldsymbol{F}_{21}^{+} \boldsymbol{K} + \boldsymbol{F}_{32} \boldsymbol{Z}_5 \end{bmatrix}, \tag{5.93}$$

$$\boldsymbol{P}_2 = \begin{bmatrix} -\boldsymbol{F}_{12}^{+} & \boldsymbol{0} \\ -\boldsymbol{F}_{21} \boldsymbol{Z}_4 \boldsymbol{F}_{11}^{+} & -\boldsymbol{F}_{23} \\ -\boldsymbol{F}_{31} \boldsymbol{Z}_4 \boldsymbol{F}_{11}^{+} & -\boldsymbol{F}_{33} \end{bmatrix}. \tag{5.94}$$

Thus, $\mathbf{U}_2$, $\mathbf{U}_3$ and $\mathbf{I}_5$ may be calculated from (5.92) by the inversion of $\boldsymbol{P}_1$. Hence, (5.65), (5.66) and (5.67) yield $\mathbf{I}_2$, $\mathbf{U}_5$ and $\mathbf{I}_3$ allowing the application of (5.74), (5.68), (5.71) and (5.76) to obtain $\mathbf{I}_4$, $\mathbf{U}_4$, $\mathbf{U}_1$ and $\mathbf{I}_6$, respectively.

The elimination of $\mathbf{U}_3$, $\mathbf{I}_3$ or $\mathbf{U}_4$, $\mathbf{I}_4$ permits the solution of the problem by the inversion of a matrix of lower order than in (5.62). The choice of the branch-voltages and branch-currents to be eliminated is made either to obtain the quantities sought directly or to solve the problem by the inversion of a matrix with lower order.

Our analyses concern networks formed by two-terminal elements and two-ports. The methods presented above may be extended in a simple way to networks containing n-ports as well, provided that the impedance or admittance parameters of the n-ports are known. If the impedance parameters of an n-port (Fig. 5.13, a), i.e. the relationships

$$\begin{bmatrix} U_1 \\ U_2 \\ \vdots \\ U_n \end{bmatrix} = \begin{bmatrix} z_{11} & z_{12} & \cdots & z_{1n} \\ z_{21} & z_{22} & \cdots & z_{2n} \\ \cdots & \cdots & \cdots & \cdots \\ z_{n1} & z_{n2} & \cdots & z_{nn} \end{bmatrix} \begin{bmatrix} I_1 \\ I_2 \\ \vdots \\ I_n \end{bmatrix} \tag{5.95}$$

are given, the equivalent connection shown in Fig. 5.13, b may be employed. If admittance parameters are known to characterize the n-port, i.e. the admittance matrix in

$$\begin{bmatrix} I_1 \\ I_2 \\ \vdots \\ I_n \end{bmatrix} = \begin{bmatrix} y_{11} & y_{12} & \cdots & y_{1n} \\ y_{21} & y_{22} & \cdots & y_{2n} \\ \cdots & \cdots & \cdots & \cdots \\ y_{n1} & y_{n2} & \cdots & y_{nn} \end{bmatrix} \begin{bmatrix} U_1 \\ U_2 \\ \vdots \\ U_n \end{bmatrix} \tag{5.96}$$

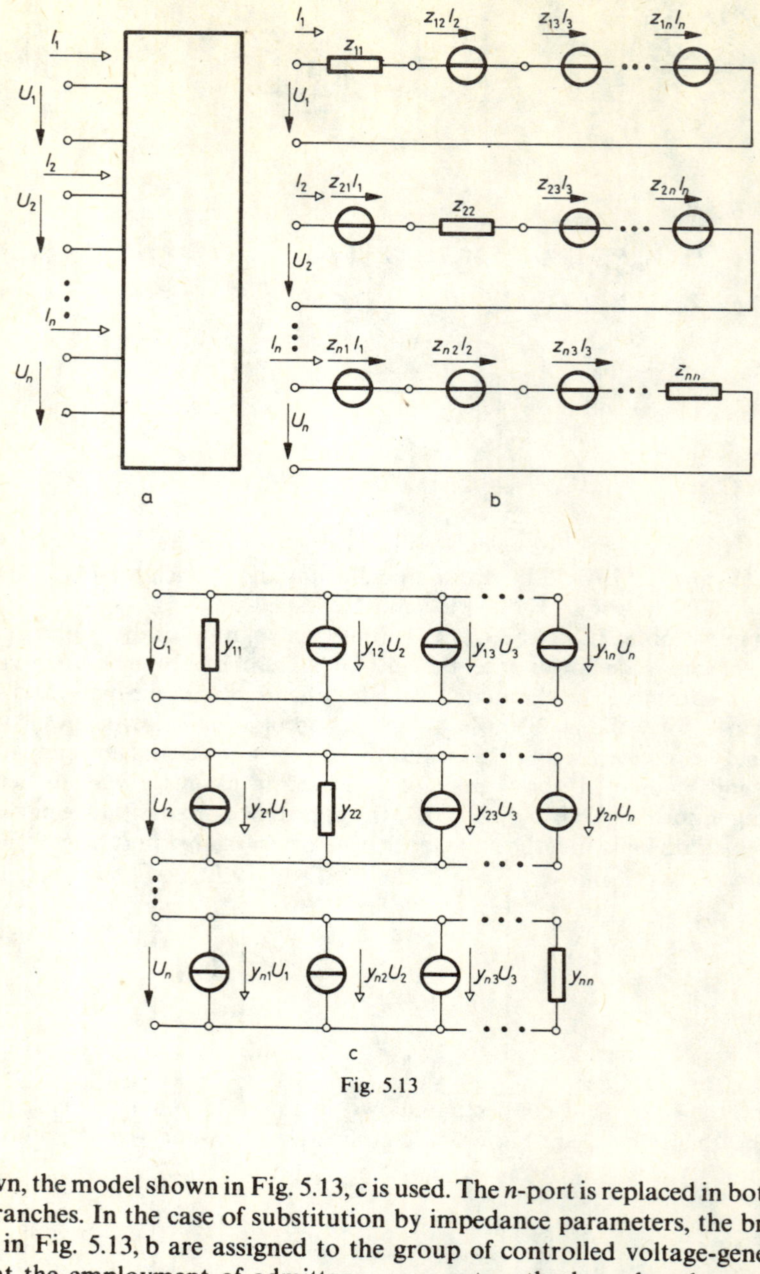

Fig. 5.13

is known, the model shown in Fig. 5.13, c is used. The n-port is replaced in both cases by n branches. In the case of substitution by impedance parameters, the branches shown in Fig. 5.13, b are assigned to the group of controlled voltage-generators, while at the employment of admittance parameters the branches drawn in Fig. 5.13, c belong to the group of controlled current-generators.

Analysis with the aid of nullators and norators

Models of two-ports, differing from those thus far used, are commonly employed for the analysis of electronic circuits. Nullators and norators are also utilized for modelling in this section [7, 13, 14, 23, 29, 37, 55, 56]. The concept of a norator has already been explained while a nullator is a fixator with both its current and voltage equal to zero. It is denoted as shown in Fig. 5.14.

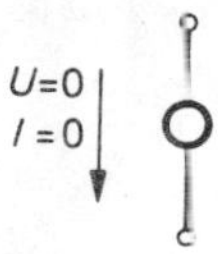

Fig. 5.14

Nullators connected in parallel or in series may be replaced by a single nullator. The series connection of a nullator and norator is equivalent to an open-circuit (Fig. 5.15, a), while their parallel connection models a short-circuit (Fig. 5.15, b). Note that the nullor shown in Table 5.4 is a two-port, with its primary side a nullator, and its secondary side a norator. Thus, both the voltage and current of the primary side are zero, while the secondary voltage and current are determined by those of the network connected to the element. All of the chain parameters of a nullor are zero (see Table 5.4). A nullor may model an ideal operational amplifier (Fig. 5.9).

Networks consisting of two-terminal elements, two-ports, electronic components and two-ports with extreme parameters may all be modelled with the aid of nullators and norators to yield networks formed by two-terminal elements only, with no mutual impedances, mutual admittances or controlled sources among them, i.e. the two-terminal branches of the network are independent sources, impedances, nullators or norators. This ensures at the same time that the graph of the model is connected.

In the following discussion a few possibilities are presented at first to model two-ports by circuits containing nullators and norators.

A necessary condition for the solvability of the network analysis problem is the equality of the number of nullators and norators in the model obtained after the replacement of series or parallel nullators by one nullator and of series or parallel norators by one norator. This may be shown in the same way as for the corresponding situation in the case of networks containing fixators and norators. In the equivalent circuits to be presented nullator–norator pairs appear.

For a loop consisting of nullators and voltage-sources only and for a cutset formed exclusively by nullators and current-sources the relevant Kirchhoff equation is not necessarily satisfied. The equations of networks containing such loops or cutsets may be contradictory.

If an impedance (admittance) belongs to a loop containing voltage-sources and nullators only in addition to this impedance (admittance), the voltage of this impedance (admittance) is determined by the other elements of the loop. If at the same time this impedance (admittance) is included in a cutset with all of its other branches being current-sources and nullators only, the latter determine the current of this impedance (admittance). In such cases the relationship between the current and voltage of the impedance (admittance) required by Ohm's law is not necessarily satisfied, i.e. the equations written may be contradictory.

If the network contains loops formed by norators, nullators and voltage-sources only, or cutsets formed exclusively by norators, nullators and current-sources, the equations written for the network are indefinite.

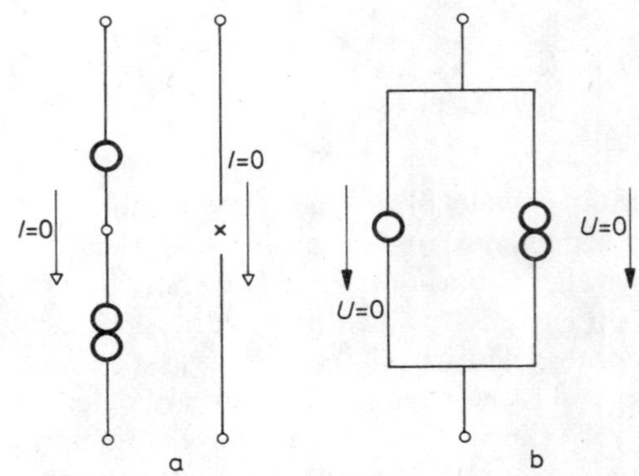

Fig. 5.15

Four possible equivalent circuits belonging to each of the characterizing equations of controlled sources are shown in Table 5.5. In each case, for two of these circuits the voltage between a pole of the primary and secondary sides is zero, e.g. primary and secondary share a common terminal while the other circuits are not so restricted. Two possible reference directions of the voltage and current on the secondary side may be chosen. One role of the choice of reference direction is to ensure that the real part of the impedance in a particular circuit is non-negative. From the point of view of analysis, the choice of the equivalent used is immaterial. In case of synthesis problems, however, not discussed here, it may be important. The equations shown in Table 5.5 are easily seen to hold on inspection of the equivalent circuits given.

On the basis of the equivalent circuits of controlled sources, models of two-ports with zero elements on the main diagonal of their parameter matrices may also be given. Each of the equivalent circuits is composed of two network parts connected to the ports by series nullators or norators. If such a network part connects to both ports by series nullators, the prescribed relationship between the primary and secondary voltages is established. If the network part connects to the ports by norators on both sides, the prescribed interdependence between the currents of the ports is ensured. In case it connects by a nullator to one port and by a norator to the

Table 5.5

Name	Characterizing equation	Equivalent circuit
Current-controlled voltage-source	$\begin{bmatrix} U_1 \\ U_2 \end{bmatrix} = \begin{bmatrix} 0 & 0 \\ Z & 0 \end{bmatrix} \begin{bmatrix} I_1 \\ I_2 \end{bmatrix}$	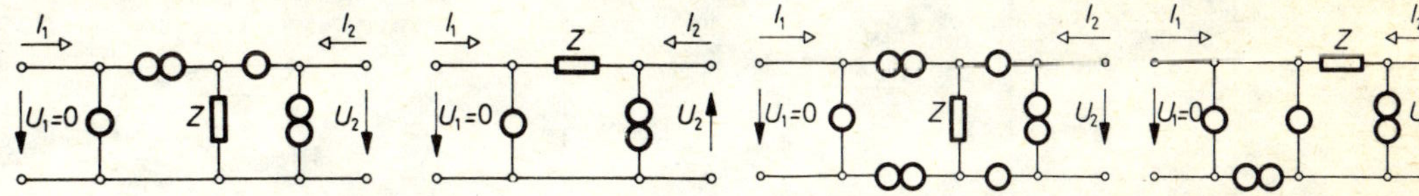
Voltage-controlled current-source	$\begin{bmatrix} I_1 \\ I_2 \end{bmatrix} = \begin{bmatrix} 0 & 0 \\ Y & 0 \end{bmatrix} \begin{bmatrix} U_1 \\ U_2 \end{bmatrix}$	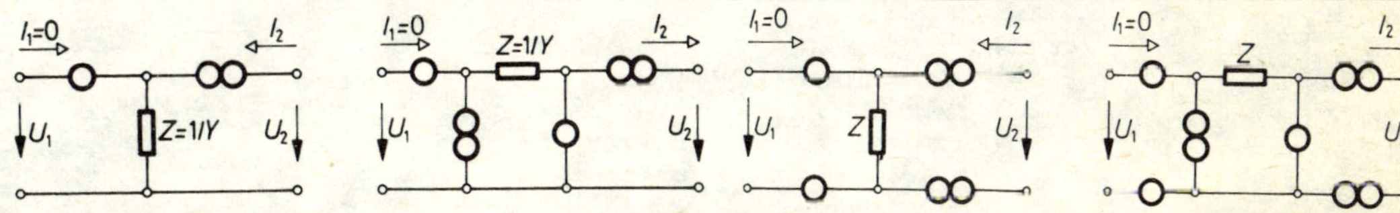
Current-controlled current-source	$\begin{bmatrix} U_1 \\ I_2 \end{bmatrix} = \begin{bmatrix} 0 & 0 \\ \mu & 0 \end{bmatrix} \begin{bmatrix} I_1 \\ U_2 \end{bmatrix}$	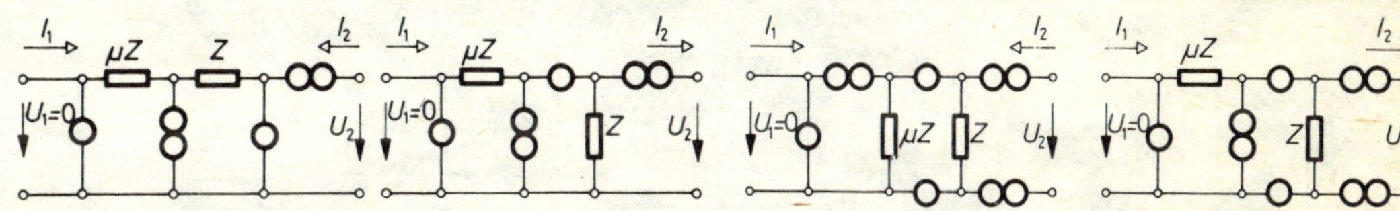
Voltage-controlled voltage-source	$\begin{bmatrix} I_1 \\ U_2 \end{bmatrix} = \begin{bmatrix} 0 & 0 \\ \alpha & 0 \end{bmatrix} \begin{bmatrix} U_1 \\ I_3 \end{bmatrix}$	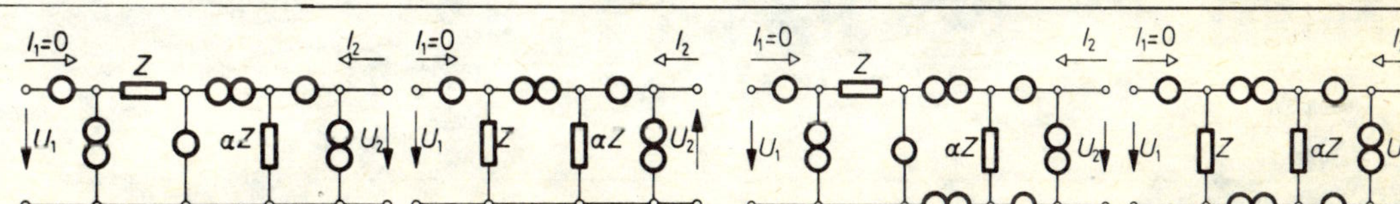

Table 5.6

Number	Equivalent circuit	Relationship between the parameters and impedances
1	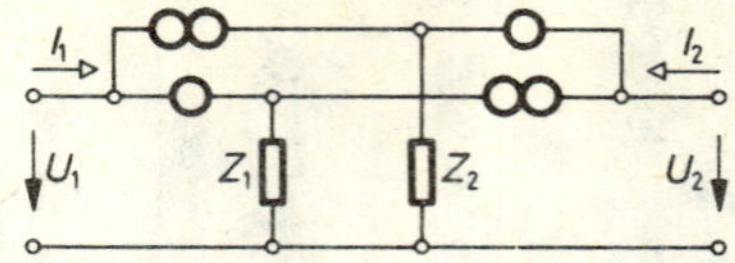	$Z_1=z_{12}=\frac{1}{y_{21}}=a_{12}$ $Z_2=z_{21}=\frac{1}{y_{12}}=\frac{1}{a_{21}}$
2	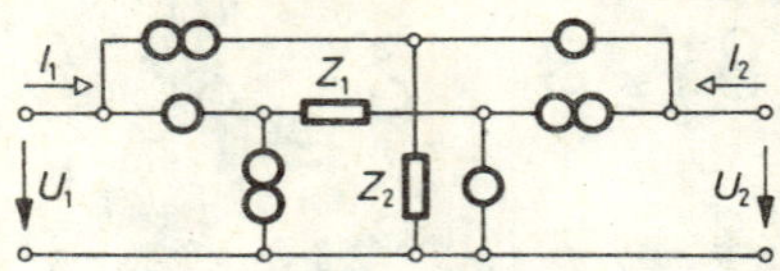	$Z_1=-z_{12}=-\frac{1}{y_{21}}=-a_{12}$ $Z_2=z_{21}=\frac{1}{y_{12}}=-\frac{1}{a_{21}}$
3	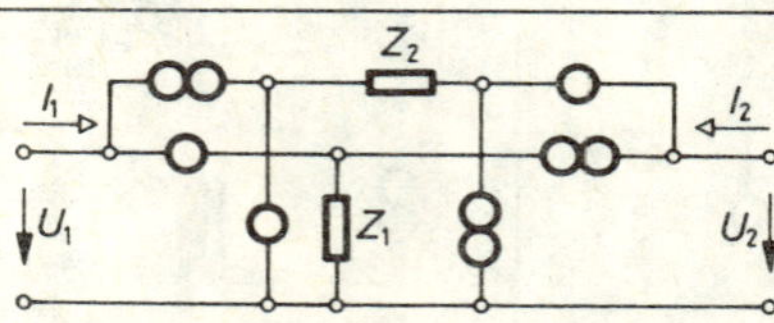	$Z_1=z_{12}=\frac{1}{y_{21}}=a_{12}$ $Z_2=-z_{21}=-\frac{1}{y_{12}}=-\frac{1}{a_{21}}$
4	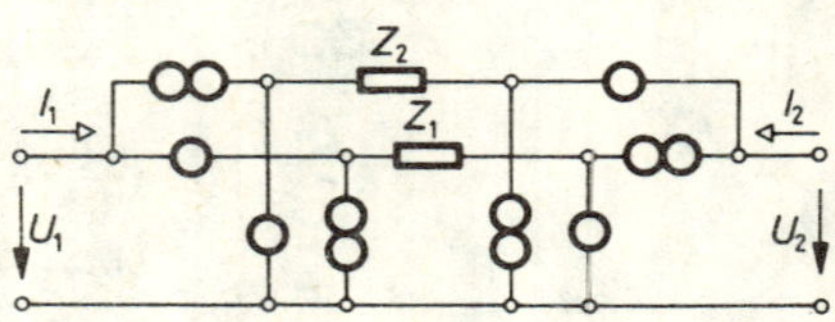	$Z_1=-z_{12}=-\frac{1}{y_{21}}=a_{12}$ $Z_2=-z_{21}=-\frac{1}{y_{12}}=-\frac{1}{a_{21}}$

$$\begin{bmatrix}U_1\\U_2\end{bmatrix}=\begin{bmatrix}0 & z_{12}\\z_{21} & 0\end{bmatrix}\begin{bmatrix}I_1\\I_2\end{bmatrix},\quad \begin{bmatrix}I_1\\I_2\end{bmatrix}=\begin{bmatrix}0 & y_{12}\\y_{21} & 0\end{bmatrix}\begin{bmatrix}U_1\\U_2\end{bmatrix},\quad \begin{bmatrix}U_1\\I_1\end{bmatrix}=\begin{bmatrix}0 & a_{12}\\a_{21} & 0\end{bmatrix}\begin{bmatrix}U_2\\I_2\end{bmatrix}$$

other, the prescribed relationship between the voltage of one port and the current of the other is guaranteed.

On the basis of the simplest equivalent circuits of controlled sources given in Table 5.5, the equivalent circuits consisting of nullators and norators of two-ports with zero elements on the main diagonals of their parameter matrices are given first. Thus the circuits shown in Table 5.6 may be regarded as the models of two-ports described by the equations

$$\begin{bmatrix}U_1\\U_2\end{bmatrix}=\begin{bmatrix}0 & z_{12}\\z_{21} & 0\end{bmatrix}\begin{bmatrix}I_1\\I_2\end{bmatrix} \tag{5.97}$$

or

$$\begin{bmatrix}I_1\\I_2\end{bmatrix}=\begin{bmatrix}0 & y_{21}\\y_{21} & 0\end{bmatrix}\begin{bmatrix}U_1\\U_2\end{bmatrix} \tag{5.98}$$

as well as

$$\begin{bmatrix} U_1 \\ I_1 \end{bmatrix} = \begin{bmatrix} 0 & a_{12} \\ a_{21} & 0 \end{bmatrix} \begin{bmatrix} U_2 \\ I_2 \end{bmatrix}. \tag{5.99}$$

The values of the impedances appearing in the circuits may be expressed in terms of the above parameters. The relationships between the impedances and the parameters are given in Table 5.6.

Four models satisfying the equation

$$\begin{bmatrix} U_1 \\ I_2 \end{bmatrix} = \begin{bmatrix} 0 & h_{12} \\ h_{21} & 0 \end{bmatrix} \begin{bmatrix} I_1 \\ U_2 \end{bmatrix} \tag{5.100}$$

are shown in Table 5.7, indicating the relationships between the impedances in the circuits and the hybrid parameters.

Table 5.7

Number	Equivalent circuit	Relationship between the parameters and impedances
1	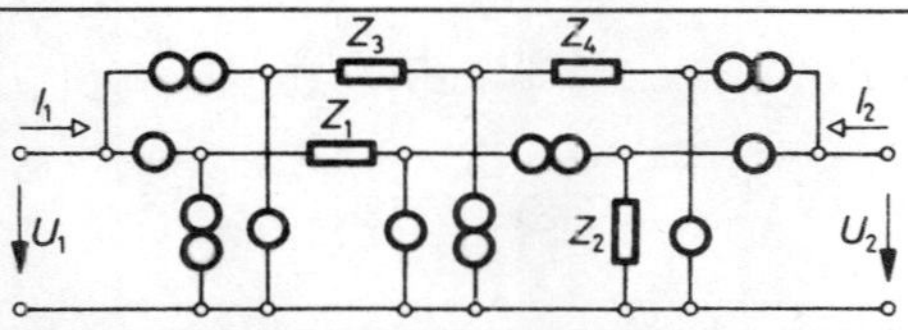	$h_{12} = \frac{Z_1}{Z_2}$ $h_{21} = \frac{Z_3}{Z_4}$
2	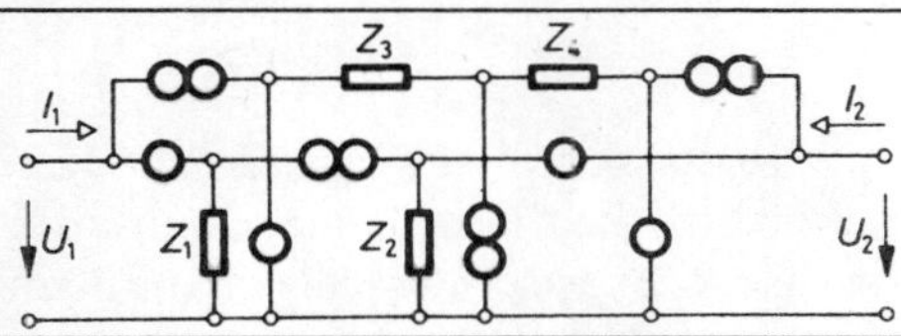	$h_{12} = -\frac{Z_1}{Z_2}$ $h_{21} = \frac{Z_3}{Z_4}$
3	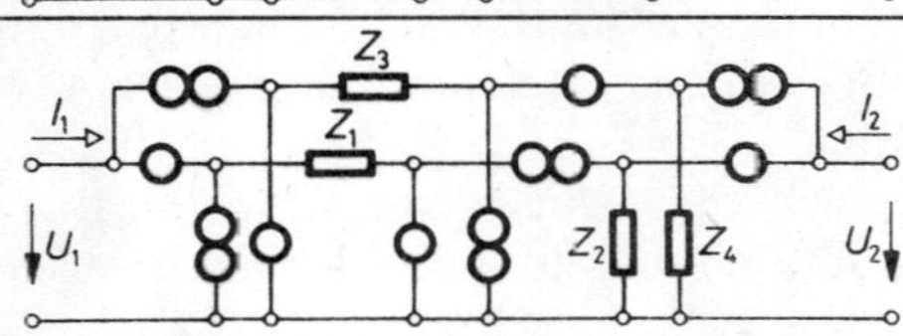	$h_{12} = \frac{Z_1}{Z_2}$ $h_{21} = -\frac{Z_3}{Z_4}$
4	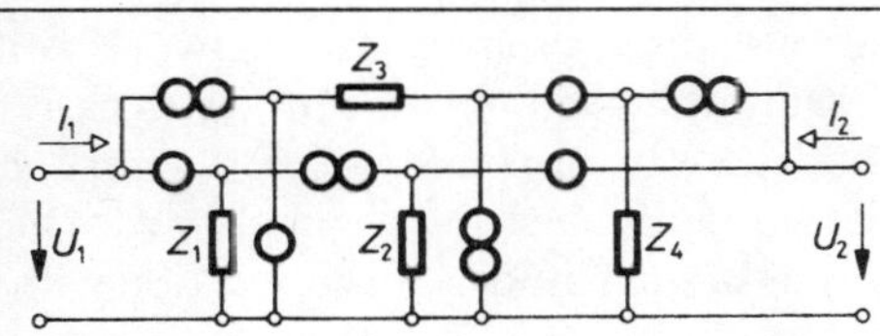	$h_{12} = -\frac{Z_1}{Z_2}$ $h_{21} = -\frac{Z_3}{Z_4}$

$$\begin{bmatrix} U_1 \\ I_2 \end{bmatrix} = \begin{bmatrix} 0 & h_{12} \\ h_{21} & 0 \end{bmatrix} \begin{bmatrix} I_1 \\ U_2 \end{bmatrix}.$$

Table 5.8

The equation characterizing the two-port	Equivalent circuit	Relationship between the parameters and impedances
$\begin{bmatrix} U_1 \\ U_2 \end{bmatrix} = \begin{bmatrix} z_{11} & z_{12} \\ z_{21} & z_{22} \end{bmatrix} \begin{bmatrix} I_1 \\ I_2 \end{bmatrix}$	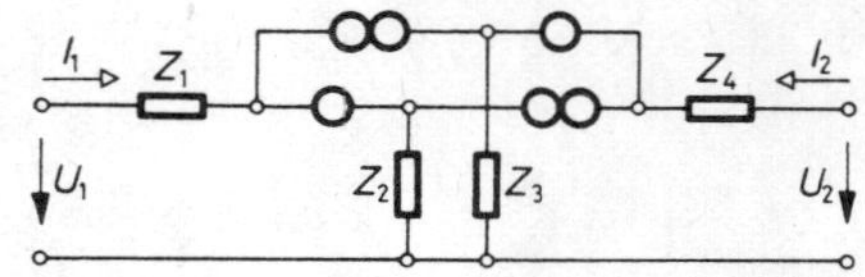	$z_{11}=Z_1$ $z_{12}=Z_2$ $z_{21}=Z_3$ $z_{22}=Z_4$
$\begin{bmatrix} I_1 \\ I_2 \end{bmatrix} = \begin{bmatrix} y_{11} & y_{12} \\ y_{21} & y_{22} \end{bmatrix} \begin{bmatrix} U_1 \\ U_2 \end{bmatrix}$	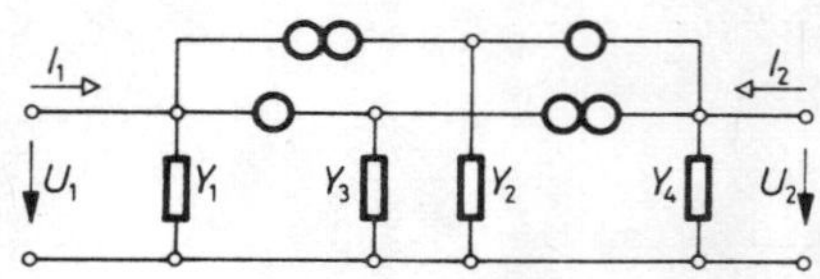	$y_{11}=Y_1$ $y_{12}=Y_2$ $y_{21}=Y_3$ $y_{22}=Y_4$
$\begin{bmatrix} U_1 \\ I_2 \end{bmatrix} = \begin{bmatrix} h_{11} & h_{12} \\ h_{21} & h_{22} \end{bmatrix} \begin{bmatrix} I_1 \\ U_2 \end{bmatrix}$	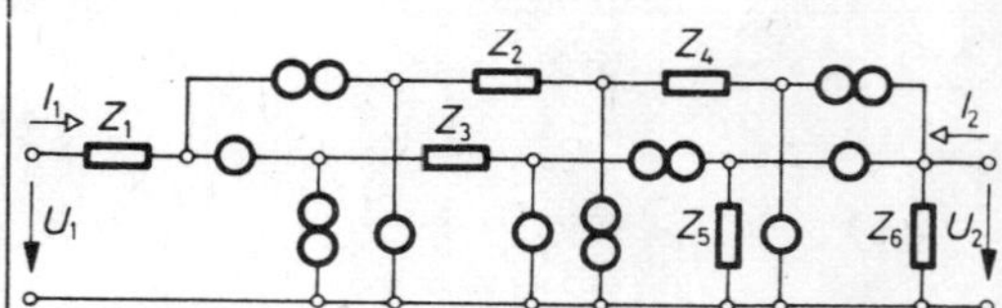	$h_{11}=Z_1$ $h_{12}=Z_3/Z_5$ $h_{21}=Z_2/Z_4$ $h_{22}=1/Z_6$

With the aid of the equivalent circuits shown in Table 5.6 and 5.7 or other analogous circuits, two-ports described by their impedance, admittance or hybrid parameters may be simply modelled. For the case of elements on the main diagonals with nonnegative real parts, series or parallel immittances connected to the primary and secondary sides yield two-ports with the prescribed parameters. As examples a few such circuits are given in Table 5.8.

On the basis of Tables 5.6 and 5.7 equivalent circuits of the ideal transformer, negative impedance converter, gyrator and nullor may also be constructed (Table 5.9).

With the aid of the equivalent circuits presented, any network consisting of independent sources, impedances, admittances and two-ports with extreme or non extreme parameters may be modelled by an equivalent network containing nullators and norators besides the two-terminal elements mentioned. The following methods are presented for the analysis of such models. In connection with one of the methods a procedure is given to determine whether the network examined leads to a contradictory or indefinite problem.

The application of the method of node voltages is initially introduced for networks containing nullators and norators [13].

Table 5.9

Name	Characterizing equation	Equivalent circuit
Ideal transformer	$\begin{bmatrix} U_1 \\ I_2 \end{bmatrix} = \begin{bmatrix} 0 & n \\ -n & 0 \end{bmatrix} \begin{bmatrix} I_1 \\ U_2 \end{bmatrix}$	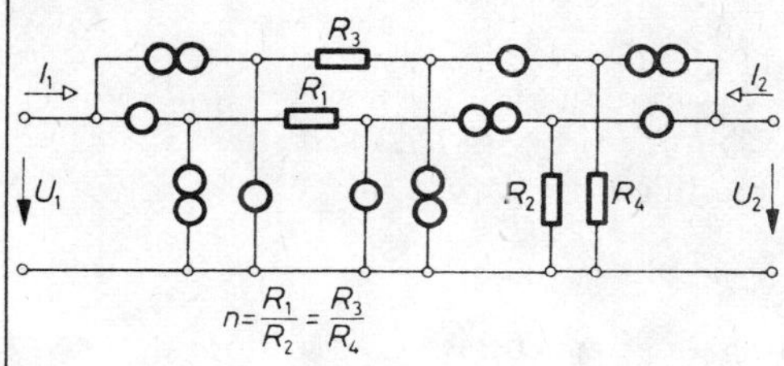
Negative impedance converter	$\begin{bmatrix} U_1 \\ I_2 \end{bmatrix} = \begin{bmatrix} 0 & k \\ k & 0 \end{bmatrix} \begin{bmatrix} I_1 \\ U_2 \end{bmatrix}$	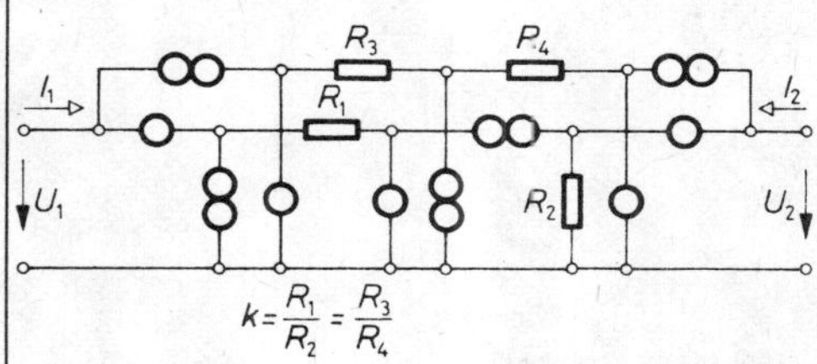
Gyrator	$\begin{bmatrix} U_1 \\ U_2 \end{bmatrix} = \begin{bmatrix} 0 & -R_1 \\ R_2 & 0 \end{bmatrix} \begin{bmatrix} I_1 \\ I_2 \end{bmatrix}$	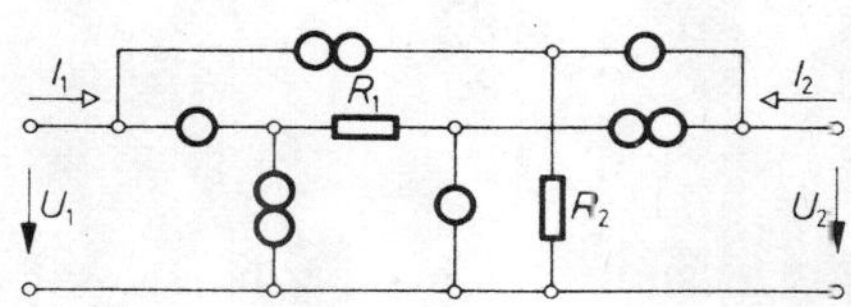
Nullor	$\begin{bmatrix} U_1 \\ I_1 \end{bmatrix} = \begin{bmatrix} 0 & 0 \\ 0 & 0 \end{bmatrix} \begin{bmatrix} U_2 \\ I_2 \end{bmatrix}$	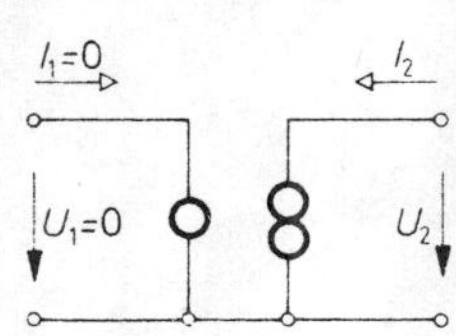

With the notation

$$\mathbf{I}_0 = A(Y\mathbf{U}_g - \mathbf{I}_g) \tag{5.101}$$

the Eq. (2.122) written for node-voltages in chapter 2 may be rewritten as

$$Y_A \mathbf{\Phi} = \mathbf{I}_0. \tag{5.102}$$

Removing the open edges containing nullators and norators from the network examined, Eq. (5.102) is written denoting by upper index 1 the quantities of this modified network:

$$Y_A^{(1)} \mathbf{\Phi}^{(1)} = \mathbf{I}_0^{(1)}. \tag{5.103}$$

This matrix equation in a more detailed form is as follows:

$$Y_{11}^{(1)}\Phi_1^{(1)} + Y_{12}^{(1)}\Phi_2^{(1)} + \ldots + Y_{1i}^{(1)}\Phi_i^{(1)} + \ldots + Y_{1j}^{(1)}\Phi_j^{(1)} + \ldots + Y_{1,n-1}^{(1)}\Phi_{n-1}^{(1)} = I_{01}^{(1)}$$

$$Y_{21}^{(1)}\Phi_1^{(1)} + Y_{22}^{(1)}\Phi_2^{(1)} + \ldots + Y_{2i}^{(1)}\Phi_i^{(1)} + \ldots + Y_{2j}^{(1)}\Phi_j^{(1)} + \ldots + Y_{2,n-1}^{(1)}\Phi_{n-1}^{(1)} = I_{02}^{(1)}$$

$$\cdots\cdots\cdots\cdots\cdots\cdots\cdots\cdots\cdots\cdots\cdots\cdots\cdots\cdots$$

$$Y_{i1}^{(1)}\Phi_1^{(1)} + Y_{i2}^{(1)}\Phi_2^{(1)} + \ldots + Y_{ii}^{(1)}\Phi_i^{(1)} + \ldots + Y_{ij}^{(1)}\Phi_j^{(1)} + \ldots + Y_{i,n-1}^{(1)}\Phi_{n-1}^{(1)} = I_{0i}^{(1)}$$

$$\cdots\cdots\cdots\cdots\cdots\cdots\cdots\cdots\cdots\cdots\cdots\cdots\cdots\cdots$$

$$Y_{n-1,1}^{(1)}\Phi_1^{(1)} + Y_{n-1,2}^{(1)}\Phi_2^{(1)} + \ldots + Y_{n-1,i}^{(1)}\Phi_i^{(1)} + \ldots + Y_{n-1,j}^{(1)}\Phi_j^{(1)} + \ldots +$$
$$+ Y_{n-1,n-1}^{(1)}\Phi_{n-1}^{(1)} = I_{0,n-1}^{(1)} \tag{5.104}$$

where n denotes the number of nodes in the network containing the nullators and norators.

The next objective is to consider how the above set of equations is modified on the reinsertion of a nullator into the network between nodes (i) and (j) (Fig. 5.16). Let the

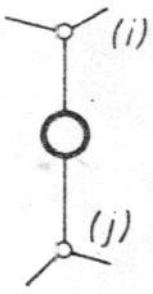

Fig. 5.16

voltages of the nodes be denoted by $\Phi'_k (k = 1, 2, \ldots, n-1)$ in this case. The voltages of nodes (i) and (j) are made equal by the nullator, and denoted by Φ'_{ij}. This means that the i-th and j-th elements in the column matrix of the node-voltages coincide, so that the set of equations (5.104) becomes redundant. This set of equations may be written as follows:

$$Y^{(1)}_{11}\Phi'_1+Y^{(1)}_{12}\Phi'_2+\ \dots\ +(Y^{(1)}_{1i}+Y^{(1)}_{1j})\Phi'_{ij}+\ \dots\ +Y^{(1)}_{1k}\Phi'_k+\ \dots\ +Y^{(1)}_{1,n-1}\Phi'_{n-1}=I^{(1)}_{01}$$

$$Y^{(1)}_{21}\Phi'_1+Y^{(1)}_{22}\Phi'_2+\ \dots\ +(Y^{(1)}_{2i}+Y^{(1)}_{2j})\Phi'_{ij}+\ \dots\ +Y^{(1)}_{2k}\Phi'_k+\ \dots\ +Y^{(1)}_{2,n-1}\Phi'_{n-1}=I^{(1)}_{02}$$

. . . .

$$Y^{(1)}_{i1}\Phi'_1+Y^{(1)}_{i2}\Phi'_2+\ \dots\ +(Y^{(1)}_{ii}+Y^{(1)}_{ij})\Phi'_{ij}+\ \dots\ +Y^{(1)}_{ik}\Phi'_k+\ \dots\ +Y^{(1)}_{i,n-1}\Phi'_{n-1}=I^{(1)}_{0i}$$

. . . .

$$Y^{(1)}_{n-1,1}\Phi'_1+Y^{(1)}_{n-1,2}\Phi'_2+\ \dots\ +(Y^{(1)}_{n-1,i}+Y^{(1)}_{n-1,j})\Phi'_{ij}+\ \dots\ +Y^{(1)}_{n-1,k}\Phi'_k+\ \dots\ +$$
$$+Y^{(1)}_{n-1,n-1}\Phi'_{n-1}=I^{(1)}_{0,n-1}, \qquad (5.105)$$

i.e. the coefficient of Φ'_{ij} here may be obtained from (5.104) by adding the coefficients of $\Phi^{(1)}_i$ and $\Phi^{(1)}_j$ in the appropriate equation. If the nullator connects node (i) to the node chosen to have zero voltage then the voltage of (i) is zero as well. In this case $Y^{(1)}_{ij}$ ($j=1, 2, \dots, n-1$) does not appear in (5.105). Thus, on the reinsertion of all of the nullators into the network, the equation

$$\mathbf{Y}^{(2)}_A\mathbf{\Phi}^{(2)}=\mathbf{I}^{(2)}_0 \qquad (5.106)$$

may be written, where upper index (2) denotes the quantities of the network obtained by the reinsertion of the nullators. The number of columns in $\mathbf{Y}^{(2)}_A$ is less by the number of nullators than the number of columns in $\mathbf{Y}^{(1)}_A$. The reinsertion of the nullators into the network does not effect the right side of (5.104), i.e. $\mathbf{I}^{(1)}_0=\mathbf{I}^{(2)}_0$.

$\mathbf{Y}^{(2)}_A$ may alternatively be obtained from $\mathbf{Y}^{(1)}_A$ with the aid of a transformation matrix $\mathbf{T}'$. $\mathbf{T}'$ is derived from the unit matrix of order $n-1$ as follows. If there is a nullator between nodes (i) and (j), and the voltage Φ'_{ij} is preferred to be the i-th element, the sum of the i-th and j-th columns of the unit matrix is the i-th column with the j-th column omitted. If the nullator connects node (i) to the zero potential node, the i-th column is omitted. Taking each nullator into account as above matrix $\mathbf{T}'$ is obtained to yield

$$\mathbf{Y}^{(2)}_A=\mathbf{Y}^{(1)}_A\mathbf{T}', \qquad (5.107)$$

i.e. the redundant set of equations under (5.106) may be written as

$$\mathbf{Y}^{(1)}_A\mathbf{T}'\mathbf{\Phi}^{(2)}=\mathbf{I}^{(1)}_0=\mathbf{I}^{(2)}_0. \qquad (5.108)$$

The next objective is to consider how this set of equations is modified if the norators are also reinserted into the network. Let a norator connect nodes (p) and (q), with the direction of the branch pointing from (p) towards (q) (Fig. 5.17). The current of the norator may be expressed with the aid of the currents of the other branches incident with nodes (p) and (q), namely as the sum of the other currents flowing out of node (p) multiplied by -1, or the sum of the currents flowing out of node (q). Adding the node equations relating to nodes (p) and (q), the current of the norator does not appear in the sum. The sum of the two node equations coincides with the continuity equation written for the cutset dividing the graph of the network into two connected subgraphs with only nodes (p) and (q) present in one of them (Fig. 5.17). Thus, the reinsertion of the norator into the network means from the point of view of the analysis, that instead of the two equations of the nodes incident with the norator, one cutset equation appears. So, in (5.105) the p-th and q-th equations are added and written instead of the p-th e.g., omitting the q-th. If the

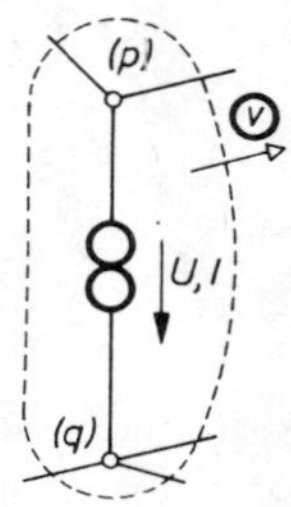

Fig. 5.17

norator is between node (p) and the zero potential node, one cutset equation appears instead of the two node equations written for node (p) and the zero potential node. This, however, is the one linearly dependent of the remaining equations. Thus the p-th equation may and should be omitted.

Let the node admittance matrix of the network obtained after the reinsertion of the norators, i.e. of the complete network be denoted by $\boldsymbol{Y}_A$, the column matrix of node-voltages by $\boldsymbol{\Phi}$, and the matrix characterizing the excitations by $\mathbf{I}_0$. Among these $\boldsymbol{Y}_A$ and $\boldsymbol{I}_0$ are obtained from $\boldsymbol{Y}_A^{(2)}$ and $\mathbf{I}_0^{(2)}$ by adding the two rows or two elements corresponding to the two nodes of each norator. This sum is written instead of one of the rows and the corresponding element, omitting the other row and element. If the norator connects a node to the zero voltage node, the row and element corresponding to it are omitted from the matrices.

$\boldsymbol{Y}_A$ and $\mathbf{I}_0$ may alternatively be obtained from the matrices $\boldsymbol{Y}_A^{(2)}$ and $\mathbf{I}_0^{(2)}=\mathbf{I}_0^{(1)}$ with the aid of a transformation matrix $\boldsymbol{T}''$. $\boldsymbol{T}''$ is derived from the unit matrix of order $n-1$ by writing the sum of the two rows corresponding to the two nodes of each norator instead of one row, omitting the other. If one node of a norator is the one with zero voltage, the row corresponding to the other node is omitted. Hence

$$\boldsymbol{Y}_A=\boldsymbol{T}''\boldsymbol{Y}_A^{(2)}, \tag{5.109}$$

$$\mathbf{I}_0=\boldsymbol{T}''\mathbf{I}_0^{(2)}=\boldsymbol{T}''\mathbf{I}_0^{(1)}, \tag{5.110}$$

and thus using (5.107) and (5.108):

$$\boldsymbol{Y}_A\boldsymbol{\Phi}=\boldsymbol{T}''\boldsymbol{Y}_A^{(2)}\boldsymbol{\Phi}=\boldsymbol{T}''\boldsymbol{Y}_A^{(1)}\boldsymbol{T}'\boldsymbol{\Phi}=\boldsymbol{T}''\mathbf{I}_0^{(2)}=\mathbf{I}_0. \tag{5.111}$$

$\boldsymbol{Y}_A$ is a square matrix with its order being less then $n-1$ by the number of nullator-norator pairs. From the last equation, the node-voltages, and hence branch-voltages and branch currents may be determined.

Next a method of analysis is presented which utilizes the loop and cutset matrices without the modification of the network containing nullators and norators [56]. Non-ideal generators are modelled by two branches, a source and an impedance, in accordance with their Thevenin or Norton equivalents. Short-circuits are regarded as voltage-sources with zero source-voltage, open-circuits as current-sources with zero source-current. A tree of the network graph is chosen for the analysis with each nullator and voltage-source of the network corresponding to tree-branches, and each norator and current-source to chords. Such a choice is possible if no loop is formed by nullators and voltage-sources, and no cutset by norators and current-sources.

Let the branches of the network be classified into six groups as follows:

1. current-sources (chords),
2. norators (chords),
3. finite admittances (chords),
4. finite impedances (tree-branches),
5. nullators (tree-branches),
6. voltage-sources (tree-branches).

If the network is to be examined to find out if there are further contradictions in the network analysis problem, a tree of the network is chosen with each norator and voltage-source of the network corresponding to tree-branches, and each nullator and current-source to chords. Now the branches of the network are classified into six groups as follows:

1. current-sources (chords),
2. nullators (chords),
3. finite admittances (chords),
4. finite impedances (tree-branches),
5. norators (tree-branches),
6. voltage-sources (tree-branches).

If such a classification of the branches is also possible, the network contains no loop formed by norators and voltage-sources only, and no cutset formed exclusively by nullators and current-sources.

If both classifications are possible, branches should be classified into groups 3 and 4 in such a way that group 3 is formed by the same branches in both cases. Thus, naturally, group 4 is also formed by the same branches. If this is possible the analysis problem has a unique solution.

The analysis may be carried out with the aid of either classifications given above. In the following example, the classification discussed first will be employed.

The number of branches in the groups is $b_1, b_2, b_3, b_4, b_5, b_6$, respectively. It is noted that the number of branches in groups 2 and 5 coincide, i.e. $b_2=b_5$.

Let the branches be given order numbers in accordance with the classification. The loops in the set of loops generated by the tree chosen are assigned order numbers in the order of the corresponding chords, while order numbers of the set of cutsets generated by the same tree are in the order of the corresponding tree-branches. With the aid of loop matrix $\boldsymbol{B}$ of the set of loops the loop equations of the network are

$$\boldsymbol{B}\mathbf{U}=\mathbf{0}, \tag{5.112}$$

where $\mathbf{U}$ is the column matrix of branch-voltages. Let $\boldsymbol{B}$ and $\mathbf{U}$ be partitioned in accordance with the six groups of branches. Thus (5.112) is written as

$$\begin{matrix} & b_1 & b_2 & b_3 & b_4 & b_5 & b_6 \\ b_1 & 1 & 0 & 0 & F_{11} & F_{12} & F_{13} \\ b_2 & 0 & 1 & 0 & F_{21} & F_{22} & F_{23} \\ b_3 & 0 & 0 & 1 & F_{31} & F_{32} & F_{33} \end{matrix} \begin{bmatrix} \mathbf{U}_1 \\ \mathbf{U}_2 \\ \mathbf{U}_3 \\ \mathbf{U}_4 \\ \mathbf{0} \\ \mathbf{U}_{g6} \end{bmatrix} = \mathbf{0}. \tag{5.113}$$

The numbers of columns in the submatrices have been indicated above matrix $\boldsymbol{B}$, and those of rows in the submatrices to the left of the matrix. It has been taken into account that $\mathbf{U}_6=\mathbf{U}_{g6}$ is the column matrix of the source-voltages of voltage-sources and $\mathbf{U}_5=\mathbf{0}$ the voltage of nullators. From (5.113):

$$\mathbf{U}_1+\boldsymbol{F}_{11}\mathbf{U}_4+\boldsymbol{F}_{13}\mathbf{U}_{g6}=\mathbf{0}, \tag{5.114}$$

$$\mathbf{U}_2+\boldsymbol{F}_{21}\mathbf{U}_4+\boldsymbol{F}_{23}\mathbf{U}_{g6}=\mathbf{0}, \tag{5.115}$$

$$\mathbf{U}_3+\boldsymbol{F}_{31}\mathbf{U}_4+\boldsymbol{F}_{33}\mathbf{U}_{g6}=\mathbf{0}. \tag{5.116}$$

Let the cutset equations be written:

$$\boldsymbol{Q}\mathbf{I}=\mathbf{0}, \tag{5.117}$$

where $\boldsymbol{Q}$ is the matrix of the set of cutsets generated by the tree chosen in accordance with the above numbering, and $\mathbf{I}$ is the column matrix of the branch-currents in the network. Let these also be partitioned according to the six groups of branches. Taking into account the above numbering of branches, loops and cutsets, (5.117) may be written as follows:

$$\begin{bmatrix} -\boldsymbol{F}_{11}^{+} & -\boldsymbol{F}_{21}^{+} & -\boldsymbol{F}_{31}^{+} & 1 & 0 & 0 \\ -\boldsymbol{F}_{12}^{+} & -\boldsymbol{F}_{22}^{+} & -\boldsymbol{F}_{32}^{+} & 0 & 1 & 0 \\ -\boldsymbol{F}_{13}^{+} & -\boldsymbol{F}_{23}^{+} & -\boldsymbol{F}_{33}^{+} & 0 & 0 & 1 \end{bmatrix} \begin{bmatrix} \mathbf{I}_{g1} \\ \mathbf{I}_2 \\ \mathbf{I}_3 \\ \mathbf{I}_4 \\ \mathbf{0} \\ \mathbf{I}_6 \end{bmatrix} = \mathbf{0}, \tag{5.118}$$

where $\mathbf{I}_{g1}$ is the column matrix of the source-currents of current-sources, and $\mathbf{I}_5=\mathbf{0}$ is the current of nullators. Hence

$$-\boldsymbol{F}_{11}^{+}\mathbf{I}_{g1}-\boldsymbol{F}_{21}^{+}\mathbf{I}_2-\boldsymbol{F}_{31}^{+}\mathbf{I}_3+\mathbf{I}_4=\mathbf{0}, \tag{5.119}$$

$$-\boldsymbol{F}_{12}^{+}\mathbf{I}_{g1}-\boldsymbol{F}_{22}^{+}\mathbf{I}_2-\boldsymbol{F}_{32}^{+}\mathbf{I}_3=\mathbf{0}, \tag{5.120}$$

$$-\boldsymbol{F}_{13}^{+}\mathbf{I}_{g1}-\boldsymbol{F}_{23}^{+}\mathbf{I}_2-\boldsymbol{F}_{33}^{+}\mathbf{I}_3+\mathbf{I}_6=\mathbf{0}. \tag{5.121}$$

The voltages and currents of the branches may be determined from the above equation e.g. as follows. $\boldsymbol{F}_{22}$ is seen from (5.113) to be a square matrix. Provided that it is nonsingular, (5.120) yields:

$$\mathbf{I}_2=-\boldsymbol{F}_{22}^{+\,-1}\boldsymbol{F}_{12}^{+}\mathbf{I}_{g1}-\boldsymbol{F}_{22}^{+\,-1}\boldsymbol{F}_{32}^{+}\mathbf{I}_3. \tag{5.122}$$

Substituting this into (5.119)

$$(\boldsymbol{F}_{21}^{+}\boldsymbol{F}_{22}^{+\,-1}\boldsymbol{F}_{32}^{+}-\boldsymbol{F}_{31}^{+})\mathbf{I}_3+\mathbf{I}_4=(\boldsymbol{F}_{11}^{+}-\boldsymbol{F}_{21}^{+}\boldsymbol{F}_{22}^{+\,-1}\boldsymbol{F}_{12}^{+})\mathbf{I}_{g1}. \tag{5.123}$$

Here $\mathbf{I}_3$ and $\mathbf{I}_4$, the currents of branches containing finite admittances and impedances, are unknown, while in (5.116) are the voltages of the same branches. These are employed for the remaining calculations.

The currents and voltages of the immitances are in the relationships

$$\mathbf{I}_3=\boldsymbol{Y}_3\mathbf{U}_3 \tag{5.124}$$

$$\mathbf{U}_4=\boldsymbol{Z}_4\mathbf{I}_4. \tag{5.125}$$

$\boldsymbol{Y}_3$ is the admittance matrix of the branches in group 3, and $\boldsymbol{Z}_4$ the impedance matrix of the branches in group 4. There are no coupled branches, couplings having been eliminated by the nullator-norator model. Thus $\boldsymbol{Y}_3$ and $\boldsymbol{Z}_4$ are diagonal matrices.

It is convenient to evaluate either $\mathbf{U}_3$ or $\mathbf{I}_4$ from the equations. The former is determined by the inversion of a matrix of order b_3, the latter by that of order b_4.

To calculate $\mathbf{U}_3$ (5.123) and (5.125) yield

$$\mathbf{U}_4=-\boldsymbol{Z}_4(\boldsymbol{F}_{21}^{+}\boldsymbol{F}_{22}^{+\,-1}\boldsymbol{F}_{32}^{+}-\boldsymbol{F}_{31}^{+})\mathbf{I}_3+\boldsymbol{Z}_4(\boldsymbol{F}_{11}^{+}-\boldsymbol{F}_{21}^{+}\boldsymbol{F}_{22}^{+\,-1}\boldsymbol{F}_{12}^{+})\mathbf{I}_{g1}. \tag{5.126}$$

This is substituted into (5.116), applying (5.124):

$$\mathbf{U}_3=[\boldsymbol{1}-\boldsymbol{F}_{31}\boldsymbol{Z}_4(\boldsymbol{F}_{21}^{+}\boldsymbol{F}_{22}^{+\,-1}\boldsymbol{F}_{32}^{+}-$$

$$-\boldsymbol{F}_{31}^{+})\boldsymbol{Y}_3]^{-1}[\boldsymbol{F}_{31}\boldsymbol{Z}_4(\boldsymbol{F}_{21}^{+}\boldsymbol{F}_{22}^{+\,-1}\boldsymbol{F}_{12}^{+}-\boldsymbol{F}_{11}^{+})\mathbf{I}_{g1}-\boldsymbol{F}_{33}\mathbf{U}_{g6}]. \tag{5.127}$$

Hence (5.124) yields $\mathbf{I}_3$, so $\mathbf{U}_4$ may be determined from (5.126) and $\mathbf{I}_4$ from (5.125).

To calculate $\mathbf{I}_4$, $\mathbf{I}_3$ is similarly expressed from (5.116) with the aid of (5.124) and (5.125):

$$\mathbf{I}_3=-\boldsymbol{Y}_3\boldsymbol{F}_{31}\boldsymbol{Z}_4\mathbf{I}_4-\boldsymbol{Y}_3\boldsymbol{F}_{33}\mathbf{U}_{g6}. \tag{5.128}$$

Substituting this into (5.123), rearrangement yields:

$$\mathbf{I}_4=[\boldsymbol{1}-(\boldsymbol{F}_{21}^{+}\boldsymbol{F}_{22}^{+\,-1}\boldsymbol{F}_{32}^{+}-\boldsymbol{F}_{31}^{+})\boldsymbol{Y}_3\boldsymbol{F}_{31}\boldsymbol{Z}_4]^{-1}-[(\boldsymbol{F}_{21}^{+}\boldsymbol{F}_{22}^{+\,-1}\boldsymbol{F}_{32}^{+}-$$
$$-\boldsymbol{F}_{31}^{+})\boldsymbol{Y}_3\boldsymbol{F}_{33}\mathbf{U}_{g6}+(\boldsymbol{F}_{11}^{+}-\boldsymbol{F}_{21}^{+}\boldsymbol{F}_{22}^{+\,-1}\boldsymbol{F}_{12}^{+})\mathbf{I}_{g1}]. \tag{5.129}$$

Hence, knowing $\mathbf{I}_4$, $\mathbf{I}_3$ may be determined from (5.128) and $\mathbf{U}_3$ and $\mathbf{U}_4$ from (5.124) and (5.125).

Thus, the voltages and currents of immitances have been determined in two ways. The remaining currents and voltages may also be calculated. Thus the currents $\mathbf{I}_2$ and voltages $\mathbf{U}_2$ of norators may be written from (5.122) and (5.115) respectively, (5.114) yields the voltages $\mathbf{U}_1$ of current-sources, while (5.121) the currents $\mathbf{I}_6$ of voltage-sources.

Examples

The analysis methods of electronic circuits are illustrated by a few examples. The setting of working points, the determination of the adjusting resistors is first dealt with, and then the analysis of certain circuits is carried out.

1. The method presented for the adjustment of working points of active elements in electronic circuits is applied to the case of the circuit with two transistors shown in Fig. 5.18, with the working points of both transistors prescribed, i.e. U_{BE1}, U_{CE1}, I_{B1}, I_{C1} as well as U_{BE2}, U_{CE2}, I_{B2}, I_{C2} are given.

The setting of working points is effected by the appropriate choice of the four resistors in the circuit. The model of the circuit formed by fixators and norators is

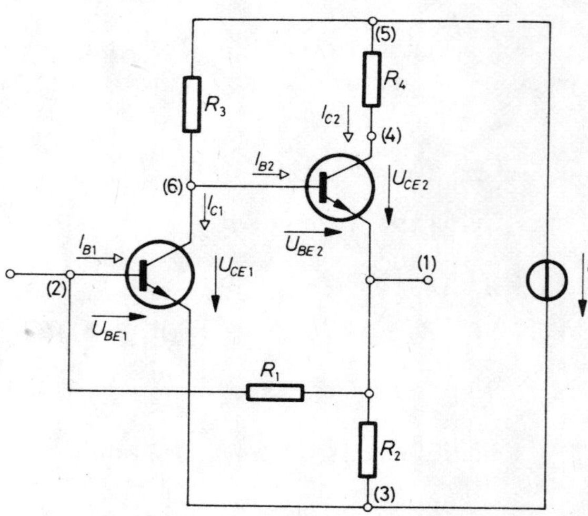

Fig. 5.18

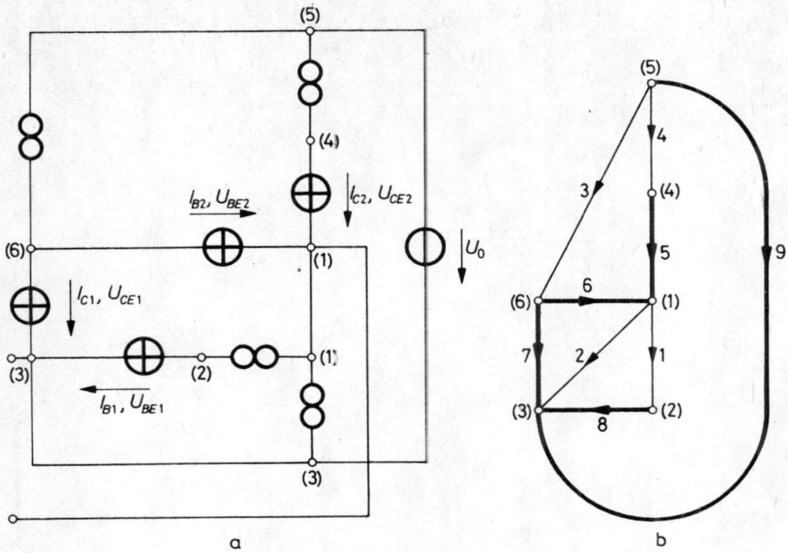

Fig. 5.19

drawn in Fig. 5.19, a, its graph being shown in Fig. 5.19, b. Branches have been numbered in the order of the classification. In this case there are no branches in the model belonging to the first, third and fourth groups. The second group is formed by branches 1, 2, 3, 4, the fifth by branches 5, 6, 7, 8, while branch 9 is to be assigned to the sixth group. The branches in the fifth and sixth groups are indeed seen to form a tree of the graph. If this were not so, the problem would not be solvable. The matrix of the fundamental set of loops generated by the tree is

$$\boldsymbol{B}=\left[\begin{array}{cccc|cccc|c} 1 & 0 & 0 & 0 & 0 & 1 & -1 & 1 & 0 \\ 0 & 1 & 0 & 0 & 0 & 1 & -1 & 0 & 0 \\ 0 & 0 & 1 & 0 & 0 & 0 & 1 & 0 & -1 \\ 0 & 0 & 0 & 1 & 1 & -1 & 1 & 0 & -1 \end{array}\right].$$

The partitioning indicated in (5.2) yields

$$\boldsymbol{F}_{22}=\begin{bmatrix} 0 & 1 & -1 & 1 \\ 0 & 1 & -1 & 0 \\ 0 & 0 & 1 & 0 \\ 1 & -1 & 1 & 0 \end{bmatrix}, \qquad \boldsymbol{F}_{23}=\begin{bmatrix} 0 \\ 0 \\ -1 \\ -1 \end{bmatrix}$$

while $\boldsymbol{F}_{11}, \boldsymbol{F}_{12}, \boldsymbol{F}_{13}, \boldsymbol{F}_{21}, \boldsymbol{F}_{31}, \boldsymbol{F}_{32}, \boldsymbol{F}_{33}$ do not exist. The column matrix of excitations is

$$\mathbf{U}_{g6} = U_0.$$

To determine $\mathbf{U}_2$ and $\mathbf{I}_2$

$$\mathbf{U}_5 = \begin{bmatrix} U_{CE2} \\ U_{BE2} \\ U_{CE1} \\ U_{BE1} \end{bmatrix}, \qquad \mathbf{I}_5 = \begin{bmatrix} I_{C2} \\ I_{B2} \\ I_{C1} \\ I_{B1} \end{bmatrix}$$

are also employed.

Substitution into (5.20) yields

$$\mathbf{U}_2 = \begin{bmatrix} U_1 \\ U_2 \\ U_3 \\ U_4 \end{bmatrix} = -\boldsymbol{F}_{23}\mathbf{U}_{g6} - \boldsymbol{F}_{22}\mathbf{U}_5 = \begin{bmatrix} 0 \\ 0 \\ U_0 \\ U_0 \end{bmatrix} - \begin{bmatrix} U_{BE2} - U_{CE1} + U_{BE1} \\ U_{BE2} - U_{CE1} \\ U_{CE1} \\ U_{CE2} - U_{BE2} + U_{CE1} \end{bmatrix} =$$

$$= \begin{bmatrix} -U_{BE2} + U_{CE1} - U_{BE1} \\ -U_{BE2} + U_{CE1} \\ U_0 - U_{CE1} \\ U_0 - U_{CE2} + U_{BE2} - U_{CE1} \end{bmatrix}.$$

Similarly, substituting into (5.22):

$$\mathbf{I}_2 = \begin{bmatrix} I_1 \\ I_2 \\ I_3 \\ I_4 \end{bmatrix} = \boldsymbol{F}_{22}^{+\,-1}\mathbf{I}_5 = \begin{bmatrix} 0 & 0 & 0 & 1 \\ 1 & 1 & 0 & -1 \\ 0 & 1 & 1 & 0 \\ 1 & 0 & 0 & 0 \end{bmatrix} \begin{bmatrix} I_{C2} \\ I_{B2} \\ I_{C1} \\ I_{B1} \end{bmatrix} =$$

$$= \begin{bmatrix} I_{B1} \\ I_{C2} + I_{B2} - I_{B1} \\ I_{B2} + I_{C1} \\ I_{C2} \end{bmatrix}.$$

From $\mathbf{U}_2 = \boldsymbol{R}_2\mathbf{I}_2$:

$$\boldsymbol{R}_2 = \langle R_1 \;\; R_2 \;\; R_3 \;\; R_4 \rangle,$$

with

$$R_1 = \frac{1}{I_{B1}}(-U_{BE2}+U_{CE1}-U_{BE1}),$$

$$R_2 = \frac{1}{I_{C2}+I_{B2}-I_{B1}}(-U_{BE2}+U_{CE1}),$$

$$R_3 = \frac{1}{I_{B2}+I_{C1}}(U_0-U_{CE1}),$$

$$R_4 = \frac{1}{I_{C2}}(U_0-U_{CE2}+U_{BE2}-U_{CE1})$$

being the values of the adjusting resistors (Fig. 5.18).

2. Next, some examples are presented for the analysis of networks containing two-ports with extreme parameters.

The voltages of the impedances Z_2 and Z_8 in the network containing an ideal transformer and gyrator, shown in Fig. 5.20, a will be calculated with the aid of chain parameters, the characteristics of the network elements being known.

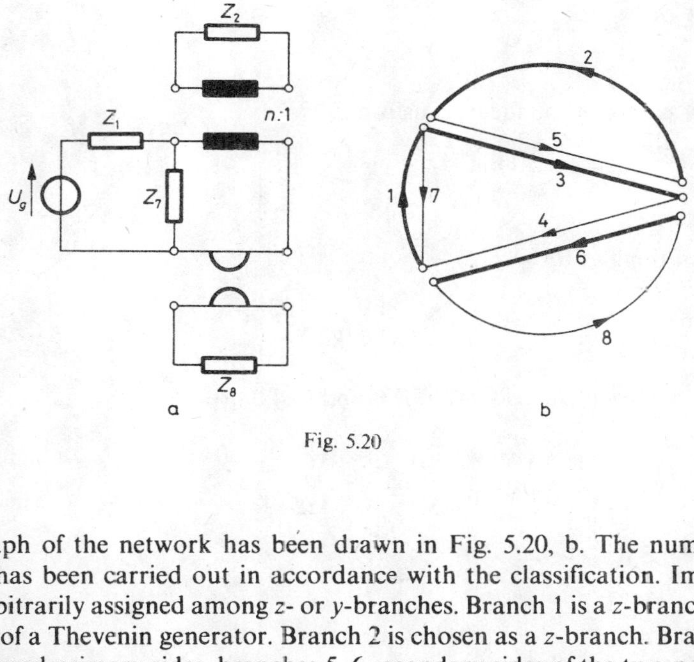

Fig. 5.20

The graph of the network has been drawn in Fig. 5.20, b. The numbering of branches has been carried out in accordance with the classification. Impedances may be arbitrarily assigned among *z*- or *y*-branches. Branch 1 is a *z*-branch because it consists of a Thevenin generator. Branch 2 is chosen as a *z*-branch. Branches 3, 4 are considered primary sides, branches 5, 6 secondary sides of the two-ports, while branches 7, 8 are *y*-branches. Branches 1, 2, 3 and 6 are chosen tree-branches. The matrix of the fundamental set of loops generated by this forest is

$$
\boldsymbol{B}=\left[\begin{array}{cc|cc|cc|cc}
1 & 0 & 1 & 1 & 0 & 0 & 0 & 0 \\
0 & 1 & 0 & 0 & 1 & 0 & 0 & 0 \\
0 & 0 & 0 & 0 & 0 & 1 & 0 & 1 \\
1 & 0 & 0 & 0 & 0 & 0 & 1 & 0
\end{array}\right],
$$

while the cutset matrix is

$$
\boldsymbol{Q}=\left[\begin{array}{cc|cc|cc|cc}
1 & 0 & 0 & -1 & 0 & 0 & -1 & 0 \\
0 & 1 & 0 & 0 & -1 & 0 & 0 & 0 \\
0 & 0 & 1 & -1 & 0 & 0 & 0 & 0 \\
0 & 0 & 0 & 0 & 0 & 1 & 0 & -1
\end{array}\right].
$$

The impedance matrix of z-branches is

$$
\boldsymbol{Z}=\langle Z_1 \;\; Z_2\rangle,
$$

the admittances matrix of y-branches, with the notation $Y_7=1/Z_7$ and $Y_8=1/Z_8$ is

$$
\boldsymbol{Y}=\langle Y_7 \;\; Y_8\rangle .
$$

The chain matrix of the ideal transformer is

$$
\boldsymbol{A}_t=\begin{bmatrix} n & 0 \\ 0 & -1/n \end{bmatrix},
$$

and that of the gyrator:

$$
\boldsymbol{A}_g=\begin{bmatrix} 0 & -R_{12} \\ 1/R_{21} & 0 \end{bmatrix}.
$$

Thus, the coefficient matrices in (5.43) and (5.44) are

$$
\boldsymbol{A}_{11}=\langle n \;\; 0\rangle, \qquad \boldsymbol{A}_{12}=\langle \;0 \;\; -R_{12}\rangle,
$$

$$
\boldsymbol{A}_{21}=\langle 0 \;\; 1/R_{21}\rangle, \qquad \boldsymbol{A}_{22}=\langle -1/n \;\; 0\rangle.
$$

Accordingly:

$$
\boldsymbol{B}_z\boldsymbol{Z}=\begin{bmatrix} 1 & 0 \\ 0 & 1 \\ 0 & 0 \\ 1 & 0 \end{bmatrix}\begin{bmatrix} Z_1 & 0 \\ 0 & Z_2 \end{bmatrix}=\begin{bmatrix} Z_1 & 0 \\ 0 & Z_2 \\ 0 & 0 \\ Z_1 & 0 \end{bmatrix},
$$

$$\boldsymbol{B}_p\boldsymbol{A}_{11}=\begin{bmatrix}1 & 1\\ 0 & 0\\ 0 & 0\\ 0 & 0\end{bmatrix}\begin{bmatrix}n & 0\\ 0 & 0\end{bmatrix}=\begin{bmatrix}n & 0\\ 0 & 0\\ 0 & 0\\ 0 & 0\end{bmatrix},$$

$$\boldsymbol{B}_p\boldsymbol{A}_{12}=\begin{bmatrix}0 & -R_{12}\\ 0 & 0\\ 0 & 0\\ 0 & 0\end{bmatrix},$$

$$\boldsymbol{Q}_p\boldsymbol{A}_{21}=\begin{bmatrix}0 & -1\\ 0 & 0\\ 1 & -1\\ 0 & 0\end{bmatrix}\begin{bmatrix}0 & 0\\ 0 & 1/R_{21}\end{bmatrix}=\begin{bmatrix}0 & -1/R_{21}\\ 0 & 0\\ 0 & -1/R_{21}\\ 0 & 0\end{bmatrix},$$

$$\boldsymbol{Q}_p\boldsymbol{A}_{22}=\begin{bmatrix}0 & 0\\ 0 & 0\\ -1/n & 0\\ 0 & 0\end{bmatrix},$$

$$\boldsymbol{Q}_y\boldsymbol{Y}=\begin{bmatrix}-1 & 0\\ 0 & 0\\ 0 & 0\\ 0 & -1\end{bmatrix}\begin{bmatrix}Y_7 & 0\\ 0 & Y_8\end{bmatrix}=\begin{bmatrix}-Y_7 & 0\\ 0 & 0\\ 0 & 0\\ 0 & -Y_8\end{bmatrix},$$

and further with the application of

$$\mathbf{I}_g=\mathbf{0} \qquad \text{and} \qquad \mathbf{U}_g=\begin{bmatrix}U_g\\ 0\end{bmatrix},$$

$$\boldsymbol{B}_z\mathbf{U}_g=\begin{bmatrix}U_g\\ 0\\ 0\\ U_g\end{bmatrix}, \qquad \boldsymbol{Q}_y\mathbf{I}_g=\begin{bmatrix}0\\ 0\\ 0\\ 0\end{bmatrix}.$$

Substituting these into (5.47) the following matrix equation is obtained:

$$\begin{bmatrix} Z_1 & 0 & n & 0 & 0 & -R_{12} & 0 & 0 \\ 0 & Z_2 & 1 & 0 & 0 & 0 & 0 & 0 \\ 0 & 0 & 0 & 1 & 0 & 0 & 0 & 1 \\ Z_1 & 0 & 0 & 0 & 0 & 0 & 1 & 0 \\ 1 & 0 & 0 & -1/R_{21} & 0 & 0 & -Y_7 & 0 \\ 0 & 1 & 0 & 0 & -1 & 0 & 0 & 0 \\ 0 & 0 & 0 & -1/R_{21} & -1/n & 0 & 0 & 0 \\ 0 & 0 & 0 & 0 & 0 & 1 & 0 & -Y_8 \end{bmatrix} \begin{bmatrix} I_1 \\ I_2 \\ U_5 \\ U_6 \\ I_5 \\ I_6 \\ U_7 \\ U_8 \end{bmatrix} = \begin{bmatrix} U_g \\ 0 \\ 0 \\ U_g \\ 0 \\ 0 \\ 0 \\ 0 \end{bmatrix}.$$

Since only I_2 and U_8 are to be determined in the problem, not all elements of the inverse of the matrix on the left side are to be calculated. The second and eighth rows of the inverse suffice, and according to the nonzero elements of the column matrix on the right side, the first and fourth elements only are to be determined. Thus

$$I_2 = \frac{n}{Z_1 + (1 + Z_1 Y_7)(n^2 Z_2 + Y_8 R_{12} R_{21})} U_g,$$

$$U_8 = \frac{-R_{21}}{Z_1 + (1 + Z_1 Y_7)(n^2 Z_2 + Y_8 R_{12} R_{21})} U_g,$$

i.e. according to $U_2 = Z_2 I_2$ the quantities sought have been determined.

3. The circuit diagram of a two-stage (Darlington) amplifier is shown in Fig. 5.21. The hybrid parameters of the transistors in the amplifier as well as the resistances and the inductance are known. In order to determine the voltage transfer coefficient U_t/U_g, the equivalent circuit shown in Table 5.1 is applied to yield the network drawn in Fig. 5.22. As a consequence of the approximation $h_{12} \approx 0$, no controlled voltage-source appears in the network. It has been taken into account that the

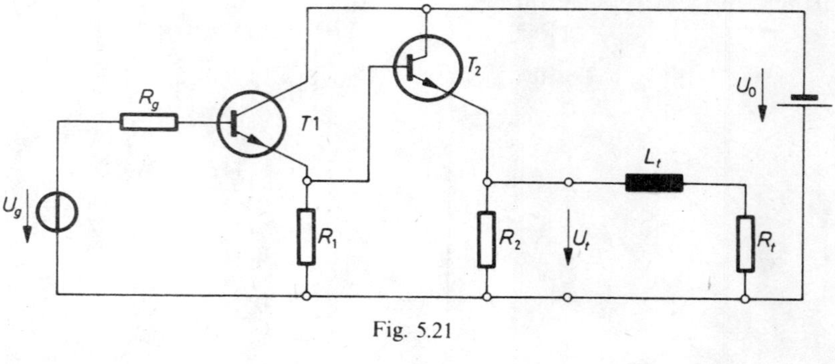

Fig. 5.21

supply voltage-sources represent short-circuit from the point of view of sinusoidal signals.

The analysis may be carried out by the application of mixed parameters to the model shown in Fig. 5.22. The graph of the model is seen in Fig. 5.23. As explained previously, it is appropriate to classify the branches of the network into four groups. There is no independent current-source in this network, thus no branch belongs to

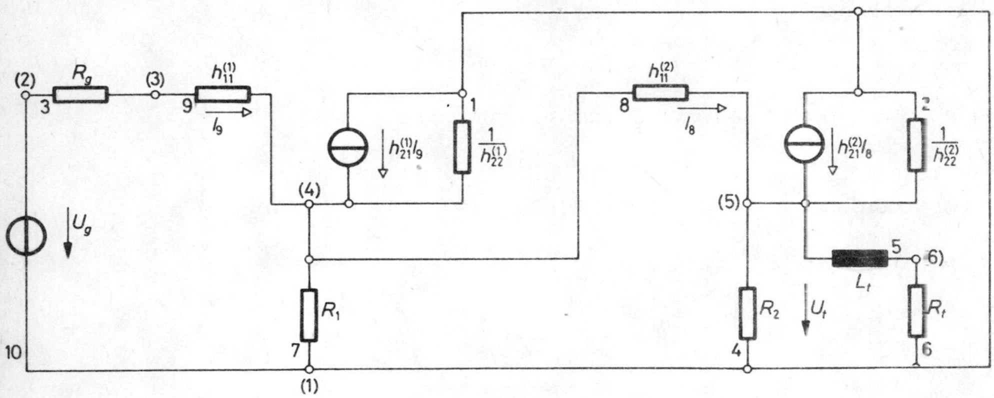

Fig. 5.22

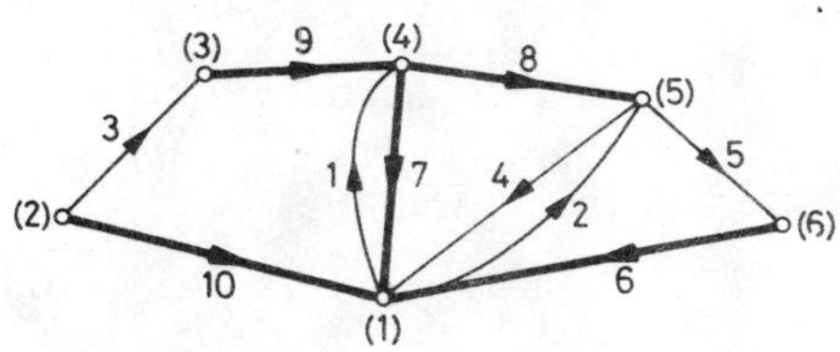

Fig. 5.23

the first group. The branches in the second group are chords, and the two controlled current-generators should be assigned here. The branches 3, 4, 5 in the graph of the network (Fig. 5.23) are also classified here. The remaining branches of the network containing impedances belong to the third group, and the voltage-source of voltage U_g forms the fourth group. In the case of such a choice the branches in the third and fourth groups make up the tree. The matrix of the fundamental set of loops generated by this tree is

$$\boldsymbol{B}=\left[\begin{array}{ccccc:cccc:c}
1 & 0 & 0 & 0 & 0 & 0 & 1 & 0 & 0 & 0 \\
0 & 1 & 0 & 0 & 0 & 0 & 1 & -1 & 0 & 0 \\
0 & 0 & 1 & 0 & 0 & 0 & 1 & 0 & 1 & -1 \\
0 & 0 & 0 & 1 & 0 & 0 & -1 & 1 & 0 & 0 \\
0 & 0 & 0 & 0 & 1 & 1 & -1 & 1 & 0 & 0
\end{array}\right],$$

i.e. $\mathbf{F}_{11}$ and $\mathbf{F}_{12}$ do not exist, and

$$\boldsymbol{F}_{21}=\begin{bmatrix} 0 & 1 & 0 & 0 \\ 0 & 1 & -1 & 0 \\ 0 & 1 & 0 & 1 \\ 0 & -1 & 1 & 0 \\ 1 & -1 & 1 & 0 \end{bmatrix}, \qquad \boldsymbol{F}_{22}=\begin{bmatrix} 0 \\ 0 \\ -1 \\ 0 \\ 0 \end{bmatrix}.$$

According to the above classification of branches:

$$\boldsymbol{Y}_2=\langle h_{22}^{(1)}\ h_{22}^{(1)}\ 1/R_g\ 1/R_4\ 1/j\omega L_t\rangle\,.$$

In accordance with (5.51) and (5.53):

$$\boldsymbol{K}=\begin{array}{c} \\ 1 \\ 2 \\ 3 \\ 4 \\ 5 \end{array}\!\!\begin{array}{c} \begin{array}{cccc} 6 & 7 & 8 & 9 \end{array} \\ \begin{bmatrix} 0 & 0 & 0 & h_{21}^{(1)} \\ 0 & 0 & h_{21}^{(1)} & 0 \\ 0 & 0 & 0 & 0 \\ 0 & 0 & 0 & 0 \\ 0 & 0 & 0 & 0 \end{bmatrix} \end{array},$$

and

$$\boldsymbol{M}=\boldsymbol{0},$$

and further

$$\boldsymbol{Z}_3=\langle R_t\ R_7\ h_{11}^{(2)}\ h_{11}^{(1)}\rangle\,.$$

Substituting the matrices written, (5.63) yields

$$\begin{bmatrix}\mathbf{U}_2\\ \mathbf{I}_3\end{bmatrix}=\begin{bmatrix} U_1 \\ U_2 \\ U_3 \\ U_4 \\ U_5 \\ I_6 \\ I_7 \\ I_8 \\ I_9 \end{bmatrix}=\left[\begin{array}{ccccc|cccc} 0 & 0 & 0 & 0 & -1/j\omega L_t & 1 & 0 & 0 & 0 \\ -h_{22}^{(1)} & -h_{22}^{(2)} & -1/R_g & 1/R_4 & 1/j\omega L_t & 0 & 1 & -h_{21}^{(2)} & -h_{21}^{(1)} \\ 0 & h_{22}^{(2)} & 0 & -1/R_4 & -1/j\omega L_t & 0 & 0 & (1-h_{21}^{(2)} & 0 \\ 0 & 0 & -1/R_g & 0 & 0 & 0 & 0 & 0 & 1 \\ 0 & 0 & 0 & 0 & 0 & 0 & R_7 & 0 & 0 \\ 0 & 0 & 0 & 0 & 0 & 0 & R_7 & -h_{11}^{(2)} & 0 \\ 0 & 0 & 0 & 0 & 0 & 0 & R_7 & 0 & h_{11}^{(1)} \\ 0 & 0 & 0 & 0 & 0 & 0 & -R_7 & h_{11}^{(2)} & 0 \\ 0 & 0 & 0 & 0 & 0 & R_t & -R_7 & h_{11}^{(2)} & 0 \end{array}\right]^{-1}\begin{bmatrix} 0 \\ 0 \\ 0 \\ 0 \\ 0 \\ 0 \\ U_g \\ 0 \\ 0 \end{bmatrix}.$$

The complex voltage transfer coefficient is the seventh element in the fourth row of the inverse matrix.

4. In the case of the analysis using a model containing controlled generators a procedure has been seen to exist which makes possible the determination of branch-voltages and branch-currents by the inversion of a matrix of lower order than in the method used in the previous example. This will be employed to calculate the complex voltage transfer coefficient U_t/U_g of the network shown in Fig. 5.24. The following quantities are presumed to be given: U_{CE1}, I_{C1}, U_{BE1}, I_{B1}, U_{CE2}, I_{C2}, U_{BE2}, I_{B2}, i.e. the working points of the two transistors, as well as the direct voltage U_0, the resistors R_g and R_5, the capacitors C_g and C_5, and the hybrid parameters $h_{11}^{(1)}, h_{12}^{(1)} \approx 0, h_{21}^{(1)}, h_{22}^{(1)}$ and $h_{11}^{(2)}, h_{12}^{(2)} \approx 0, h_{21}^{(2)}, h_{22}^{(2)}$ of the two transistors. The calculation is carried out for angular frequency ω.

The prescribed values of the working points can be adjusted by the resistors R_1, R_2, R_3, R_4 with the aid of the results of example 1.

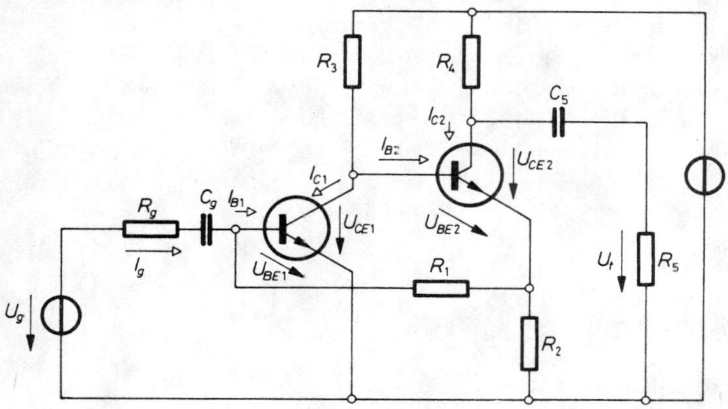

Fig. 5.24

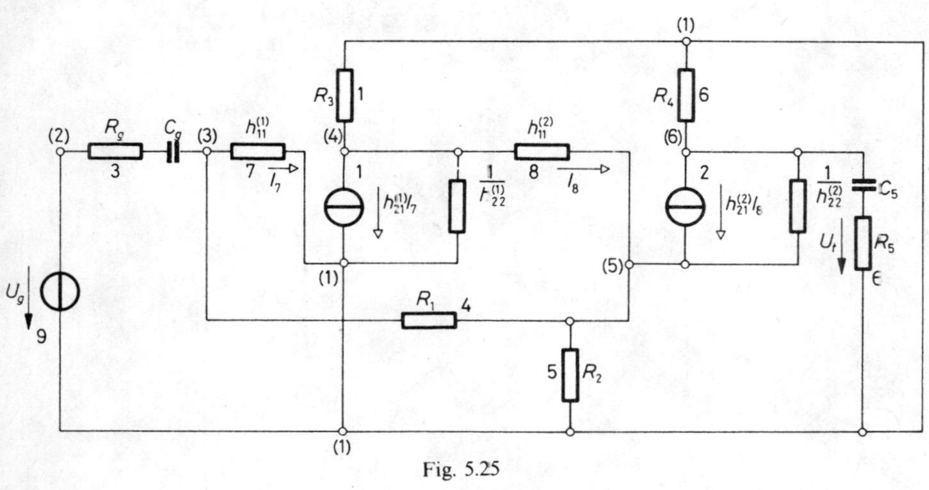

Fig. 5.25

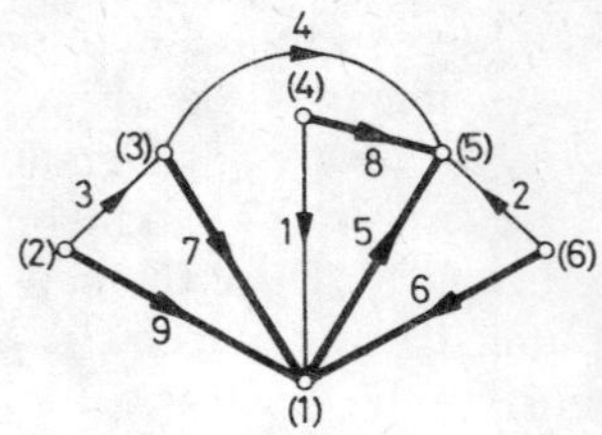

Fig. 5.26

The model of the network shown in Fig. 5.25 is obtained by modelling the transistors by equivalent circuits containing current controlled current-generators according to Table 5.1, and by regarding the direct current generator as a short-circuit. Its graph has been drawn in Fig. 5.26.

To apply the method mentioned, branches are classified into six groups. No branch of the connection belongs to the first group. Branches 1, 2 belong to the second group, 3, 4 to the third, 5, 6 to the fourth, 7, 8 to the fifth and 9 to the sixth group. This means at the same time that the tree consisting of branches 5, 6, 7, 8, 9 of the graph has been selected for the calculation.

The matrices necessary for the calculation are:

$$\boldsymbol{Y}_2=\begin{bmatrix} h_{22}^{(1)}+\dfrac{1}{R_3} & 0 \\ 0 & h_{22}^{(2)} \end{bmatrix},$$

according to (5.65) and (5.66)

$$\boldsymbol{K}=\begin{array}{c} \\ 1 \\ 2 \end{array}\begin{array}{c} \begin{array}{cc} 7 & 8 \end{array} \\ \begin{bmatrix} h_{21}^{(1)} & 0 \\ 0 & h_{21}^{(2)} \end{bmatrix} \end{array}, \quad \text{and} \quad \boldsymbol{M}=\boldsymbol{0},$$

and further

$$\boldsymbol{Z}_5=\begin{bmatrix} h_{11}^{(1)} & 0 \\ 0 & h_{11}^{(2)} \end{bmatrix},$$

$$\boldsymbol{Y}_3=\begin{bmatrix} \dfrac{1}{R_g}+j\omega C_g & 0 \\ 0 & \dfrac{1}{R_1} \end{bmatrix},$$

$$\boldsymbol{Z}_4=\begin{bmatrix} R_2 & 0 \\ 0 & R_4\times\left(R_5+\dfrac{1}{j\omega C_5}\right) \end{bmatrix}.$$

The matrix of the fundamental set of loops generated by the tree chosen is

$$
\boldsymbol{B}=\left[\begin{array}{cc|cc|cc|cc|c}
1 & 0 & 0 & 0 & 1 & 0 & 0 & -1 & 0\\
0 & 1 & 0 & 0 & -1 & -1 & 0 & 0 & 0\\
\hline
0 & 0 & 1 & 0 & 0 & 0 & 1 & 0 & -1\\
0 & 0 & 0 & 1 & -1 & 0 & -1 & 0 & 0
\end{array}\right],
$$

i.e. $\boldsymbol{F}_{11}$, $\boldsymbol{F}_{12}$, $\boldsymbol{F}_{13}$ do not exist, and

$$
\boldsymbol{F}_{21}=\begin{bmatrix}1 & 0\\ -1 & -1\end{bmatrix},\qquad \boldsymbol{F}_{22}=\begin{bmatrix}0 & -1\\ 0 & 0\end{bmatrix},\qquad \boldsymbol{F}_{23}=\begin{bmatrix}0\\ 0\end{bmatrix},
$$

$$
\boldsymbol{F}_{31}=\begin{bmatrix}0 & 0\\ -1 & 0\end{bmatrix},\qquad \boldsymbol{F}_{32}=\begin{bmatrix}1 & 0\\ -1 & 0\end{bmatrix},\qquad \boldsymbol{F}_{33}=\begin{bmatrix}-1\\ 0\end{bmatrix}.
$$

Hence, according to (5.84) and (5.85):

$$
\boldsymbol{N}_1=\begin{bmatrix}
1 & 0 & R_2 & 0 & 0 & -h^{(2)}_{11}\\
0 & 1 & -R_2 & -\left[R_4\times\left(R_5+\dfrac{1}{j\omega C_5}\right)\right] & 0 & 0\\
-h^{(1)}_{22}-\dfrac{1}{R_3} & h^{(2)}_{22} & 1+\dfrac{R_2}{R_1} & 0 & \dfrac{1}{R_1}h^{(1)}_{11}-h^{(1)}_{21} & h^{(2)}_{21}\\
0 & h^{(2)}_{22} & 0 & 1 & 0 & h^{(2)}_{21}\\
0 & 0 & \dfrac{R_2}{R_1} & 0 & 1+h^{(1)}_{11}\left(\dfrac{1}{R_g}+j\omega C_g+\dfrac{1}{R}\right) & 0\\
h^{(1)}_{22}+\dfrac{1}{R_3} & 0 & 0 & 0 & h^{(1)}_{21} & 1
\end{bmatrix},
$$

$$
\boldsymbol{N}_2=\begin{bmatrix}0\\ 0\\ 0\\ 0\\ \dfrac{1}{R_g}+j\omega C_g\\ 0\end{bmatrix},
$$

i.e. from (5.83), using $\mathbf{U}_g = U_g$:

$$\begin{bmatrix} \mathbf{U}_2 \\ \mathbf{I}_4 \\ \mathbf{I}_5 \end{bmatrix} = \begin{bmatrix} U_1 \\ U_2 \\ I_5 \\ I_6 \\ I_7 \\ I_8 \end{bmatrix} = \mathbf{N}_1^{-1}\mathbf{N}_2 U_g .$$

Hence I_6 can be determined. According to Fig. 5.25 $U_t = I_6 R_5 R_4 \Big/ \left(R_4 + R_5 + \right.$ $\left. + \frac{1}{j\omega C_5} \right)$, and thus U_t/U_g can be calculated.

5. The voltage U_0 in the band-filter element of second order shown in Fig. 5.27 is to be determined. The operational amplifier is substituted by a nullor (Fig. 5.28), and on the basis of the model thus obtained the method of node-voltages is applied to calculate the sought voltage as a function of angular frequency ω.

Removing the nullators and norators from the model (Fig. 5.29), the equation of node-voltages is written:

$$\begin{bmatrix} G_7 + j\omega C_2 & -j\omega C_2 & 0 & 0 & 0 & -G_7 \\ -j\omega C_2 & G_1 + G_3 + j\omega(C_1 + C_2) & -j\omega C_1 & 0 & -G_3 & 0 \\ 0 & -j\omega C_1 & G_2 + j\omega C_1 & 0 & 0 & 0 \\ 0 & 0 & 0 & G_4 + G_5 & -G_5 & 0 \\ 0 & -G_3 & 0 & -G_5 & G_3 + G_5 + G_6 & -G_6 \\ -G_7 & 0 & 0 & 0 & -G_6 & G_6 + G_7 \end{bmatrix} \times$$

$$\times \begin{bmatrix} \Phi_1^{(1)} \\ \Phi_2^{(1)} \\ \Phi_3^{(1)} \\ \Phi_4^{(1)} \\ \Phi_5^{(1)} \\ \Phi_6^{(1)} \end{bmatrix} = \begin{bmatrix} 0 \\ U_{be} G_1 \\ 0 \\ 0 \\ 0 \\ 0 \end{bmatrix}$$

with the node (0) chosen as the zero potential node, and the notation $G_1 = 1/R_1$ ($i = 1, 2, \ldots, 7$) employed. The reinsertion of the two nullators between nodes (3) and

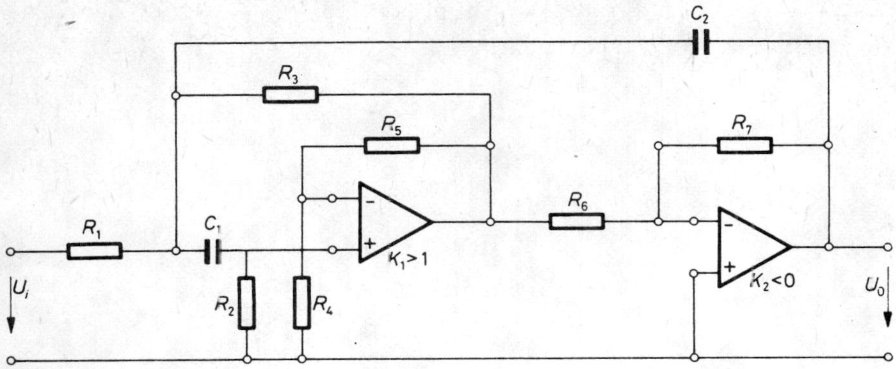

Fig. 5.27

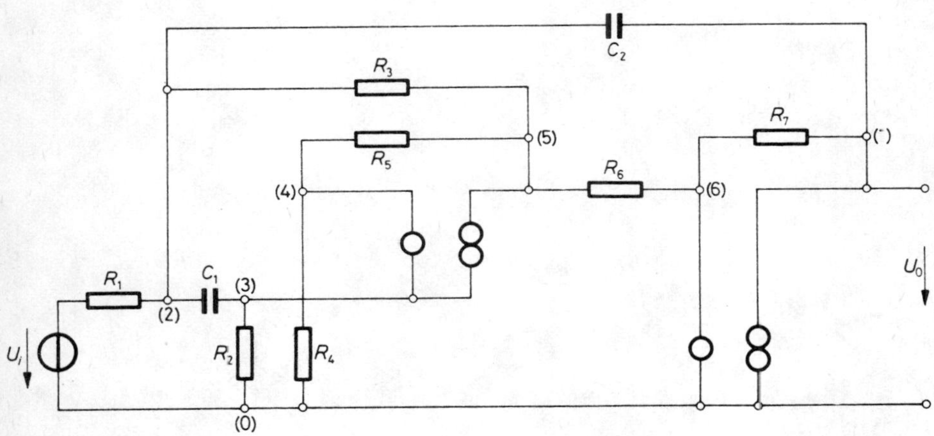

Fig. 5.28

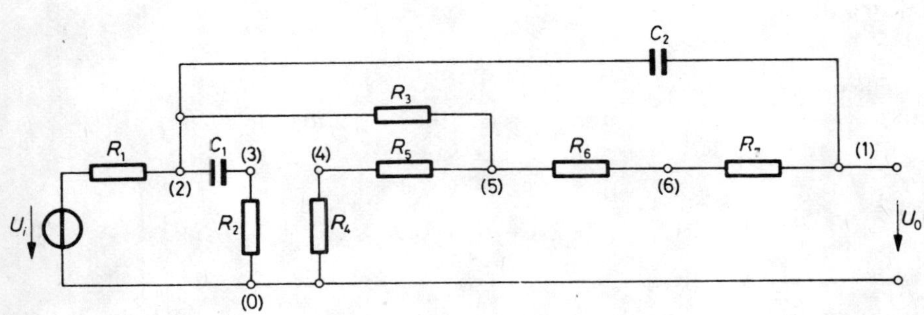

Fig. 5.29

(4) as well as between nodes (0) and (6) is taken into account by premultiplying the node admittance matrix by the matrix

$$T' = \begin{bmatrix} 1 & 0 & 0 & 0 \\ 0 & 1 & 0 & 0 \\ 0 & 0 & 1 & 0 \\ 0 & 0 & 1 & 0 \\ 0 & 0 & 0 & 1 \\ 0 & 0 & 0 & 0 \end{bmatrix},$$

while the reinsertion of the two norators between nodes (3) and (5) as well as between nodes (0) and (6) necessitates post-multiplication by the matrix

$$T'' = \begin{bmatrix} 0 & 1 & 0 & 0 & 0 & 0 \\ 0 & 0 & 1 & 0 & 1 & 0 \\ 0 & 0 & 0 & 1 & 0 & 0 \\ 0 & 0 & 0 & 0 & 0 & 1 \end{bmatrix}.$$

Therefore the equation corresponding to (5.106) for the network obtained after the reinsertion of the nullators is

$$\begin{bmatrix} G_7 + j\omega C_2 & -j\omega C_2 & 0 & 0 \\ -j\omega C_2 & G_1 + G_3 + j\omega(C_1 + C_2) & -j\omega C_1 & -G_3 \\ 0 & -j\omega C_1 & G_2 + j\omega C_1 & 0 \\ 0 & 0 & G_4 + G_5 & -G_5 \\ 0 & -G_3 & -G_5 & G_3 + G_5 + G_6 \\ -G_7 & 0 & 0 & -G_6 \end{bmatrix} \begin{bmatrix} \Phi_1^{(2)} \\ \Phi_2^{(2)} \\ \Phi_{34}^{(2)} \\ \Phi_5^{(2)} \end{bmatrix} = \begin{bmatrix} 0 \\ U_i G_1 \\ 0 \\ 0 \\ 0 \\ 0 \end{bmatrix}.$$

After the reinsertion of the norators the following equation holds:

$$\begin{bmatrix} -j\omega C_2 & G_1 + G_3 + j\omega(C_1 + C_2) & -j\omega C_1 & -G_3 \\ 0 & -(G_3 + j\omega C_1) & G_2 - G_5 + j\omega C_1 & G_3 + G_5 + G_6 \\ 0 & 0 & G_4 + G_5 & -G_5 \\ -G_7 & 0 & 0 & -G_6 \end{bmatrix} \begin{bmatrix} \Phi_1 \\ \Phi_2 \\ \Phi_{34} \\ \Phi_5 \end{bmatrix} = \begin{bmatrix} U_i(\\ 0 \\ 0 \\ 0 \end{bmatrix}$$

Hence

$$\begin{aligned}U_0=\Phi_1=U_iG_1G_6(G_4+G_5)(G_3+j\omega C_1)[G_7\{[G_1+\\ +G_3+j\omega(C_1+C_2)]\,[G_5(G_5-G_2-j\omega C_1)-\\ -(G_4+G_5)(G_3+G_5+G_6)]+\\ +(G_3+j\omega C_1)(j\omega C_1G_5+G_3G_4+G_3G_5)\}-\\ -j\omega C_2G_6(G_3+j\omega C_1)(G_4+G_5)]^{-1}\end{aligned}$$

is the sought voltage.

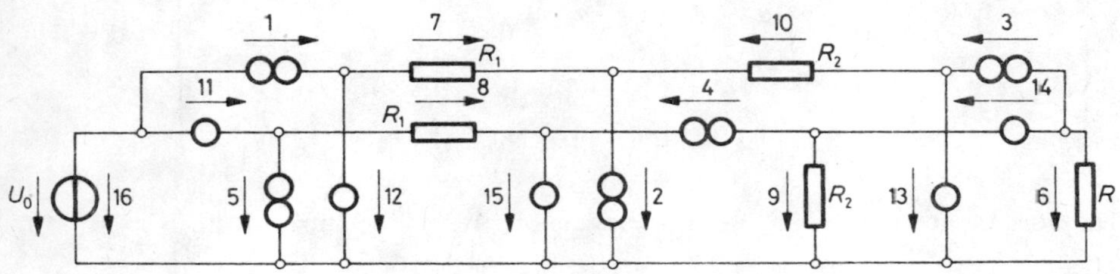

Fig. 5.30

6. The equivalent circuit of a negative impedance converter terminated by resistance R, and excited by source-voltage U_0 is shown in Fig. 5.30. The current of the voltage-source is to be determined.

The branches have been numbered for the analysis as indicated in Fig. 5.30. Branches 6, 7, . . ., 10 contain impedances. There being 9 nodes in the network, two branches containing impedance are also necessary to form the tree beside the voltage-source and the five nullator branches. Branches 9 and 10 are chosen for this purpose. Thus branches 1, . . ., 5 containing norators and 6, 7, 8 formed by impedances are chords, 9, 10 containing impedances, 11, . . ., 15 formed by nullators and branch 16 containing a voltage-source are tree-branches (Fig. 5.31). The matrix of the fundamental set of loops generated by this tree is:

$$\boldsymbol{B}=\left[\begin{array}{ccccc|ccc|cc|ccccc|c}
1 & 0 & 0 & 0 & 0 & 0 & 0 & 0 & 0 & 0 & 0 & 1 & 0 & 0 & 0 & -1\\
0 & 1 & 0 & 0 & 0 & 0 & 0 & 0 & 0 & 1 & 0 & 0 & -1 & 0 & 0 & 0\\
0 & 0 & 1 & 0 & 0 & 0 & 0 & 0 & -1 & 0 & 0 & 0 & 1 & -1 & 0 & 0\\
0 & 0 & 0 & 1 & 0 & 0 & 0 & 0 & -1 & 0 & 0 & 0 & 0 & 0 & 1 & 0\\
0 & 0 & 0 & 0 & 1 & 0 & 0 & 0 & 0 & 0 & 1 & 0 & 0 & 0 & 0 & -1\\
\hline
0 & 0 & 0 & 0 & 0 & 1 & 0 & 0 & -1 & 0 & 0 & 0 & 0 & -1 & 0 & 0\\
0 & 0 & 0 & 0 & 0 & 0 & 1 & 0 & 0 & -1 & 0 & -1 & 1 & 0 & 0 & 0\\
0 & 0 & 0 & 0 & 0 & 0 & 0 & 1 & 0 & 0 & 1 & 0 & 0 & 0 & 1 & -1
\end{array}\right],$$

F_{11}, F_{12}, F_{13} do not exist, and

$$F_{21}=\begin{bmatrix}0 & 0\\ 0 & 1\\ -1 & 0\\ -1 & 0\\ 0 & 0\end{bmatrix}, \qquad F_{22}=\begin{bmatrix}0 & 1 & 0 & 0 & 0\\ 0 & 0 & -1 & 0 & 0\\ 0 & 0 & 1 & -1 & 0\\ 0 & 0 & 0 & 0 & 1\\ 1 & 0 & 0 & 0 & 0\end{bmatrix}, \qquad F_{23}=\begin{bmatrix}-1\\ 0\\ 0\\ 0\\ -1\end{bmatrix},$$

$$F_{31}=\begin{bmatrix}-1 & 0\\ 0 & -1\\ 0 & 0\end{bmatrix}, \qquad F_{32}=\begin{bmatrix}0 & 0 & 0 & -1 & 0\\ 0 & -1 & 1 & 0 & 0\\ 1 & 0 & 0 & 0 & 1\end{bmatrix}, \qquad F_{33}=\begin{bmatrix}0\\ 0\\ -1\end{bmatrix}.$$

The admittance matrix of the branches belonging to the third group is

$$Y_3=\left\langle \frac{1}{R} \quad \frac{1}{R_1} \quad \frac{1}{R_1}\right\rangle,$$

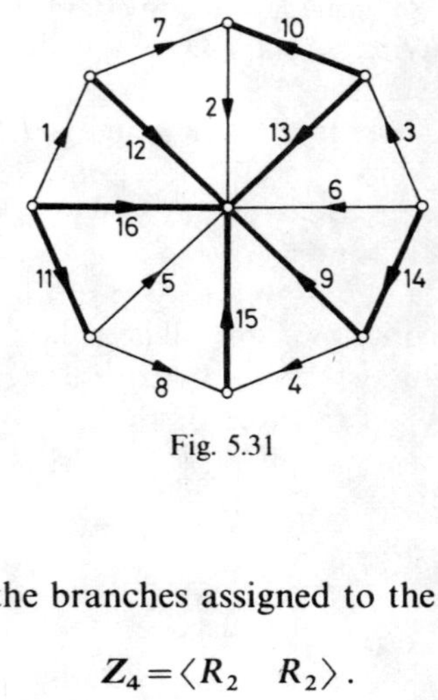

Fig. 5.31

the impedance matrix of the branches assigned to the fourth group is

$$Z_4=\langle R_2 \quad R_2\rangle .$$

Hence according to (5.129):

$$\mathbf{I}_4=\begin{bmatrix}I_9\\ \\ I_{10}\end{bmatrix}=\begin{bmatrix}1 & 0\\ \\ \frac{R_2}{R} & 1\end{bmatrix}^{-1}\begin{bmatrix}\frac{U_0}{R_1}\\ \\ 0\end{bmatrix}=\begin{bmatrix}\frac{U_0}{R_1}\\ \\ -\frac{R_2}{R_1}\frac{U_0}{R}\end{bmatrix}.$$

Substitution into (5.128) yields

$$\mathbf{I}_3 = \begin{bmatrix} I_6 \\ I_7 \\ I_8 \end{bmatrix} = \frac{U_0}{R} \begin{bmatrix} R_2/R_1 \\ -(R_2/R_1)^2 \\ R/R_1 \end{bmatrix}.$$

From (5.122):

$$\mathbf{I}_2 = \begin{bmatrix} I_1 \\ I_2 \\ I_3 \\ I_4 \\ I_5 \end{bmatrix} = U_0/R \begin{bmatrix} -(R_2/R_1)^2 \\ -R_2/R_1 - (R_2/R_1)^2 \\ -R_2/R_1 \\ -R/R_1 \\ -R/R_1 \end{bmatrix}.$$

Substituting $\mathbf{I}_2$ and $\mathbf{I}_3$ into (5.121):

$$\mathbf{I}_6 = I_{16} = \left(\frac{R_2}{R_1}\right)^2 \frac{U_0}{R} = \frac{1}{k^2} \frac{U_0}{R}$$

as expected.

7. The output voltage of a second order low-pass filter element (Fig. 5.32) is calculated in the following with the aid of the method employing the loop-matrix. The operational amplifier is substituted by a nullor. (Fig. 5.33).

The graph of the network is shown in Fig. 5.34. The tree-branches are drawn by thick lines. There being no current-source in the network, $b_1 = 0$, in accordance with the number of nullators and norators $b_2 = b_5 = 1$. There is one voltage-source in the network, so $b_6 = 1$. The number of nodes is 5, thus there are 4 tree-branches. Therefore $b_3 = 3$, $b_4 = 2$. The matrix of the fundamental set of loops generated by the tree selected is

$$\boldsymbol{B} = \left[\begin{array}{c|ccc|cc|c|c} 1 & 0 & 0 & 0 & 1 & -1 & -1 & 0 \\ \hline 0 & 1 & 0 & 0 & 1 & -1 & 0 & 0 \\ 0 & 0 & 1 & 0 & 0 & -1 & -1 & 0 \\ 0 & 0 & 0 & 1 & 0 & 1 & 1 & -1 \end{array}\right],$$

i.e. $\boldsymbol{F}_{11}$, $\boldsymbol{F}_{12}$, $\boldsymbol{F}_{13}$ do not exist, and

$$\boldsymbol{F}_{21} = [1 \;\; -1], \qquad \boldsymbol{F}_{22} = -1, \qquad \boldsymbol{F}_{23} = 0,$$

$$\boldsymbol{F}_{31} = \begin{bmatrix} 1 & -1 \\ 0 & -1 \\ 0 & 1 \end{bmatrix}, \qquad \boldsymbol{F}_{32} = \begin{bmatrix} 0 \\ -1 \\ 1 \end{bmatrix}, \qquad \boldsymbol{F}_{33} = \begin{bmatrix} 0 \\ 0 \\ -1 \end{bmatrix}.$$

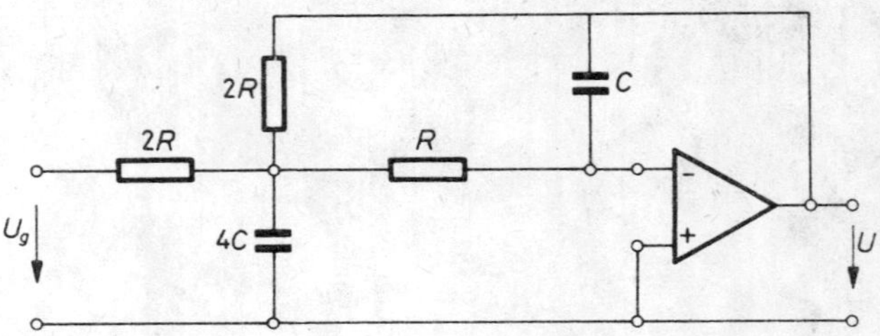

Fig. 5.32

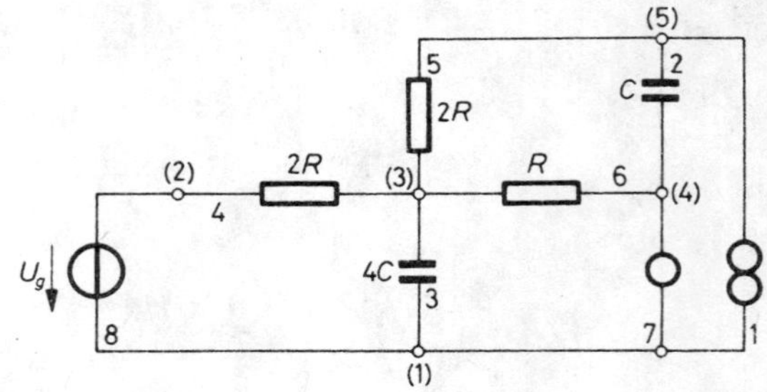

Fig. 5.33

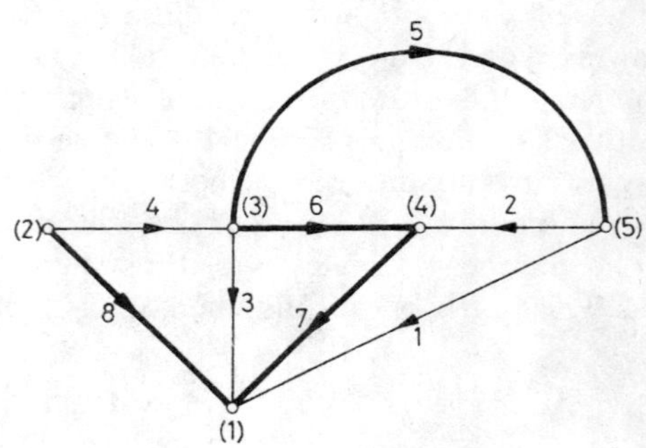

Fig. 5.34

The column matrix of excitations is

$$\mathbf{U}_{g6} = U_g .$$

The admittance matrix of the branches belonging to the third group is

$$\boldsymbol{Y}_3 = \langle j\omega C \;\; j\omega 4C \;\; 1/2R \rangle .$$

The impedance matrix of the branches assigned to the fourth group is

$$\boldsymbol{Z}_4 = \langle 2R \;\; R \rangle .$$

Using the matrices written, (5.127) yields

$$\mathbf{U}_3 = \begin{bmatrix} U_2 \\ U_3 \\ U_4 \end{bmatrix} = \begin{bmatrix} U \\ U_3 \\ U_4 \end{bmatrix} = \begin{bmatrix} 1+j\omega 3CR & -j\omega 8CR & 1 \\ j\omega CR & 1 & 0 \\ -j\omega CR & 0 & 1 \end{bmatrix}^{-1} \begin{bmatrix} 0 \\ 0 \\ U_g \end{bmatrix},$$

i.e.

$$U = \frac{U_g}{\omega^2 8C^2R^2 - j\omega 4CR - 1}$$

is the sought voltage.

CHAPTER 6

STATE EQUATIONS

The Kirchhoff equations of a network written in the time domain constitute a set of differential equations which can be solved by calculations which involve only time functions, without the use of Laplace, Fourier or other transformations. Such an approach to network equations is presented in this chapter.

In the course of our analysis the connections of the network, the characteristics of the network elements and the initial conditions necessary for the determination of unknown signals appearing in the network are assumed to be known. A set of differential equations may be written, involving the signals occurring in the branches of the network, enabling the unknown signals to be calculated. Some of the unknowns can be eliminated to yield a minimal number in a set of differential equations which have at most the first derivatives of these unknowns and in terms of which the remaining unknowns may be expressed. The minimal set of unknowns which remain in the equations are the *state variables* of the network. The state variables may be summarized in a column matrix called the *state vector*. Thus, the state vector satisfies a set of differential equations of first order called *state equations*, and the remaining unknown responses occurring in the network are expressible with the aid of its elements, the state variables. State equations may also be written for physical systems other than electrical networks.

The state variables of electrical networks may be the charges of capacitors and fluxes of inductors in the network. For the analysis of linear networks, it is convenient to regard quantities proportional to these, namely the voltages of capacitors and currents of inductors as state variables. In the following these latter are chosen as the state variables.

It should be noted that calculations concerning state equations are of great importance in the case of nonlinear networks, this procedure being appropriate for the examination of transient processes. In the course of our discussion, however, as in the previous chapters, only linear networks are dealt with.

A loop of the network formed by capacitors, independent and controlled voltage-sources, short-circuits, fixators and nullators is called a capacitive loop. The voltage of any capacitor in each capacitive loop can be expressed in terms of the voltages of the other elements of the loop, i.e. the voltage of one capacitor in the capacitive loop is not independent and does not constitute a state variable.

A cutset of the network formed by inductors, independent and controlled current-sources, open-circuits, fixators, nullators is called an inductive cutset. The current of any inductor in each inductive cutset can be expressed in terms of the currents of the other elements of the cutset, i.e. the current of one inductor in the inductive cutset is not independent and does not represent a state variable.

The number of state variables is equal to the sum of the number of capacitors and inductors in the network decreased by one for each capacitive loop and inductive cutset. Therefore, the number of state variables equals the number of independent initial conditions in the networks.

To determine the presence of capacitive loops and inductive cutsets, coupled two-terminal elements and n-ports are modelled by controlled generators or sources as explained in Chapter 5, and to determine the number of state variables the network so obtained is investigated for the presence of capacitive loops and inductive cutsets. If the occurrence of a capacitive loop or inductive cutset can be avoided by the replacement of parallel capacitors or series inductors by a single element, it is preferable to make such replacements before writing state equations.

The variables (currents, voltages) of the network are classified into three groups as follows:

1. Excitations (source currents, source-voltages) with their time-functions given. These are arranged in a column matrix denoted by $\mathbf{r}$. The number of excitations, i.e. of the elements in $\mathbf{r}$ is p.
2. The number of state variables is q. The state vector formed by state variables is denoted by $\mathbf{x}$.
3. The currents and voltages of the network, not being state variables or excitations.

In case of network analysis problems the determination of some voltages or currents in groups 2 and 3 is often required. These variables belonging to groups 2 and 3 form a column matrix with s elements, denoted by $\mathbf{y}$. This is the vector of responses.

The state equations of a linear, time-invariant system with the above notations are

$$\dot{\mathbf{x}} = \boldsymbol{D}\mathbf{x} + \boldsymbol{E}\mathbf{r}\,, \tag{6.1}$$

where $\boldsymbol{D}$ and $\boldsymbol{E}$ are matrices dependent upon the characteristics of passive elements in the network, and upon their interconnections, and independent of excitations and time.* $\boldsymbol{D}$ is a square matrix of order q, the number of rows in $\boldsymbol{E}$ is q, that of columns is p. $\dot{\mathbf{x}}$ denotes the time derivative of $\mathbf{x}$.

* In the state equation (6.1) the symbols $\boldsymbol{A}$ and $\boldsymbol{B}$ are commonly used in the literature instead of $\boldsymbol{D}$ and $\boldsymbol{E}$. The symbol $\mathbf{u}$ is commonly used for inputs instead of $\mathbf{r}$.

The solution of (6.1) is composed of two terms. The sum of the general solution $\mathbf{X}(t)$ of the homogeneous equation

$$\dot{\mathbf{X}} - \boldsymbol{D}\mathbf{X} = \mathbf{0} \qquad (6.2)$$

and a particular solution $\mathbf{x}_1(t)$ of Eq. (6.1) yields the general solution of (6.1):

$$\mathbf{x}(t) = \mathbf{X}(t) + \mathbf{x}_1(t) \,. \qquad (6.3)$$

The time-variation of $\mathbf{X}(t)$ is independent of the excitation signals $\mathbf{r}(t)$. If $\mathbf{r}(t)$ is periodic (e.g. constant or sinusoidally varying with time), $\mathbf{X}(t)$ is the transient term of the solution, with its time variation determined by the time constants of the network, while the steady-state part of the solution, $\mathbf{x}_1(t)$ varies periodically, similarly to the excitations. Given the solution $\mathbf{x}$ of the state equations, i.e. of the state variables, the column matrix $\mathbf{y}$ of the responses is obtained from

$$\mathbf{y} = \boldsymbol{H}\mathbf{x} + \boldsymbol{S}\mathbf{r} \,. \qquad (6.4)$$

The number of rows in $\boldsymbol{H}$ and $\boldsymbol{S}$ is s, the number of columns in $\boldsymbol{H}$ is q, that in $\boldsymbol{S}$ is p.*

In the following, the establishment of state equations in networks not including capacitive loops or inductive cutsets is discussed. In such networks the voltage of each capacitor and the current of each inductor constitutes a state variable. A method is presented to determine the presence of capacitive loops or inductive cutsets in the network. Initially, networks consisting of independent sources, independent generators, resistors, capacitors, and self and mutual inductors are discussed from the point of view of writing state equations, followed by the examination of models containing controlled sources in addition to the above elements. Finally, a method for the solution of state equations is also touched upon.

Writing the state equations of networks without controlled sources

To write the state equations [16] of a network consisting of independent sources, generators, resistors, capacitors and inductors, the branches of the network are classified into six groups. Generators are represented by their Thevenin or Norton equivalents, regarding the sources and the passive two-poles in them as distinct branches to permit the following classification. One edge of the graph corresponds to each source, inductor, capacitor or resistor of the network. A tree of the graph is chosen with each branch containing a current-source, inductor or open-circuit corresponding to a chord, while each branch formed by a capacitor, voltage-source or short-circuit is associated with a tree-branch. Further branches containing resistors may equally be chords or tree-branches, and their inclusion in the appropriate groups is arranged so as to permit the selection of the tree mentioned above. Thus, the six groups of branches are:

* The symbols $\boldsymbol{C}$ and $\boldsymbol{D}$ are commonly used in the literature instead of $\boldsymbol{H}$ and $\boldsymbol{S}$.

1. current-sources (chords);
2. finite conductances (chords);
3. inductors (chords);
4. capacitors (tree-branches);
5. finite resistances (tree-branches);
6. voltage-sources (tree-branches).

If the network contains no capacitive loop or inductive cutset, such a classification of branches is possible. Conversely, if capacitors and voltage-sources form a loop, not all of the edges associated with them may be tree-branches. Similarly, if the branches in groups 1 and 3 may not all be chords, an inductive cutset is present in the network.

Thus, the branches of a network not including capacitive loops or inductive cutsets may be classified into the groups stated above. Let the branches of the network be assigned order numbers in the order of the classification. The numbers and orientations of the loops in the fundamental set of loops generated by the tree are chosen to coincide with the order number and orientation of the chord in the loop. The cutsets in the fundamental set of cutsets generated by the same tree are numbered in the order of the tree-branches in the cutsets with the orientation of the cutset in the tree-branches coinciding with that of the branch. Column matrices are formed by the time-functions of voltages and currents of the branches in the network. The respective notations are $\mathbf{u}(t)$ and $\mathbf{i}(t)$.

The Kirchhoff equations of the network in the time-domain are:

$$\boldsymbol{B}\mathbf{u}(t)=\mathbf{0}, \tag{6.5}$$

$$\boldsymbol{Q}\mathbf{i}(t)=\mathbf{0}. \tag{6.6}$$

Partitioning these in accordance with branch-classification:

$$\begin{bmatrix} \mathit{1} & \mathit{0} & \mathit{0} & \boldsymbol{F}_{11} & \boldsymbol{F}_{12} & \boldsymbol{F}_{13} \\ \mathit{0} & \mathit{1} & \mathit{0} & \boldsymbol{F}_{21} & \boldsymbol{F}_{22} & \boldsymbol{F}_{23} \\ \mathit{0} & \mathit{0} & \mathit{1} & \boldsymbol{F}_{31} & \boldsymbol{F}_{32} & \boldsymbol{F}_{33} \end{bmatrix} \begin{bmatrix} \mathbf{u}_1 \\ \mathbf{u}_2 \\ \mathbf{u}_3 \\ \mathbf{u}_4 \\ \mathbf{u}_5 \\ \mathbf{u}_6 \end{bmatrix} = \mathbf{0}, \tag{6.7}$$

$$\begin{bmatrix} -\boldsymbol{F}_{11}^{+} & -\boldsymbol{F}_{21}^{+} & -\boldsymbol{F}_{31}^{+} & \mathit{1} & \mathit{0} & \mathit{0} \\ -\boldsymbol{F}_{12}^{+} & -\boldsymbol{F}_{22}^{+} & -\boldsymbol{F}_{32}^{+} & \mathit{0} & \mathit{1} & \mathit{0} \\ -\boldsymbol{F}_{13}^{+} & -\boldsymbol{F}_{23}^{+} & -\boldsymbol{F}_{33}^{+} & \mathit{0} & \mathit{0} & \mathit{1} \end{bmatrix} \begin{bmatrix} \mathbf{i}_1 \\ \mathbf{i}_2 \\ \mathbf{i}_3 \\ \mathbf{i}_4 \\ \mathbf{i}_5 \\ \mathbf{i}_6 \end{bmatrix} = \mathbf{0}. \tag{6.8}$$

The following notation is introduced for futher analysis:

$$\mathbf{i}_1 = \mathbf{i}_0, \tag{6.9}$$

$$\mathbf{u}_2 = \mathbf{u}_G, \qquad \mathbf{i}_2 = \mathbf{i}_G, \tag{6.10}$$

$$\mathbf{u}_3 = \mathbf{u}_L, \qquad \mathbf{i}_3 = \mathbf{i}_L, \tag{6.11}$$

$$\mathbf{u}_4 = \mathbf{u}_C, \qquad \mathbf{i}_4 = \mathbf{i}_C, \tag{6.12}$$

$$\mathbf{u}_5 = \mathbf{u}_R, \qquad \mathbf{i}_5 = \mathbf{i}_R, \tag{6.13}$$

$$\mathbf{u}_6 = \mathbf{u}_0. \tag{6.14}$$

Hence (6.7) and (6.8) yield

$$\mathbf{u}_1 + \boldsymbol{F}_{11}\mathbf{u}_C + \boldsymbol{F}_{12}\mathbf{u}_R + \boldsymbol{F}_{13}\mathbf{u}_0 = \mathbf{0}, \tag{6.15}$$

$$\mathbf{u}_G + \boldsymbol{F}_{21}\mathbf{u}_C + \boldsymbol{F}_{22}\mathbf{u}_R + \boldsymbol{F}_{23}\mathbf{u}_0 = \mathbf{0}, \tag{6.16}$$

$$\mathbf{u}_L + \boldsymbol{F}_{31}\mathbf{u}_C + \boldsymbol{F}_{32}\mathbf{u}_R + \boldsymbol{F}_{33}\mathbf{u}_0 = \mathbf{0}, \tag{6.17}$$

$$-\boldsymbol{F}_{11}^{+}\mathbf{i}_0 - \boldsymbol{F}_{21}^{+}\mathbf{i}_G - \boldsymbol{F}_{31}^{+}\mathbf{i}_L + \mathbf{i}_C = \mathbf{0}, \tag{6.18}$$

$$-\boldsymbol{F}_{12}^{+}\mathbf{i}_0 - \boldsymbol{F}_{22}^{+}\mathbf{i}_G - \boldsymbol{F}_{32}^{+}\mathbf{i}_L + \mathbf{i}_R = \mathbf{0}, \tag{6.19}$$

$$-\boldsymbol{F}_{13}^{+}\mathbf{i}_0 - \boldsymbol{F}_{23}^{+}\mathbf{i}_G - \boldsymbol{F}_{33}^{+}\mathbf{i}_L + \mathbf{i}_6 = \mathbf{0}. \tag{6.20}$$

The currents and voltages of groups 2 and 5 are in the relation

$$\mathbf{i}_G = \boldsymbol{G}\mathbf{u}_G, \tag{6.21}$$

$$\mathbf{u}_R = \boldsymbol{R}\mathbf{i}_R, \tag{6.22}$$

where $\boldsymbol{G}$ and $\boldsymbol{R}$ are the diagonal matrices formed by the conductances of branches in group 2 and the resistances of branches in group 5, respectively.

The equation

$$\mathbf{u}_L = \boldsymbol{L}\dot{\mathbf{i}}_L \tag{6.23}$$

holds for the branches in group 3. The main diagonal of $\boldsymbol{L}$ consists of the self inductances in the branches of group 3, while the k-th element of the i-th row in the matrix is the mutual inductance between the i-th and k-th branches in group 3. Time-derivatives are denoted by dots.

The relationship between the voltages and currents of branches in group 4 is written as

$$\mathbf{i}_C = \boldsymbol{C}\dot{\mathbf{u}}_C, \tag{6.24}$$

where $\boldsymbol{C}$ is the diagonal matrix formed by the capacitances of the branches in the group.

$\mathbf{u}_C$ and $\mathbf{i}_L$ being state variables, these equations are to be rearranged to obtain a state equation of the form

$$\begin{bmatrix} \dot{\mathbf{i}}_L \\ \dot{\mathbf{u}}_C \end{bmatrix} = \boldsymbol{D}\begin{bmatrix} \mathbf{i}_L \\ \mathbf{u}_C \end{bmatrix} + \boldsymbol{E}\begin{bmatrix} \mathbf{i}_0 \\ \mathbf{u}_0 \end{bmatrix}. \tag{6.25}$$

From (6.16), using (6.21):

$$\mathbf{i}_G = -\boldsymbol{GF}_{21}\mathbf{u}_C - \boldsymbol{GF}_{22}\mathbf{u}_R - \boldsymbol{GF}_{23}\mathbf{u}_0, \tag{6.26}$$

and from (6.19) in accordance with (6.22):

$$\mathbf{u}_R = \boldsymbol{RF}^+_{12}\mathbf{i}_0 + \boldsymbol{RF}^+_{22}\mathbf{i}_G + \boldsymbol{RF}^+_{32}\mathbf{i}_L. \tag{6.27}$$

Writing this into (6.26), rearrangement yields

$$\mathbf{i}_G = -(\boldsymbol{1} + \boldsymbol{GF}_{22}\boldsymbol{RF}^+_{22})^{-1}(\boldsymbol{GF}_{22}\boldsymbol{RF}^+_{12}\mathbf{i}_0 + \boldsymbol{GF}_{22}\boldsymbol{RF}^+_{32}\mathbf{i}_L + \boldsymbol{GF}_{21}\mathbf{u}_C + \boldsymbol{GF}_{23}\mathbf{u}_0). \tag{6.28}$$

On substitution of (6.26) into (6.27)

$$\mathbf{u}_R = (\boldsymbol{1} + \boldsymbol{RF}^+_{22}\boldsymbol{GF}_{22})^{-1}(\boldsymbol{RF}^+_{12}\mathbf{i}_0 + \boldsymbol{RF}^+_{32}\mathbf{i}_L - \boldsymbol{RF}^+_{22}\boldsymbol{GF}_{21}\mathbf{u}_C - \boldsymbol{RF}^+_{22}\boldsymbol{GF}_{23}\mathbf{u}_0) \tag{6.29}$$

is obtained. Writing this into (6.17), and (6.28) into (6.18), the following is derived;

$$\mathbf{u}_L + \boldsymbol{F}_{31}\mathbf{u}_C + \boldsymbol{F}_{32}(\boldsymbol{1} + \boldsymbol{RF}^+_{22}\boldsymbol{GF}_{22})^{-1}\boldsymbol{R}(\boldsymbol{F}^+_{12}\mathbf{i}_0 + \boldsymbol{F}^+_{32}\mathbf{i}_L - \\ - \boldsymbol{F}^+_{22}\boldsymbol{GF}_{21}\mathbf{u}_C - \boldsymbol{F}^+_{22}\boldsymbol{GF}_{23}\mathbf{u}_0) + \boldsymbol{F}_{33}\mathbf{u}_0 = \boldsymbol{0}, \tag{6.30}$$

$$-\boldsymbol{F}^+_{11}\mathbf{i}_0 + \boldsymbol{F}^+_{21}(\boldsymbol{1} + \boldsymbol{GF}_{22}\boldsymbol{RF}^+_{22})^{-1}\boldsymbol{G}(\boldsymbol{F}_{22}\boldsymbol{RF}^+_{12}\mathbf{i}_0 + \\ + \boldsymbol{F}_{22}\boldsymbol{RF}^+_{32}\mathbf{i}_L + \boldsymbol{F}_{21}\mathbf{u}_C + \boldsymbol{F}_{23}\mathbf{u}_0) - \boldsymbol{F}^+_{31}\mathbf{i}_L + \mathbf{i}_C = \boldsymbol{0}. \tag{6.31}$$

$\dot{\mathbf{i}}_L$ and $\dot{\mathbf{u}}_C$ may be expressed from (6.23) and (6.24). Let us substitute (6.30) and (6.31) into these. Thus the state equations

$$\begin{bmatrix} \dot{\mathbf{i}}_L \\ \dot{\mathbf{u}}_C \end{bmatrix} = \begin{bmatrix} \boldsymbol{L}^{-1} & \boldsymbol{0} \\ \boldsymbol{0} & \boldsymbol{C}^{-1} \end{bmatrix} \left\{ \begin{bmatrix} \boldsymbol{D}_{11} & \boldsymbol{D}_{12} \\ \boldsymbol{D}_{21} & \boldsymbol{D}_{22} \end{bmatrix} \begin{bmatrix} \mathbf{i}_L \\ \mathbf{u}_C \end{bmatrix} + \begin{bmatrix} \boldsymbol{E}_{11} & \boldsymbol{E}_{12} \\ \boldsymbol{E}_{21} & \boldsymbol{E}_{22} \end{bmatrix} \begin{bmatrix} \mathbf{i}_0 \\ \mathbf{u}_0 \end{bmatrix} \right\} \tag{6.32}$$

are obtained, where the notations

$$\boldsymbol{D}_{11} = -\boldsymbol{F}_{32}(\boldsymbol{1} + \boldsymbol{RF}^+_{22}\boldsymbol{GF}_{22})^{-1}\boldsymbol{RF}^+_{32}, \tag{6.33}$$

$$\boldsymbol{D}_{12} = -\boldsymbol{F}_{31} + \boldsymbol{F}_{32}(\boldsymbol{1} + \boldsymbol{RF}^+_{22}\boldsymbol{GF}_{22})^{-1}\boldsymbol{RF}^+_{22}\boldsymbol{GF}_{21}, \tag{6.34}$$

$$\boldsymbol{D}_{21} = -\boldsymbol{F}^+_{21}(\boldsymbol{1} + \boldsymbol{GF}_{22}\boldsymbol{RF}^+_{22})^{-1}\boldsymbol{GF}_{22}\boldsymbol{RF}^+_{32} + \boldsymbol{F}^+_{31}, \tag{6.35}$$

$$\boldsymbol{D}_{22} = -\boldsymbol{F}^+_{21}(\boldsymbol{1} + \boldsymbol{GF}_{22}\boldsymbol{RF}^+_{22})^{-1}\boldsymbol{GF}_{21}, \tag{6.36}$$

$$\boldsymbol{E}_{11} = -\boldsymbol{F}_{32}(\boldsymbol{1} + \boldsymbol{RF}^+_{22}\boldsymbol{GF}_{22})^{-1}\boldsymbol{RF}^+_{12}, \tag{6.37}$$

$$\boldsymbol{E}_{12} = -\boldsymbol{F}_{33} + \boldsymbol{F}_{32}(\boldsymbol{1} + \boldsymbol{RF}^+_{22}\boldsymbol{GF}_{22})^{-1}\boldsymbol{RF}^+_{22}\boldsymbol{GF}_{23}, \tag{6.38}$$

$$\boldsymbol{E}_{21} = \boldsymbol{F}^+_{11} - \boldsymbol{F}^+_{21}(\boldsymbol{1} + \boldsymbol{GF}_{22}\boldsymbol{RF}^+_{22})^{-1}\boldsymbol{GF}_{22}\boldsymbol{RF}^+_{12}, \tag{6.39}$$

$$\boldsymbol{E}_{22} = -\boldsymbol{F}^+_{21}(\boldsymbol{1} + \boldsymbol{GF}_{22}\boldsymbol{RF}^+_{22})^{-1}\boldsymbol{GF}_{23} \tag{6.40}$$

have been employed.

A knowledge of the state variables and excitations permits the calculation of the currents of the branches in group 2 from (6.28) and of the voltages of the branches in group 5 from (6.29), and hence (6.15) and (6.16) yield $\mathbf{u}_1$ and $\mathbf{u}_G$, (6.23) $\mathbf{u}_L$, (6.24) $\mathbf{i}_C$, (6.19) $\mathbf{i}_R$, and (6.20) $\mathbf{i}_6$, i.e. the column matrix of the responses can be expressed in the form of (6.4).

Writing the state equations of networks containing controlled sources

To write the state equation of networks containing coupled two-poles and two-ports, our discussion is restricted to the case where the substitution of the two-ports by controlled generators yields a model containing controlled sources with the time-functions of their controlled signals being linearly proportional to that of the controlling signals. Two methods will be presented for writing the state equations. For one of them, two-ports are modelled by controlled generators, while nullators and norators are employed for modelling of the other, and the state equations are written for the model thus obtained. In the course of the selection of the tree of the graph, the possible presence of capacitive loops or inductive cutsets in the network may be determined.

Using models containing controlled sources

For the first method to be discussed the two-ports of the network (with the exception of coupled inductors) are modelled by controlled generators as shown in Chapter 5 [60]. Modelling is to be carried out on the basis of the relationships between the time-functions of the currents and voltages of the two-ports.

Controlled and independent generators are considered to consist of a source and a passive two-terminal element. The substitution of the two-ports is carried out in accordance with Table 5.1 in Chapter 5 to obtain the primary and secondary sides of the two-ports consisting of a source and of a passive two-terminal element or elements. In degenerate cases controlled sources with zero source-current or zero source-voltage are regarded as zero conductance or zero resistance branches. Some degenerate cases are shown in Fig. 6.1.

After modelling the two-ports as explained previously, the branches of the network obtained are independent and controlled sources, inductors, capacitors, and resistors.

Additional controlled sources may be present in the network, not as a result of modelling the two-ports as mentioned above. In such cases the controlling signals may not be currents or voltages of resistors (or short- or open-circuits), but, for example, may be voltages of independent current-sources, currents of inductors, etc. In the case of a controlling voltage an open-circuit is added in parallel to the relevant element, while for a controlling current a short-circuit in series with the appropriate element is added to the network. In such situations, the voltage of the open-circuit, and the current of the short-circuit will be regarded as the controlling signals. Thus,

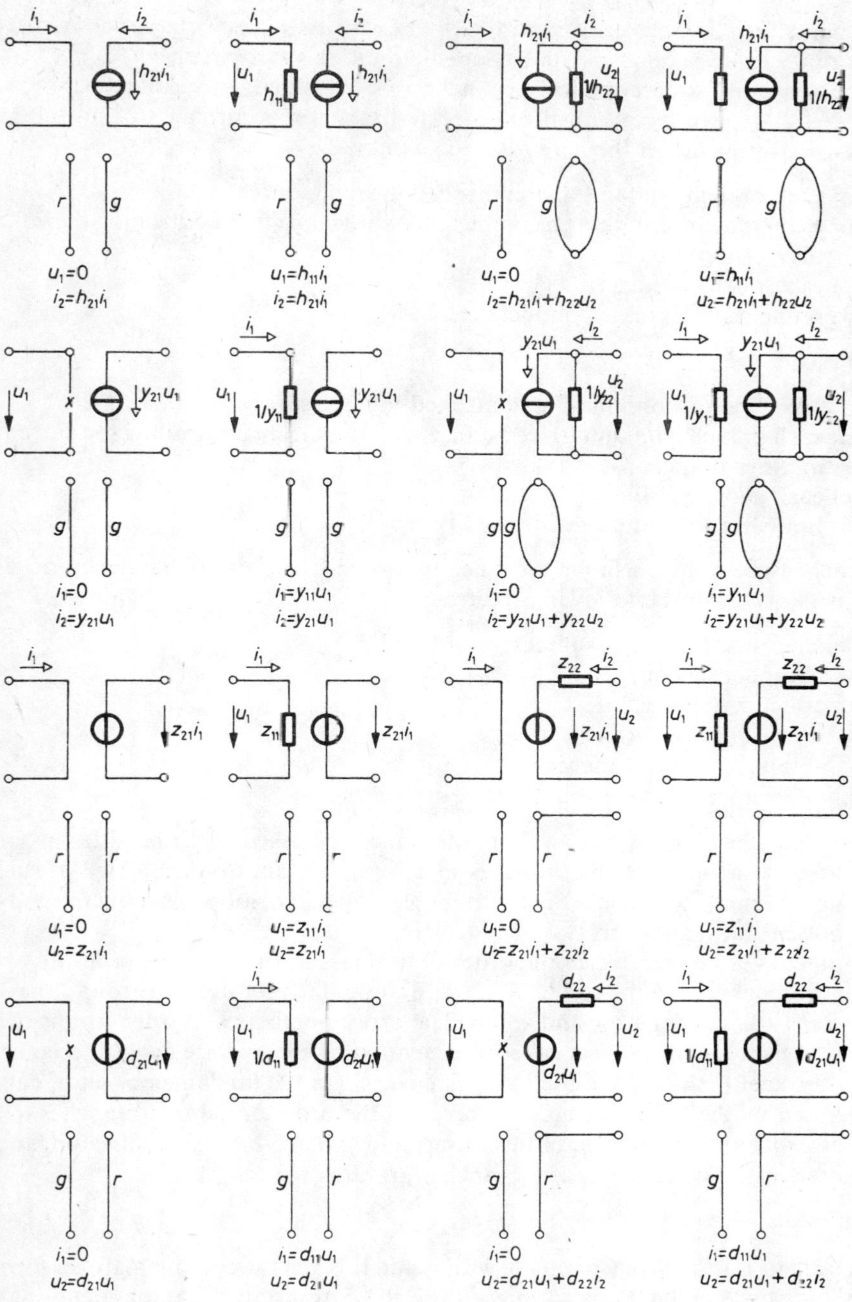

u1=0
i2=h21i1
u1=h11i1
i2=h21i1
u1=0
i2=h21i1+h22u2
u1=h11i1
u2=h21i1+h22u2
i1=0
i2=y21u1
i1=y11u1
i2=y21u1
i1=0
i2=y21u1+y22u2
i1=y11u1
i2=y21u1+y22u2
u1=0
u2=z21i1
u1=z11i1
u2=z21i1
u1=0
u2=z21i1+z22i2
u1=z11i1
u2=z21i1+z22i2
i1=0
u2=d21u1
i1=d11u1
u2=d21u1
i1=0
u2=d21u1+d22i2
i1=d11u1
u2=d21u1+d22i2

Fig. 6.1

it is possible to consider each controlled source as a two-port with the passive part at its primary side being a resistance, open-circuit or short-circuit.

To write the state equations, branches are classified into groups. Controlled sources and resistances constitute g- or r-branches, according to the following scheme. g-branches in the network are as follows:

(a) each branch containing a controlled current-source;
(b) each branch containing a conductance with its voltage controlling the voltage of another branch;
(c) each open-circuit;
(d) branches with finite conductance.

r-branches are as follows:

(a) each branch containing a controlled voltage-source;
(b) each branch containing a resistance with its current controlling the current of another branch;
(c) each short-circuit;
(d) branches with finite resistance.

Branches with nonzero, finite resistance are g- or r-branches, so that the branches of the network form the following six groups:

1. independent current-sources (chords);
2. g-branches (chords);
3. inductors (chords);
4. capacitors (tree-branches);
5. r-branches (tree-branches);
6. independent voltage-sources (tree-branches).

Thus, branches with nonzero, finite resistance are assigned to be either g- or r-branches so as to have the branches in groups 4, 5 and 6 form a tree. If such a classification of the branches is not possible, capacitive loops or inductive cutsets are present in the network.

Branches are numbered in the order of the classification, i.e. the order numbers of the branches in group 1 are $1, 2, \ldots, b_1$, and similarly for those in group 2 they are $b_1+1, b_1+2, \ldots, b_1+b_2$, and so on. The order numbers and orientations of the loops in the fundamental set of loops generated by the tree are chosen to coincide with those of the chords in the loop. The cutsets in the fundamental set of cutsets generated by the same tree are numbered in the order of the tree-branches in the cutsets, with the orientation of the cutsets and the tree-branches coinciding along the branches. The currents of g-branches are thus written as

$$\mathbf{i}_2 = \boldsymbol{G}_2\mathbf{u}_2 + \boldsymbol{K}\mathbf{i}_5, \tag{6.41}$$

similarly to Eq. (5.65) of Chapter 5, with $\mathbf{i}_2$ and $\mathbf{i}_5$ being the column matrices formed by the currents of the branches in groups 2 and 5, respectively, and $\mathbf{u}_2$ denoting the column matrix formed by the voltages of the branches in group 2. The elements on the main diagonal of G_2 are the conductances of the branches in group 2, while the elements off the main diagonal are the conductance parameters characterizing the relationships between the source-currents of voltage controlled current-sources and their controlling voltages. K is constructed of the proportionality constants between the source-currents of current controlled current-sources in group 2 and their controlling currents. The rows and columns of K correspond to the branches of groups 2 and 5, respectively.

The voltages of r-branches forming group 5 are described by

$$\mathbf{u}_5 = \boldsymbol{M}\mathbf{u}_2 + \boldsymbol{R}_5\mathbf{i}_5 \tag{6.42}$$

similarly to Eq. (5.66) of Chapter 5, with $\mathbf{u}_5$ denoting the column matrix formed by the voltages of r-branches. The elements of M yield the relationships between the source-voltages of voltage controlled voltage-sources and their controlling voltages. The rows and columns of M are ordered in accordance with the branches of groups 5 and 2, respectively. R_5 is a square matrix with the elements on its main diagonal being the resistances of r-branches, and the elements off the main diagonal being the resistance parameters describing the relationships between the source-voltages of current controlled voltage-sources and their controlling currents.

The relationship between the column matrices $\mathbf{u}_3$ and $\mathbf{i}_3$, formed by the voltages and currents of the branches assigned to group 3 is

$$\mathbf{u}_3 = \boldsymbol{L}\dot{\mathbf{i}}_3 \,. \tag{6.43}$$

The self inductances of the branches appear on the main diagonal of L, while the appropriate mutual inductances occur off the main diagonal.

The currents and voltages of the branches in group 4 are related by

$$\mathbf{i}_4 = \boldsymbol{C}\dot{\mathbf{u}}_4 \,. \tag{6.44}$$

C is a diagonal matrix, its elements being the capacitances of capacitors. $\mathbf{i}_4$ and $\mathbf{u}_4$ are the column matrices formed by the currents and voltages of the branches in the group.

Introducing the notations $\mathbf{i}_1$, $\mathbf{i}_6$ and $\mathbf{u}_1$, $\mathbf{u}_6$ for the column matrices formed by the currents and voltages of branches in groups 1 and 6, and partitioning the matrices of the fundamental sets of loops and cutsets generated by the tree chosen in accordance with the classification of the branches, Kirchhoff's loop and cutset equations can be written for the network as follows:

$$\begin{bmatrix} \boldsymbol{1} & \boldsymbol{0} & \boldsymbol{0} & \boldsymbol{F}_{11} & \boldsymbol{F}_{12} & \boldsymbol{F}_{13} \\ \boldsymbol{0} & \boldsymbol{1} & \boldsymbol{0} & \boldsymbol{F}_{21} & \boldsymbol{F}_{22} & \boldsymbol{F}_{23} \\ \boldsymbol{0} & \boldsymbol{0} & \boldsymbol{1} & \boldsymbol{F}_{31} & \boldsymbol{F}_{32} & \boldsymbol{F}_{33} \end{bmatrix} \begin{bmatrix} \mathbf{u}_1 \\ \mathbf{u}_2 \\ \mathbf{u}_3 \\ \mathbf{u}_4 \\ \mathbf{u}_5 \\ \mathbf{u}_6 \end{bmatrix} = \mathbf{0} \tag{6.45}$$

and

$$\begin{bmatrix} -F^{+}_{11} & -F^{+}_{21} & -F^{+}_{31} & 1 & 0 & 0 \\ -F^{+}_{12} & -F^{+}_{22} & -F^{+}_{32} & 0 & 1 & 0 \\ -F^{+}_{13} & -F^{+}_{23} & -F^{+}_{33} & 0 & 0 & 1 \end{bmatrix} \begin{bmatrix} \mathbf{i}_1 \\ \mathbf{i}_2 \\ \mathbf{i}_3 \\ \mathbf{i}_4 \\ \mathbf{i}_5 \\ \mathbf{i}_6 \end{bmatrix} = \mathbf{0}. \tag{6.46}$$

The following equations are obtained from these:

$$\mathbf{u}_1 + F_{11}\mathbf{u}_4 + F_{12}\mathbf{u}_5 + F_{13}\mathbf{u}_6 = \mathbf{0}, \tag{6.47}$$

$$\mathbf{u}_2 + F_{21}\mathbf{u}_4 + F_{22}\mathbf{u}_5 + F_{23}\mathbf{u}_6 = \mathbf{0}, \tag{6.48}$$

$$\mathbf{u}_3 + F_{31}\mathbf{u}_4 + F_{32}\mathbf{u}_5 + F_{33}\mathbf{u}_6 = \mathbf{0}, \tag{6.49}$$

$$-F^{+}_{11}\mathbf{i}_1 - F^{+}_{21}\mathbf{i}_2 - F^{+}_{31}\mathbf{i}_3 + \mathbf{i}_4 = \mathbf{0}, \tag{6.50}$$

$$-F^{+}_{12}\mathbf{i}_1 - F^{+}_{22}\mathbf{i}_2 - F^{+}_{32}\mathbf{i}_3 + \mathbf{i}_5 = \mathbf{0}, \tag{6.51}$$

$$-F^{+}_{13}\mathbf{i}_1 - F^{+}_{23}\mathbf{i}_2 - F^{+}_{33}\mathbf{i}_3 + \mathbf{i}_6 = \mathbf{0}. \tag{6.52}$$

In the equations $\mathbf{i}_3$ and $\mathbf{u}_4$ are the column matrices of the state variables, while $\mathbf{i}_1$ and $\mathbf{u}_6$ are those of the excitations. To write the state equations, the variables $\mathbf{u}_1$, $\mathbf{u}_2$, $\mathbf{u}_3$, $\mathbf{u}_5$, $\mathbf{i}_2$, $\mathbf{i}_4$, $\mathbf{i}_5$, $\mathbf{i}_6$ have to be eliminated from the above equations. To this end $\mathbf{u}_2$ is expressed from (6.48), $\mathbf{i}_5$ from (6.51), and substituted into (6.41) and (6.42):

$$\mathbf{i}_2 = -G_2F_{21}\mathbf{u}_4 - G_2F_{22}\mathbf{u}_5 - G_2F_{23}\mathbf{u}_6 + KF^{+}_{12}\mathbf{i}_1 + KF^{+}_{22}\mathbf{i}_2 + KF^{+}_{32}\mathbf{i}_3, \tag{6.53}$$

$$\mathbf{u}_5 = -MF_{21}\mathbf{u}_4 - MF_{22}\mathbf{u}_5 - MF_{23}\mathbf{u}_6 + R_5F^{+}_{12}\mathbf{i}_1 + R_5F^{+}_{22}\mathbf{i}_2 + R_5F^{+}_{32}\mathbf{i}_3. \tag{6.54}$$

$\mathbf{u}_5$ is written from (6.54) into (6.53). So

$$\begin{aligned} &[1 + G_2F_{22}(1 + MF_{22})^{-1}R_5F^{+}_{22} - KF^{+}_{22}]\mathbf{i}_2 = KF^{+}_{12} - GF_{22}[(1 + \\ &\quad + MF_{22})^{-1}R_5F^{+}_{12}]\,\mathbf{i}_1 + [KF^{+}_{32} - G_2F_{22}(1 + MF_{22})^{-1}R_5F^{+}_{32}]\mathbf{i}_3 + \\ &\quad + G_2[F_{22}(1 + MF_{22})^{-1}MF_{21} - F_{21}]\mathbf{u}_4 + \\ &\quad + G_2[F_{22}(1 + MF_{22})^{-1}MF_{23} - F_{23}]\mathbf{u}_6 \end{aligned} \tag{6.55}$$

is obtained. Expressing $\mathbf{i}_2$ from (6.53) and substituting into (6.54):

$$\begin{aligned} &[1 + MF_{22} + R_5F^{+}_{22}(1 - KF^{+}_{22})^{-1}G_2F_{22}]\mathbf{u}_5 = R_5[F^{+}_{12} + F^{+}_{22}(1 - \\ &\quad - KF^{+}_{22})^{-1}KF^{+}_{12}]\mathbf{i}_1 + R_5[F^{+}_{22}(1 - KF^{+}_{22})^{-1}KF^{+}_{32} + F^{+}_{32}]\mathbf{i}_3 - \\ &\quad - [MF_{21} + R_5F^{+}_{22}(1 - KF^{+}_{22})^{-1}G_2F_{21}]\mathbf{u}_4 - \\ &\quad - [MF_{23} + R_5F^{+}_{22}(1 - KF^{+}_{22})^{-1}G_2F_{23}]\mathbf{u}_6. \end{aligned} \tag{6.56}$$

Hence, $\mathbf{u}_5$ can be obtained, and substituted into (6.49), using (6.43). Similarly, expressing $\mathbf{i}_2$ from (6.55) and substituting into (6.50), using (6.44), results in the following state equations being obtained:

$$\begin{bmatrix}\dot{\mathbf{i}}_3\\ \dot{\mathbf{u}}_4\end{bmatrix}=\begin{bmatrix}\boldsymbol{L}^{-1} & \boldsymbol{0}\\ \boldsymbol{0} & \boldsymbol{C}^{-1}\end{bmatrix}\left\{\begin{bmatrix}\boldsymbol{D}_{11} & \boldsymbol{D}_{12}\\ \boldsymbol{D}_{21} & \boldsymbol{D}_{22}\end{bmatrix}\begin{bmatrix}\mathbf{i}_3\\ \mathbf{u}_4\end{bmatrix}+\begin{bmatrix}\boldsymbol{E}_{11} & \boldsymbol{E}_{12}\\ \boldsymbol{E}_{21} & \boldsymbol{E}_{22}\end{bmatrix}\begin{bmatrix}\mathbf{i}_1\\ \mathbf{u}_6\end{bmatrix}\right\}, \quad (6.57)$$

where

$$\boldsymbol{D}_{11}=-\boldsymbol{F}_{32}[\boldsymbol{1}+\boldsymbol{M}\boldsymbol{F}_{22}+\boldsymbol{R}_5\boldsymbol{F}_{22}^{+}(\boldsymbol{1}-\boldsymbol{K}\boldsymbol{F}_{22}^{+})^{-1}\boldsymbol{G}_2\boldsymbol{F}_{22}]^{-1}[\boldsymbol{R}_5\boldsymbol{F}_{22}^{+}(\boldsymbol{1}-$$
$$-\boldsymbol{K}\boldsymbol{F}_{22}^{+})^{-1}\boldsymbol{K}\boldsymbol{F}_{32}^{+}+\boldsymbol{R}_5\boldsymbol{F}_{32}^{+}], \quad (6.58)$$

$$\boldsymbol{D}_{12}=\boldsymbol{F}_{32}[\boldsymbol{1}+\boldsymbol{M}\boldsymbol{F}_{22}+\boldsymbol{R}_5\boldsymbol{F}_{22}^{+}(\boldsymbol{1}-\boldsymbol{K}\boldsymbol{F}_{22}^{+})^{-1}\boldsymbol{G}_2\boldsymbol{F}_{22}]^{-1}[\boldsymbol{M}\boldsymbol{F}_{21}+$$
$$+\boldsymbol{R}_5\boldsymbol{F}_{22}^{+}(\boldsymbol{1}-\boldsymbol{K}\boldsymbol{F}_{22}^{+})^{-1}\boldsymbol{G}_2\boldsymbol{F}_{21}]-\boldsymbol{F}_{31}, \quad (6.59)$$

$$\boldsymbol{D}_{21}=\boldsymbol{F}_{31}^{+}+\boldsymbol{F}_{21}^{+}[\boldsymbol{1}+\boldsymbol{G}_2\boldsymbol{F}_{22}(\boldsymbol{1}+\boldsymbol{M}\boldsymbol{F}_{22})^{-1}\boldsymbol{R}_5\boldsymbol{F}_{22}^{+}-\boldsymbol{K}\boldsymbol{F}_{22}^{+}]^{-1}[\boldsymbol{K}\boldsymbol{F}_{32}^{+}-$$
$$-\boldsymbol{G}_2\boldsymbol{F}_{22}(\boldsymbol{1}+\boldsymbol{M}\boldsymbol{F}_{22})^{-1}\boldsymbol{R}_5\boldsymbol{F}_{32}^{+}], \quad (6.60)$$

$$\boldsymbol{D}_{22}=\boldsymbol{F}_{21}^{+}[\boldsymbol{1}+\boldsymbol{G}_2\boldsymbol{F}_{22}(\boldsymbol{1}+\boldsymbol{M}\boldsymbol{F}_{22})^{-1}\boldsymbol{R}_5\boldsymbol{F}_{22}^{+}-\boldsymbol{K}\boldsymbol{F}_{22}^{+}]^{-1}[\boldsymbol{G}_2\boldsymbol{F}_{22}^{+}(\boldsymbol{1}+$$
$$+\boldsymbol{M}\boldsymbol{F}_{22})^{-1}\boldsymbol{M}\boldsymbol{F}_{21}-\boldsymbol{G}_2\boldsymbol{F}_{21}], \quad (6.61)$$

$$\boldsymbol{E}_{11}=-\boldsymbol{F}_{32}[\boldsymbol{1}+\boldsymbol{M}\boldsymbol{F}_{22}+\boldsymbol{R}_5\boldsymbol{F}_{22}^{+}(\boldsymbol{1}-\boldsymbol{K}\boldsymbol{F}_{22}^{+})^{-1}\boldsymbol{G}_2\boldsymbol{F}_{22}]^{-1}[\boldsymbol{R}_5\boldsymbol{F}_{12}^{+}+$$
$$+\boldsymbol{R}_5\boldsymbol{F}_{22}^{+}(\boldsymbol{1}-\boldsymbol{K}\boldsymbol{F}_{22}^{+})^{-1}\boldsymbol{K}\boldsymbol{F}_{12}^{+}]. \quad (6.62)$$

$$\boldsymbol{E}_{12}=\boldsymbol{F}_{32}[\boldsymbol{1}+\boldsymbol{M}\boldsymbol{F}_{22}+\boldsymbol{R}_5\boldsymbol{F}_{22}^{+}(\boldsymbol{1}-\boldsymbol{K}\boldsymbol{F}_{22}^{+})^{-1}\boldsymbol{G}_2\boldsymbol{F}_{22}]^{-1}[\boldsymbol{M}\boldsymbol{F}_{23}+$$
$$+\boldsymbol{R}_5\boldsymbol{F}_{22}^{+}(\boldsymbol{1}-\boldsymbol{K}\boldsymbol{F}_{22}^{+})^{-1}\boldsymbol{G}_2\boldsymbol{F}_{23}]-\boldsymbol{F}_{33}, \quad (6.63)$$

$$\boldsymbol{E}_{21}=\boldsymbol{F}_{11}^{+}+\boldsymbol{F}_{21}^{+}[\boldsymbol{1}+\boldsymbol{G}_2\boldsymbol{F}_{22}(\boldsymbol{1}+\boldsymbol{M}\boldsymbol{F}_{22})^{-1}\boldsymbol{R}_5\boldsymbol{F}_{22}^{+}-\boldsymbol{K}\boldsymbol{F}_{22}^{+}]^{-1}[\boldsymbol{K}\boldsymbol{F}_{12}^{+}-$$
$$-\boldsymbol{G}_2\boldsymbol{F}_{22}(\boldsymbol{1}+\boldsymbol{M}\boldsymbol{F}_{22})^{-1}\boldsymbol{R}_5\boldsymbol{F}_{12}^{+}], \quad (6.64)$$

$$\boldsymbol{E}_{22}=\boldsymbol{F}_{21}^{+}[\boldsymbol{1}+\boldsymbol{G}_2\boldsymbol{F}_{22}(\boldsymbol{1}+\boldsymbol{M}\boldsymbol{F}_{22})^{-1}\boldsymbol{R}_5\boldsymbol{F}_{22}^{+}-\boldsymbol{K}\boldsymbol{F}_{22}^{+}]^{-1}[\boldsymbol{G}_2\boldsymbol{F}_{22}(\boldsymbol{1}+$$
$$+\boldsymbol{M}\boldsymbol{F}_{22})^{-1}\boldsymbol{M}\boldsymbol{F}_{23}-\boldsymbol{G}_2\boldsymbol{F}_{23}]. \quad (6.65)$$

If there are no inductors in the network, i.e. the state variables are the voltages of the capacitors, the blocks, $\boldsymbol{D}_{11}$, $\boldsymbol{D}_{12}$, $\boldsymbol{D}_{21}$, $\boldsymbol{E}_{11}$ and $\boldsymbol{E}_{12}$ do not exist and the first term on the right-hand side of (6.57) is $\boldsymbol{C}^{-1}\boldsymbol{D}_{22}\mathbf{u}_4$. Conversely, if there are no capacitors in the network, i.e. only the currents of the inductors are state variables, $\boldsymbol{D}_{12}$, $\boldsymbol{D}_{21}$, $\boldsymbol{D}_{22}$, $\boldsymbol{E}_{21}$ and $\boldsymbol{E}_{22}$ do not exist and the first term on the right-hand side of (6.57) is $\boldsymbol{L}^{-1}\boldsymbol{D}_{11}\mathbf{i}_3$.

Knowing the state variables, $\mathbf{i}_2$ may be calculated from (6.55), $\mathbf{u}_5$ from (6.56), $\mathbf{u}_3$ from (6.49), and $\mathbf{i}_4$ from (6.50). Using these, (6.47), (6.48), (6.51) and (6.52) yield $\mathbf{u}_1$, $\mathbf{u}_2$, $\mathbf{i}_5$ and $\mathbf{i}_6$, respectively.

Comparing the coefficients in equations (6.58)...(6.65) with those in equations (6.33)...(6.40), the latter are seen to be obtained from the former on substitution of $\boldsymbol{K}=\boldsymbol{0}$ and $\boldsymbol{M}=\boldsymbol{0}$, i.e. they may be considered special cases of the former.

Using models containing nullators and norators

The possibility of modelling two-ports by passive two-terminal elements, nullators and norators has been shown in Chapter 5. The following method is presented for writing state equations, applicable when two-ports in the network (with the exception of mutual inductances) are modelled by the use of nullators and norators [8, 59].

To write the equations of the network, each voltage-source, current-source, resistor, inductor, capacitor, nullator and norator is regarded as a distinct branch. A tree of the network graph is chosen for the analysis, with each voltage-source, short-circuit, nullator and capacitor corresponding to tree-branches, and each current-source, open-circuit, norator and inductor associated with chords. Of the nonzero finite resistances or conductances those connected in series with a capacitor are regarded as chords, while those connected parallel to an inductor are chosen as tree-branches, while the remainder are selected as either tree-branches or chords in such a manner that the tree-branches mentioned above and those formed by resistances constitute a tree or forest of the network graph. If there is no capacitive loop or inductive cutset in the network, the classification of branches can be carried out as previously explained.

The branches of the network are classified into the following eight groups:

1. chords containing current-sources (numbering b_1);
2. chords containing norators (numbering b_2);
3. chords containing inductors (numbering b_3);
4. chords with finite conductance (numbering b_4);
5. tree-branches with finite resistance (numbering b_5);
6. tree-branches containing capacitors (numbering b_6);
7. tree-branches containing nullators (numbering b_7);
8. tree-branches containing voltage-sources (numbering b_8).

The numbers of nullators and norators in the model must be equal, i.e. $b_2 = b_7$.

Let the branches be numbered in the order of the classification, i.e. the order numbers of the branches assigned to group 1 are $1, 2, \ldots, b_1$, those in group 2 are $b_1+1, b_1+2, \ldots, b_1+b_2$, etc. The order numbers and orientations of the loops in the fundamental set of loops generated by the tree chosen correspond to those of the chord in the loop. The cutsets of the fundamental set of cutsets generated by the same tree are numbered in the order of the tree-branches in the cutsets, with the directions of the tree-branches and the cutsets corresponding.

The column matrices of the network branch-voltages and branch-currents as well as the matrices of the sets of loops and cutsets mentioned are partitioned in accordance with the branch classification. With their aid, the independent loop and cutset equations of the network take the form

$$\begin{bmatrix} 1 & 0 & 0 & 0 & F_{11} & F_{12} & F_{13} & F_{14} \\ 0 & 1 & 0 & 0 & F_{21} & F_{22} & F_{23} & F_{24} \\ 0 & 0 & 1 & 0 & F_{31} & F_{32} & F_{33} & F_{34} \\ 0 & 0 & 0 & 1 & F_{41} & F_{42} & F_{43} & F_{44} \end{bmatrix} \begin{bmatrix} \mathbf{u}_1 \\ \mathbf{u}_2 \\ \mathbf{u}_3 \\ \mathbf{u}_4 \\ \mathbf{u}_5 \\ \mathbf{u}_6 \\ \mathbf{0} \\ \mathbf{u}_8 \end{bmatrix} = \mathbf{0}, \qquad (6.66)$$

$$\begin{bmatrix} -F_{11}^{+} & -F_{21}^{+} & -F_{31}^{+} & -F_{41}^{+} & 1 & 0 & 0 & 0 \\ -F_{12}^{+} & -F_{22}^{+} & -F_{32}^{+} & -F_{42}^{+} & 0 & 1 & 0 & 0 \\ -F_{13}^{+} & -F_{23}^{+} & -F_{33}^{+} & -F_{43}^{+} & 0 & 0 & 1 & 0 \\ -F_{14}^{+} & -F_{24}^{+} & -F_{34}^{+} & -F_{44}^{+} & 0 & 0 & 0 & 1 \end{bmatrix} \begin{bmatrix} \mathbf{i}_1 \\ \mathbf{i}_2 \\ \mathbf{i}_3 \\ \mathbf{i}_4 \\ \mathbf{i}_5 \\ \mathbf{i}_6 \\ \mathbf{0} \\ \mathbf{i}_8 \end{bmatrix} = \mathbf{0}, \qquad (6.67)$$

where it has been taken into account that the voltages and currents of nullators are zero. Hence:

$$\mathbf{u}_1 + F_{11}\mathbf{u}_5 + F_{12}\mathbf{u}_6 + F_{14}\mathbf{u}_8 = \mathbf{0}, \qquad (6.68)$$

$$\mathbf{u}_2 + F_{21}\mathbf{u}_5 + F_{22}\mathbf{u}_6 + F_{24}\mathbf{u}_8 = \mathbf{0}, \qquad (6.69)$$

$$\mathbf{u}_3 + F_{31}\mathbf{u}_5 + F_{32}\mathbf{u}_6 + F_{34}\mathbf{u}_8 = \mathbf{0}, \qquad (6.70)$$

$$\mathbf{u}_4 + F_{41}\mathbf{u}_5 + F_{42}\mathbf{u}_6 + F_{44}\mathbf{u}_8 = \mathbf{0}, \qquad (6.71)$$

$$-F_{11}^{+}\mathbf{i}_1 - F_{21}^{+}\mathbf{i}_2 - F_{31}^{+}\mathbf{i}_3 - F_{41}^{+}\mathbf{i}_4 + \mathbf{i}_5 = \mathbf{0}, \qquad (6.72)$$

$$-F_{12}^{+}\mathbf{i}_1 - F_{22}^{+}\mathbf{i}_2 - F_{32}^{+}\mathbf{i}_3 - F_{42}^{+}\mathbf{i}_4 + \mathbf{i}_6 = \mathbf{0}, \qquad (6.73)$$

$$-F_{13}^{+}\mathbf{i}_1 - F_{23}^{+}\mathbf{i}_2 - F_{33}^{+}\mathbf{i}_3 - F_{43}^{+}\mathbf{i}_4 = \mathbf{0}, \qquad (6.74)$$

$$-F_{14}^{+}\mathbf{i}_1 - F_{24}^{+}\mathbf{i}_2 - F_{34}^{+}\mathbf{i}_3 - F_{44}^{+}\mathbf{i}_4 + \mathbf{i}_8 = \mathbf{0}. \qquad (6.75)$$

In these equations $\mathbf{i}_3$ and $\mathbf{u}_6$ are the column matrices of state variables, $\mathbf{i}_1$ and $\mathbf{u}_8$ are those of source-currents and source-voltages. To write the state equations, other variables must be eliminated from these equations. To this end the known relationships between currents and voltages of each respective group are used, namely

$$\mathbf{u}_3 = \boldsymbol{L}\dot{\mathbf{i}}_3\,, \tag{6.76}$$

where $\boldsymbol{L}$ contains the self-inductances of the branches in group 3 on its main diagonal, with the mutual inductances between the branches of the group written off the main diagonal.

$$\mathbf{i}_4 = \boldsymbol{G}_4\mathbf{u}_4\,. \tag{6.77}$$

$\boldsymbol{G}_4$ is the diagonal matrix formed by the conductances of the branches assigned to group 4.

$$\mathbf{u}_5 = \boldsymbol{R}_5\mathbf{i}_5\,. \tag{6.78}$$

$\boldsymbol{R}_5$ is the diagonal matrix formed by the resistances of the branches assigned to group 5.

$$\mathbf{i}_6 = \boldsymbol{C}\dot{\mathbf{u}}_6\,. \tag{6.79}$$

$\boldsymbol{C}$ is the diagonal matrix constructed of the capacitances of capacitors.

The numbers of nullators and norators being equal, $\boldsymbol{F}_{23}$ is square, and provided that it is nonsingular, $\mathbf{i}_2$, the currents of norators can be expressed from (6.74):

$$\mathbf{i}_2 = -{\boldsymbol{F}_{23}^{+}}^{-1}\boldsymbol{F}_{13}^{+}\mathbf{i}_1 - {\boldsymbol{F}_{23}^{+}}^{-1}\boldsymbol{F}_{33}^{+}\mathbf{i}_3 - {\boldsymbol{F}_{23}^{+}}^{-1}\boldsymbol{F}_{43}^{+}\mathbf{i}_4\,. \tag{6.80}$$

Substituting this into (6.72) and (6.73):

$$(\boldsymbol{F}_{21}^{+}{\boldsymbol{F}_{23}^{+}}^{-1}\boldsymbol{F}_{33}^{+} - \boldsymbol{F}_{31}^{+})\mathbf{i}_3 + (\boldsymbol{F}_{21}^{+}{\boldsymbol{F}_{23}^{+}}^{-1}\boldsymbol{F}_{43}^{+} - \boldsymbol{F}_{41}^{+})\mathbf{i}_4 + \mathbf{i}_5 =$$
$$= (\boldsymbol{F}_{11}^{+} - \boldsymbol{F}_{21}^{+}{\boldsymbol{F}_{23}^{+}}^{-1}\boldsymbol{F}_{13}^{+})\mathbf{i}_1\,, \tag{6.81}$$

$$(\boldsymbol{F}_{22}^{+}{\boldsymbol{F}_{23}^{+}}^{-1}\boldsymbol{F}_{33}^{+} - \boldsymbol{F}_{32}^{+})\mathbf{i}_3 + (\boldsymbol{F}_{22}^{+}{\boldsymbol{F}_{23}^{+}}^{-1}\boldsymbol{F}_{43}^{+} - \boldsymbol{F}_{42}^{+})\mathbf{i}_4 + \mathbf{i}_6 =$$
$$= (\boldsymbol{F}_{12}^{+} - \boldsymbol{F}_{22}^{+}{\boldsymbol{F}_{23}^{+}}^{-1}\boldsymbol{F}_{13}^{+})\mathbf{i}_1\,. \tag{6.82}$$

(6.71) yields, with the application of (6.77):

$$\mathbf{i}_4 = -\boldsymbol{G}_4\boldsymbol{F}_{41}\mathbf{u}_5 - \boldsymbol{G}_4\boldsymbol{F}_{42}\mathbf{u}_6 - \boldsymbol{G}_4\boldsymbol{F}_{44}\mathbf{u}_8\,. \tag{6.83}$$

From (6.81), taking (6.78) into account:

$$\mathbf{u}_5 = \boldsymbol{R}_5(\boldsymbol{F}_{11}^{+} - \boldsymbol{F}_{21}^{+}{\boldsymbol{F}_{23}^{+}}^{-1}\boldsymbol{F}_{13}^{+})\mathbf{i}_1 + \boldsymbol{R}_5(\boldsymbol{F}_{31}^{+} - \boldsymbol{F}_{21}^{+}{\boldsymbol{F}_{23}^{+}}^{-1}\boldsymbol{F}_{33}^{+})\mathbf{i}_3 +$$
$$+ \boldsymbol{R}_5(\boldsymbol{F}_{41}^{+} - \boldsymbol{F}_{21}^{+}{\boldsymbol{F}_{23}^{+}}^{-1}\boldsymbol{F}_{43}^{+})\mathbf{i}_4\,. \tag{6.84}$$

Let us substitute this into (6.83). Thus,

$$\mathbf{i}_4 = [\boldsymbol{1} + \boldsymbol{G}_4\boldsymbol{F}_{41}\boldsymbol{R}_5(\boldsymbol{F}_{41}^{+} - \boldsymbol{F}_{21}^{+}{\boldsymbol{F}_{23}^{+}}^{-1}\boldsymbol{F}_{43}^{+})]^{-1}[\boldsymbol{G}_4\boldsymbol{F}_{41}\boldsymbol{R}_5(\boldsymbol{F}_{21}^{+}{\boldsymbol{F}_{23}^{+}}^{-1}\boldsymbol{F}_{33}^{+} - \boldsymbol{F}_{31}^{+})\mathbf{i}_3 -$$
$$- \boldsymbol{G}_4\boldsymbol{F}_{42}\mathbf{u}_6 + \boldsymbol{G}_4\boldsymbol{F}_{41}\boldsymbol{R}_5(\boldsymbol{F}_{21}^{+}\mathrm{F}_{23}^{+\,-1}\boldsymbol{F}_{13}^{+} - \boldsymbol{F}_{11}^{+})\mathbf{i}_1 - \boldsymbol{G}_4\boldsymbol{F}_{44}\mathbf{u}_8] \tag{6.85}$$

is obtained. Substituting (6.83) into (6.84):

$$\mathbf{u}_5 = [\boldsymbol{1} + \boldsymbol{R}_5(\boldsymbol{F}_{41}^{+} - \boldsymbol{F}_{21}^{+}{\boldsymbol{F}_{23}^{+}}^{-1}\boldsymbol{F}_{43}^{+})\boldsymbol{G}_4\boldsymbol{F}_{41}]^{-1}[\boldsymbol{R}_5(\boldsymbol{F}_{31}^{+} - \boldsymbol{F}_{21}^{+}{\boldsymbol{F}_{23}^{+}}^{-1}\boldsymbol{F}_{33}^{+})\mathbf{i}_3 +$$
$$+ \boldsymbol{R}_5(\boldsymbol{F}_{21}^{+}{\boldsymbol{F}_{23}^{+}}^{-1}\boldsymbol{F}_{43}^{+} - \boldsymbol{F}_{41}^{+})\boldsymbol{G}_4\boldsymbol{F}_{42}\mathbf{u}_6 + \boldsymbol{R}_5(\boldsymbol{F}_{11}^{+} - \boldsymbol{F}_{21}^{+}{\boldsymbol{F}_{23}^{+}}^{-1}\boldsymbol{F}_{13}^{+})\mathbf{i}_1 +$$
$$+ \boldsymbol{R}_5(\boldsymbol{F}_{21}^{+}{\boldsymbol{F}_{23}^{+}}^{-1}\boldsymbol{F}_{43}^{+} - \boldsymbol{F}_{41})\boldsymbol{G}_4\boldsymbol{F}_{44}\mathbf{u}_8]\,. \tag{6.86}$$

From (6.70), after the substitution of (6.76) and (6.86), and from (6.82), taking (6.85) and (6.79) into account the state equations of the network are

$$\begin{bmatrix}\dot{\mathbf{i}}_3\\ \dot{\mathbf{u}}_6\end{bmatrix}=\begin{bmatrix}L^{-1} & 0\\ 0 & C^{-1}\end{bmatrix}\left\{\begin{bmatrix}D_{11} & D_{12}\\ D_{21} & D_{22}\end{bmatrix}\begin{bmatrix}\mathbf{i}_3\\ \mathbf{u}_6\end{bmatrix}+\begin{bmatrix}E_{11} & E_{12}\\ E_{21} & E_{22}\end{bmatrix}\begin{bmatrix}\mathbf{i}_1\\ \mathbf{u}_8\end{bmatrix}\right\}, \tag{6.87}$$

where

$$D_{11}=F_{31}[1+R_5(F_{41}^{+}-F_{21}^{+}{F_{23}^{+}}^{-1}F_{43}^{+})G_4F_{41}]^{-1}R_5(F_{21}^{+}{F_{23}^{+}}^{-1}F_{33}^{+}-F_{31}^{+}), \tag{6.88}$$

$$D_{12}=F_{31}[1+R_5(F_{41}^{+}-F_{21}^{+}{F_{23}^{+}}^{-1}F_{43}^{+})G_4F_{41}]^{-1}R_5(F_{41}^{+}-$$
$$-F_{21}^{+}{F_{23}^{+}}^{-1}F_{43}^{+})G_4F_{42}-F_{32}, \tag{6.89}$$

$$D_{21}=F_{32}^{+}-F_{22}^{+}{F_{23}^{+}}^{-1}F_{33}^{+}+(F_{42}^{+}-F_{22}^{+}{F_{23}^{+}}^{-1}F_{43}^{+})[1+G_4F_{41}R_5(F_{41}^{+}-$$
$$-F_{21}^{+}{F_{23}^{+}}^{-1}F_{43}^{+})]^{-1}G_4F_{41}R_5(F_{21}^{+}{F_{23}^{+}}^{-1}F_{33}^{+}-F_{31}^{+}), \tag{6.90}$$

$$D_{22}=(F_{22}^{+}{F_{23}^{+}}^{-1}F_{43}^{+}-F_{42}^{+})[1+G_4F_{41}R_5(F_{41}^{+}-F_{21}^{+}{F_{23}^{+}}^{-1}F_{43}^{+})]^{-1}G_4F_{42}, \tag{6.91}$$

$$E_{11}=F_{31}[1+R_5(F_{41}^{+}-F_{21}^{+}{F_{23}^{+}}^{-1}F_{43}^{+})G_4F_{41}]^{-1}R_5(F_{21}^{+}{F_{23}^{+}}^{-1}F_{13}^{+}-F_{11}^{+}), \tag{6.92}$$

$$E_{12}=F_{31}[1+R_5(F_{41}^{+}-F_{21}^{+}{F_{23}^{+}}^{-1}F_{43}^{+})G_4F_{41}]^{-1}R_5(F_{41}^{+}-$$
$$-F_{21}^{+}{F_{23}^{+}}^{-1}F_{43}^{+})G_4F_{44}-F_{34}, \tag{6.93}$$

$$E_{21}=F_{12}^{+}-F_{22}^{+}{F_{23}^{+}}^{-1}F_{13}^{+}+(F_{42}^{+}-F_{22}^{+}{F_{23}^{+}}^{-1}F_{43}^{+})[1+G_4F_{41}R_5(F_{41}^{+}-$$
$$-F_{21}^{+}{F_{23}^{+}}^{-1}F_{43}^{+})]^{-1}G_4F_{41}R_5(F_{21}^{+}{F_{23}^{+}}^{-1}F_{13}^{+}-F_{11}^{+}), \tag{6.94}$$

$$E_{22}=(F_{22}^{+}{F_{23}^{+}}^{-1}F_{43}^{+}-F_{42}^{+})[1+G_4F_{41}R_5(F_{41}^{+}-F_{21}^{+}{F_{23}^{+}}^{-1}F_{43}^{+})]^{-1}G_4F_{44}. \tag{6.95}$$

Knowing the state variables, (6.85) and (6.86) yield $\mathbf{i}_4$ and $\mathbf{u}_5$, and from these $\mathbf{i}_5$, $\mathbf{u}_1$, $\mathbf{u}_2$, $\mathbf{u}_3$, $\mathbf{u}_4$, $\mathbf{i}_2$, $\mathbf{i}_6$ and $\mathbf{i}_8$ are calculated according to (6.81), (6.68), (6.69), (6.70), (6.71), (6.80), (6.73) and (6.75), respectively.

The solution of state equations

State equations constitute a set of first-order differential equations. Several methods are known for the solution of such sets of equations. The following analytical method is applicable to the case of linear equations.

If D is a square matrix, the general solution of state equation (6.1) is known to be the sum of the general solution of the homogeneous set (6.2), and a particular solution of (6.1). The solution of the homogeneous set of equations (6.2) may be written as*

$$\mathbf{X}=e^{Dt}\mathbf{X}_0, \tag{6.96}$$

* The interpretation of the matrix function appearing here has been discussed in Chapter 4.

where $\mathbf{X}_0 = \mathbf{X}(t=0)$. The time derivative of (6.96) is

$$\dot{\mathbf{X}} = \boldsymbol{D}e^{\boldsymbol{D}t}\mathbf{X}_0 . \tag{6.97}$$

Substituting this and (6.96) into (6.2), (6.96) is indeed seen to solve the homogeneous equation.

As has been shown, the calculation of matrix function (6.96) requires the determination of the eigenvalues $\lambda_i (i=1, 2, \ldots, q)$ of matrix $\boldsymbol{D}$. The eigenvalues are the solutions of Eq. (4.104) in chapter 4, which is the characteristic equation

$$\det | \boldsymbol{D} - \lambda_i \boldsymbol{1} | = 0 . \tag{6.98}$$

Knowing these roots, matrix Lagrange-polynomials

$$\boldsymbol{L}_k = \prod_{\substack{j=1 \\ j \neq k}}^{q} \frac{\boldsymbol{D} - \lambda_j \boldsymbol{1}}{\lambda_k - \lambda_j} \tag{6.99}$$

can be written in accordance with (4.108) of chapter 4, and hence (4.115) yields the form

$$\boldsymbol{X} = e^{\boldsymbol{D}t}\boldsymbol{X}_0 = \sum_{k=1}^{q} \boldsymbol{L}_k e^{\lambda_k t}\boldsymbol{X}_0 \tag{6.100}$$

for the solution of the homogeneous equation. The eigenvalues λ_k $(k=1, 2, \ldots, q)$ of the matrix $\boldsymbol{D}$ appearing in the state equations of the network are quantities characterizing the network. If λ_k is real, its reciprocal multiplied by -1 is a time-constant of the network. If λ_k is a complex quantity, the reciprocal of its real part multiplied by -1 is a time-constant, while its imaginary part is the angular frequency of a free oscillation of the network:*

$$\lambda_k = -\frac{1}{T_k} \pm j\omega_k . \tag{6.101}$$

A particular solution of the inhomogeneous equation can be obtained by several methods, for example, by the method of variation of parameters. Varying the coefficient $\mathbf{X}_0$ appearing in the solution of the homogeneous equation, a solution of the inhomogeneous equation is sought in the form

$$\mathbf{x}_1(t) = e^{\boldsymbol{D}t}\mathbf{X}_0(t) \tag{6.102}$$

Hence:

$$\dot{\mathbf{x}}_1(t) = \boldsymbol{D}e^{\boldsymbol{D}t}\mathbf{X}_0(t) + e^{\boldsymbol{D}t}\dot{\mathbf{X}}_0(t) . \tag{6.103}$$

Substituting into the differential equation (6.1):

$$\boldsymbol{D}e^{\boldsymbol{D}t}\mathbf{X}_0(t) + e^{\boldsymbol{D}t}\dot{\mathbf{X}}_0(t) = \boldsymbol{D}e^{\boldsymbol{D}t}\mathbf{X}_0(t) + \boldsymbol{E}\mathbf{r}(t) , \tag{6.104}$$

yielding

$$\dot{\mathbf{X}}_0(t) = e^{-\boldsymbol{D}t}\boldsymbol{E}\mathbf{r}(t) , \tag{6.105}$$

* In other words, the λ_k are the complex frequencies of the natural modes of the network.

i.e.

$$\mathbf{X}_0(t) = \int_0^t e^{-\boldsymbol{D}\tau} \boldsymbol{E}\mathbf{r}(\tau)\,\mathrm{d}\tau = \int_0^t \sum_{k=1}^{q} \boldsymbol{L}_k e^{-\lambda_k \tau} \boldsymbol{E}\mathbf{r}(\tau)\,\mathrm{d}\tau\,, \tag{6.106}$$

and

$$\mathbf{x}_1(t) = e^{\boldsymbol{D}t} \int_0^t e^{-\boldsymbol{D}\tau} \boldsymbol{E}\mathbf{r}(\tau)\,\mathrm{d}\tau\,. \tag{6.107}$$

Thus, the solution of the state equations is of the form

$$\mathbf{X}(t) = e^{\boldsymbol{D}t}\mathbf{X}_0 + e^{\boldsymbol{D}t} \int_0^t e^{-\boldsymbol{D}\tau} \boldsymbol{E}\mathbf{r}(\tau)\,\mathrm{d}\tau\,. \tag{6.108}$$

The constant $\mathbf{X}_0$ appearing here can be determined from initial conditions. Namely, for $t=0$:

$$\mathbf{X}(0) = \mathbf{X}_0\,. \tag{6.109}$$

Substituting this into (6.108), the state vector has been determined.

Examples

1. In the network shown in Fig. 6.2, the voltage-source supplies a voltage U_0, constant in time. With the switch open, the network has reached a steady state, and the switch is then closed at $t=0$. The voltages of the two capacitors will be calculated by the solution of the state equation of the network. The initial conditions may be determined by inspection from the circuit: at $t=0$, $u_3(0)=u_4(0)=U_0/2$.

To write the state equations, the tree shown in Fig. 6.3, b of the network graph (Fig. 6.3, a) is selected. There are no current-sources and inductors in the network, so for the application of the method presented to write the state equations of networks without controlled sources, groups 1 and 3 of the branches contain no edge. Branches 3 and 4 belong the group 4, branch 5 to group 5, and branch 6 to group 6. Since these form a tree of the graph, the branches corresponding to the resistors $R_1 = 1/G_1$ and $R_2 = 1/G_2$, are classified into the second group. The matrix of the fundamental set of loops generated by the tree thus chosen is:

$$\boldsymbol{B} = \left[\begin{array}{cc|cc|c|c} 1 & 0 & 1 & 0 & 1 & -1 \\ 0 & 1 & 0 & 1 & -1 & 0 \end{array}\right],$$

i.e. $\boldsymbol{F}_{11}, \boldsymbol{F}_{12}, \boldsymbol{F}_{13}, \boldsymbol{F}_{31}, \boldsymbol{F}_{32}, \boldsymbol{F}_{33}$ do not exist, and

$$\boldsymbol{F}_{21} = \begin{bmatrix} 1 & 0 \\ 0 & 1 \end{bmatrix}, \quad \boldsymbol{F}_{22} = \begin{bmatrix} 1 \\ -1 \end{bmatrix}, \quad \boldsymbol{F}_{23} = \begin{bmatrix} -1 \\ 0 \end{bmatrix}.$$

The additional matrices necessary for the calculation are:

$$\boldsymbol{G} = \langle G_1 \;\; G_2 \rangle\,,$$

$$\boldsymbol{C} = \langle C \;\; C \rangle\,,$$

$$\mathbf{u}_0 = U_0\,.$$

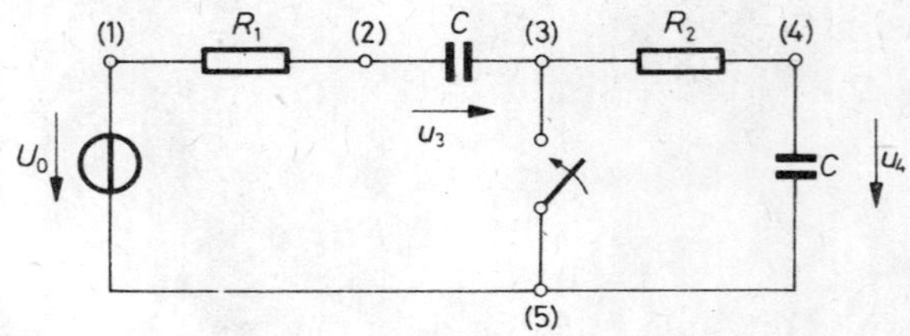

Fig. 6.2

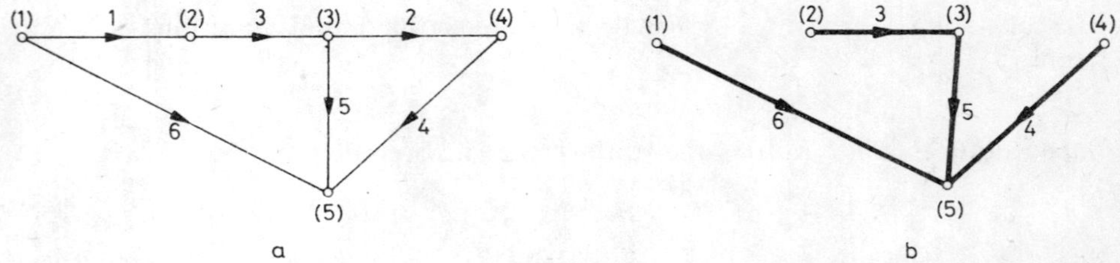

Fig. 6.3

The matrix $\boldsymbol{R}$ has no elements. Hence, the coefficients appearing in the state equations (6.32) are

$$\boldsymbol{C}^{-1} = <\frac{1}{C}\ \frac{1}{C}>,$$

$$\boldsymbol{D}_{22} = -\begin{bmatrix} G_1 & 0 \\ 0 & G_2 \end{bmatrix},$$

$$\boldsymbol{E}_{22} = -\begin{bmatrix} G_1 & 0 \\ 0 & G_2 \end{bmatrix}\begin{bmatrix} -1 \\ 0 \end{bmatrix} = \begin{bmatrix} G_1 \\ 0 \end{bmatrix}.$$

Submatrices $\boldsymbol{D}_{11}$, $\boldsymbol{D}_{12}$, $\boldsymbol{D}_{21}$, $\boldsymbol{E}_{11}$, $\boldsymbol{E}_{12}$, $\boldsymbol{E}_{21}$ do not exist, i.e.

$$\dot{\mathbf{u}}_C = \begin{bmatrix} \dot{u}_3 \\ \dot{u}_4 \end{bmatrix} = -\frac{1}{C}\left\{\begin{bmatrix} G_1 & 0 \\ 0 & G_2 \end{bmatrix}\begin{bmatrix} u_3 \\ u_4 \end{bmatrix} - \begin{bmatrix} G_1 \\ 0 \end{bmatrix} U_0\right\}.$$

Its solution, in accordance with (6.108), is

$$\mathbf{u}_C = \begin{bmatrix} u_3 \\ u_4 \end{bmatrix} = e^{\boldsymbol{D}t}\mathbf{X}_0 + e^{\boldsymbol{D}t}\int_0^t e^{-\boldsymbol{D}t}\boldsymbol{E}\mathbf{r}(\tau)\,\mathrm{d}\tau,$$

where

$$\boldsymbol{D} = \begin{bmatrix} -G_1/C & 0 \\ 0 & -G_2/C \end{bmatrix},$$

from the initial conditions:

$$\mathbf{X}_0=\begin{bmatrix}u_3(0)\\u_4(0)\end{bmatrix}=\begin{bmatrix}1\\1\end{bmatrix}U_0/2,$$

and

$$\boldsymbol{E}=\begin{bmatrix}G_1/C\\0\end{bmatrix},$$

$$\mathbf{r}=U_0.$$

The eigenvalues of $\boldsymbol{D}$ are the solutions of the equation

$$\left(-\frac{G_1}{C}-\lambda\right)\left(-\frac{G_2}{C}-\lambda\right)=0,$$

i.e.

$$\lambda_1=-\frac{G_1}{C},\qquad \lambda_2=-\frac{G_2}{C}.$$

Hence, the Lagrange-polynomials of matrix $\boldsymbol{D}$, according to (4.108) of Chapter 4 are:

$$\boldsymbol{L}_1=\frac{\boldsymbol{D}-\lambda_2\boldsymbol{1}}{\lambda_1-\lambda_2}=\begin{bmatrix}1&0\\0&0\end{bmatrix},$$

$$\boldsymbol{L}_2=\frac{\boldsymbol{D}-\lambda_1\boldsymbol{1}}{\lambda_2-\lambda_1}=\begin{bmatrix}0&0\\0&1\end{bmatrix}.$$

Thus

$$e^{\boldsymbol{D}t}=\boldsymbol{L}_1e^{\lambda_1t}+\boldsymbol{L}_2e^{\lambda_2t}=\begin{bmatrix}e^{\lambda_1t}&0\\0&e^{\lambda_2t}\end{bmatrix}.$$

The solution of the state equations, in accordance with (6.108), is:

$$\mathbf{X}(t)=e^{\boldsymbol{D}t}\int_0^t e^{-\boldsymbol{D}\tau}\boldsymbol{E}\mathbf{r}(\tau)\,\mathrm{d}\tau+e^{\boldsymbol{D}t}\mathbf{X}_0=$$

$$=U_0\begin{bmatrix}e^{\lambda_1t}&0\\0&e^{\lambda_2t}\end{bmatrix}\int_0^t\begin{bmatrix}e^{-\lambda_1\tau}&0\\0&e^{-\lambda_2\tau}\end{bmatrix}\begin{bmatrix}G_1/C\\0\end{bmatrix}\mathrm{d}\tau+$$

$$+\frac{U_0}{2}\begin{bmatrix}e^{\lambda_1t}&0\\0&e^{\lambda_2t}\end{bmatrix}\begin{bmatrix}1\\1\end{bmatrix}=U_0\begin{bmatrix}e^{\lambda_1t}&0\\0&e^{\lambda_2t}\end{bmatrix}\begin{bmatrix}e^{-\lambda_1t}-1\\0\end{bmatrix}+$$

$$+\frac{U_0}{2}\begin{bmatrix}e^{\lambda_1t}\\e^{\lambda_2t}\end{bmatrix}=\frac{U_0}{2}\begin{bmatrix}2-e^{-\frac{G_1}{C}t}\\e^{-\frac{G_2}{C}t}\end{bmatrix}.$$

2. The circuit diagram of a tuned differential amplifier is shown in Fig. 6.4. The output voltage $u_0(t)$ is to be determined in terms of the input voltages $u_{g1}(t)$ and

$u_{g2}(t)$ with the aid of the state equations. The model of the amplifier drawn in Fig. 6.5, a is employed to this end, which corresponds to an ideal amplifier. The transistors are characterized by admittance parameters (assuming $y_{11}=y_{12}=y_{22}=0$). In case of small signal analysis, DC supply sources can be substituted by short-circuits. Since no current flows on the wire connecting the center top of the inductor with the common point of the controlled current-sources, this branch may be omitted from the model. Thus, the equivalent circuit shown in Fig. 6.5, b is obtained, with its graph drawn in Fig. 6.6.

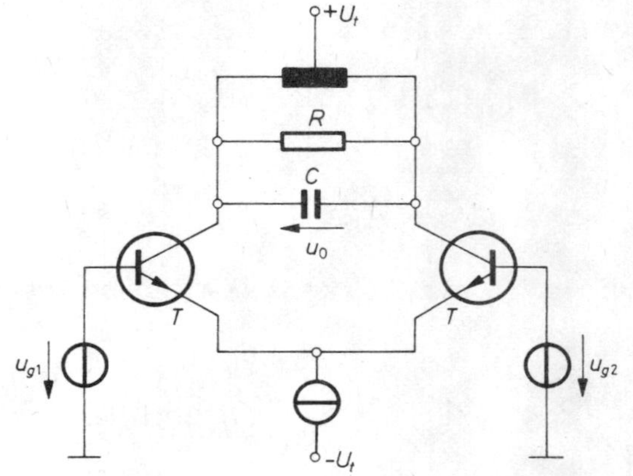

Fig. 6.4

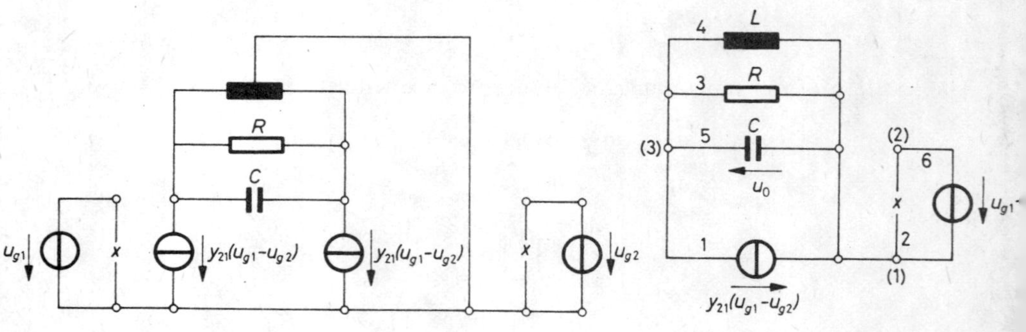

Fig. 6.5

The branches are classified as follows:

No branch belongs to groups 1 and 5. Branches 1, 2, 3 belong to group 2, branch 4 to group 3, branch 5 to group 4, and group 6 is formed by branch 6. The matrix of the fundamental set of loops generated by the tree chosen is

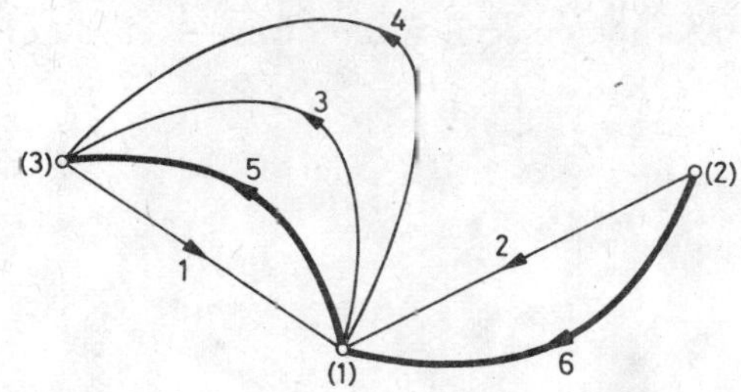

Fig. 6.6

$$B=\left[\begin{array}{ccc|c|c|c} 1 & 0 & 0 & 0 & 1 & 0 \\ 0 & 1 & 0 & 0 & 0 & -1 \\ 0 & 0 & 1 & 0 & -1 & 0 \\ \hline 0 & 0 & 0 & 1 & -1 & 0 \end{array}\right].$$

i.e. $\boldsymbol{F}_{11}$, $\boldsymbol{F}_{12}$, $\boldsymbol{F}_{13}$, $\boldsymbol{F}_{22}$, $\boldsymbol{F}_{32}$ do not exist, and

$$\boldsymbol{F}_{21}=\begin{bmatrix} 1 \\ 0 \\ -1 \end{bmatrix}, \qquad \boldsymbol{F}_{23}=\begin{bmatrix} 0 \\ -1 \\ 0 \end{bmatrix},$$

$$\boldsymbol{F}_{31}=-1, \qquad \boldsymbol{F}_{33}=0.$$

The further matrices characterizing the network are

$$\boldsymbol{G}_2=\begin{bmatrix} 0 & y_{21} & 0 \\ 0 & 0 & 0 \\ 0 & 0 & 1/R \end{bmatrix}$$

$$\boldsymbol{L}=L, \qquad \boldsymbol{C}=C.$$

Hence, according to (6.58)...(6.65):

$$\boldsymbol{D}_{11}=0, \qquad \boldsymbol{D}_{12}=1, \qquad \boldsymbol{D}_{21}=-1, \qquad \boldsymbol{D}_{22}=-1/R$$

$$\boldsymbol{E}_{12}=0, \qquad \boldsymbol{E}_{22}=y_{21}.$$

$\boldsymbol{E}_{11}$ and $\boldsymbol{E}_{21}$ do not exist there being no independent current-source in the network.

According to the above formulas the state equations of the network are:

$$\begin{bmatrix} \dot{i}_4 \\ \dot{u}_5 \end{bmatrix}=\begin{bmatrix} 0 & 1/L \\ -1/C & -1/(CR) \end{bmatrix}\begin{bmatrix} i_4 \\ u_5 \end{bmatrix}+\begin{bmatrix} 0 \\ y_{21}/C \end{bmatrix}(u_{g1}-u_{g2}).$$

The solution necessitates the determination of the eigenvalues of the matrix

$$\boldsymbol{D}=\begin{bmatrix} 0 & 1/L \\ -1/C & -1/(CR) \end{bmatrix},$$

that is the roots of the equation

$$\lambda(1/(CR)+\lambda)+1/(CL)=0.$$

These are:

$$\lambda_1=-1/(2CR)+\sqrt{1/(2CR)^2-1/(CL)}$$

$$\lambda_2=-1/(2CR)-\sqrt{1/(2CR)^2-1/(CL)}.$$

The corresponding matrix Lagrange polynomials are:

$$\boldsymbol{L}_1=\frac{1}{\lambda_1-\lambda_2}\begin{bmatrix} -\lambda_2 & 1/L \\ -1/C & -1/(CR)-\lambda_2 \end{bmatrix}$$

$$\boldsymbol{L}_2=\frac{1}{\lambda_2-\lambda_1}\begin{bmatrix} -\lambda_1 & 1/L \\ -1/C & -1/(CR)-\lambda_1 \end{bmatrix}.$$

The initial values of the state variables are:

$$\mathbf{X}_0=\begin{bmatrix} i_4(0) \\ u_5(0) \end{bmatrix}.$$

Hence

$$e^{\boldsymbol{D}t}\mathbf{X}_0=\frac{1}{\lambda_1-\lambda_2}\left\{\begin{bmatrix} -\lambda_2 & 1/L \\ -1/C & -1/(CR)-\lambda_2 \end{bmatrix}e^{\lambda_1 t}-\right.$$

$$\left.-\begin{bmatrix} -\lambda_1 & 1/L \\ -1/C & -1/(CR)-\lambda_1 \end{bmatrix}e^{\lambda_2 t}\right\}\begin{bmatrix} i_4(0) \\ u_5(0) \end{bmatrix}.$$

The second term on the right-hand side of the state equation is:

$$\mathbf{E}\mathbf{r}(t)=\begin{bmatrix} 0 \\ y_{21}/C \end{bmatrix}(u_{g1}-u_{g2})=y_{21}/C\begin{bmatrix} 0 & 0 \\ 1 & -1 \end{bmatrix}\begin{bmatrix} u_{g1}(t) \\ u_{g2}(t) \end{bmatrix}.$$

Thus, the solution (6.108) can be written as the function of $u_{g1}(t)$ and $u_{g2}(t)$, and $u_5(t)=u_0(t)$ is the sought voltage.

3. The oscillation frequency of the oscillator containing operational amplifiers shown in Fig. 6.7 can be determined with the aid of the state equation. To write the state equations, the operational amplifiers are modelled by nullors (Fig. 6.8). The graph of the network thus obtained has been drawn in Fig. 6.9.

At the classification of edges no branch is seen to belong to groups 1, 3, and 8. Branches 1, 2, 3 belong to group 2, branches 4, 5, 6, 7, 8 to group 4, branch 9 to group 5, branches 10, 11 to group 6, and branches 12, 13, 14 form group 7. The branches of

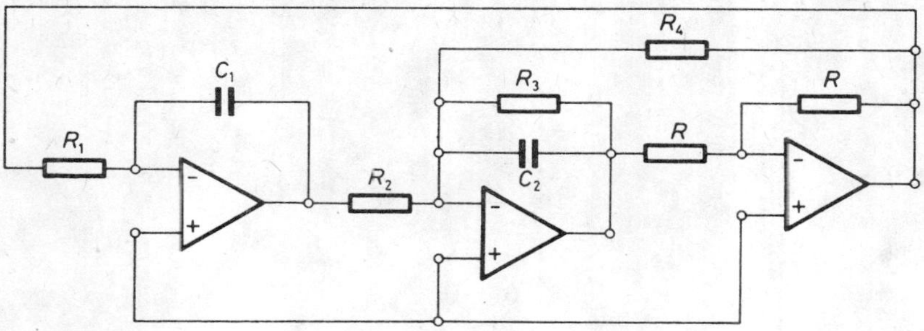

Fig. 6.7

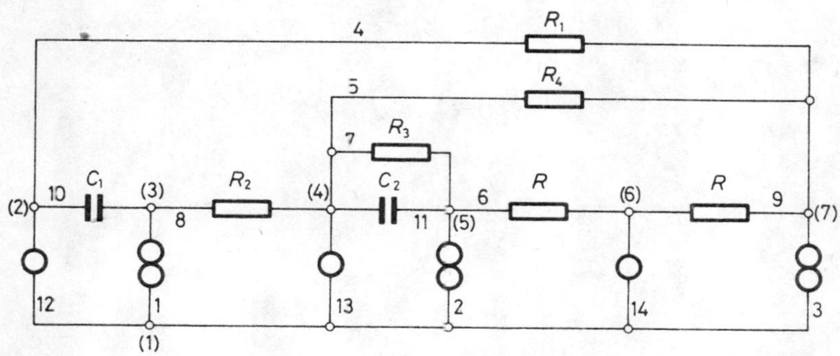

Fig. 6.8

the tree thus obtained have been indicated by thick lines in Fig. 6.9. The matrix of the fundamental set of loops generated by the tree is

$$
\boldsymbol{B}=\left[\begin{array}{ccc|cccc|c|c|cc|ccc}
1 & 0 & 0 & 0 & 0 & 0 & 0 & 0 & 0 & 1 & 0 & -1 & 0 & 0 \\
0 & 1 & 0 & 0 & 0 & 0 & 0 & 0 & 0 & 0 & 1 & 0 & -1 & 0 \\
0 & 0 & 1 & 0 & 0 & 0 & 0 & 0 & 1 & 0 & 0 & 0 & 0 & -1 \\
\hline
0 & 0 & 0 & 1 & 0 & 0 & 0 & 0 & 1 & 0 & 0 & 1 & 0 & -1 \\
0 & 0 & 0 & 0 & 1 & 0 & 0 & 0 & -1 & 0 & 0 & 0 & -1 & 1 \\
0 & 0 & 0 & 0 & 0 & 1 & 0 & 0 & 0 & 0 & -1 & 0 & 1 & -1 \\
0 & 0 & 0 & 0 & 0 & 0 & 1 & 0 & 0 & 0 & -1 & 0 & 0 & 0 \\
0 & 0 & 0 & 0 & 0 & 0 & 0 & 1 & 0 & 1 & 0 & -1 & 1 & 0
\end{array}\right]
$$

i.e. $\boldsymbol{F}_{11}$, $\boldsymbol{F}_{12}$, $\boldsymbol{F}_{13}$, $\boldsymbol{F}_{14}$, $\boldsymbol{F}_{24}$, $\boldsymbol{F}_{31}$, $\boldsymbol{F}_{32}$, $\boldsymbol{F}_{33}$, $\boldsymbol{F}_{34}$, $\boldsymbol{F}_{44}$ do not exist, and

$$\boldsymbol{F}_{21}=\begin{bmatrix}0\\0\\1\end{bmatrix},\qquad \boldsymbol{F}_{22}=\begin{bmatrix}1&0\\0&1\\0&0\end{bmatrix},\qquad \boldsymbol{F}_{23}=\begin{bmatrix}-1&0&0\\0&-1&0\\0&0&-1\end{bmatrix},$$

$$\boldsymbol{F}_{41}=\begin{bmatrix}1\\-1\\0\\0\\0\end{bmatrix},\qquad \boldsymbol{F}_{42}=\begin{bmatrix}0&0\\0&0\\0&-1\\0&-1\\1&0\end{bmatrix},\qquad \boldsymbol{F}_{43}=\begin{bmatrix}1&0&-1\\0&-1&1\\0&1&-1\\0&0&0\\-1&1&0\end{bmatrix}.$$

The further matrices characterizing the network are:

$$\boldsymbol{G}_4=\langle G_1\ \ G_4\ \ G\ \ G_3\ \ G_2\rangle,$$

$$\boldsymbol{R}_5=R,\qquad \boldsymbol{C}=\langle C_1\ \ C_2\rangle,$$

where the notation $G=1/R$, $G_i=1/R_i$ $(i=1, 2, 3, 4)$ has been employed.

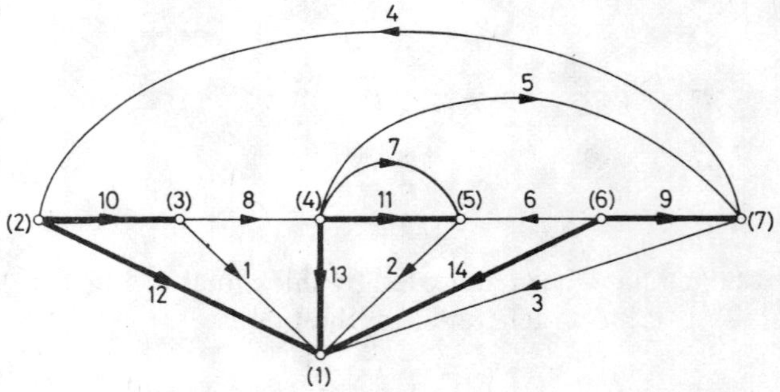

Fig. 6.9

Since there is no inductor in the network, $\boldsymbol{D}_{11}$, $\boldsymbol{D}_{12}$, $\boldsymbol{D}_{21}$ do not exist. On the basis of (6.91):

$$\boldsymbol{D}_{22}=\begin{bmatrix}0&G_1\\-G_2&-G_3+G_4\end{bmatrix}$$

and thus, according to (6.87):

$$\dot{\boldsymbol{u}}_6=\begin{bmatrix}\dot{u}_{10}\\\dot{u}_{11}\end{bmatrix}=\begin{bmatrix}1/C_1&0\\0&1/C_2\end{bmatrix}\begin{bmatrix}0&G_1\\-G_2&-G_3+G_4\end{bmatrix}\begin{bmatrix}u_{10}\\u_{11}\end{bmatrix}$$

i.e.

$$\begin{bmatrix} \dot{u}_{10} \\ \dot{u}_{11} \end{bmatrix} = \boldsymbol{D} \begin{bmatrix} u_{10} \\ u_{11} \end{bmatrix} = \begin{bmatrix} 0 & G_1/C_1 \\ -G_2/C_2 & (G_4-G_3)/C_2 \end{bmatrix} \begin{bmatrix} u_{10} \\ u_{11} \end{bmatrix}.$$

The eigenvalues of matrix $\boldsymbol{D}$, and thus those of the network are given by the solutions of the equation

$$\lambda\left(\frac{G_3-G_4}{C_2}+\lambda\right)+\frac{G_1 G_2}{C_1 C_2}=0.$$

These are

$$\lambda_{12}=\frac{G_4-G_3}{2C_2}\pm\sqrt{\left(\frac{G_4-G_3}{2C_2}\right)^2-\frac{G_1 G_2}{C_1 C_2}}=-\frac{1}{T}\pm j\omega.$$

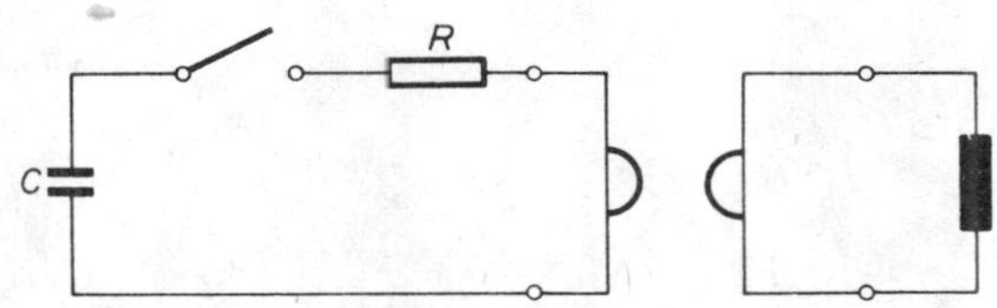

Fig. 6.10

The solution is imaginary if $G_3=G_4$, i.e. $R_3=R_4$. In this case the oscillation angular frequency of the network is $\omega=1/\sqrt{R_1 R_2 C_1 C_2}$.

4. A capacitor, of capacitance C, initially charged to voltage U_0 is connected to a gyrator terminated by an inductor by closing a switch at $t=0$ (Fig. 6.10). The inductor current is assumed to be initially zero. The capacitor is discharged so producing a time-varying current and voltage in the resistor R and inductor L.

The time-variation of the inductor current and the capacitor voltage will be determined by solving the state equations of the network. In order to write the state equations, the gyrator is modelled by a circuit containing nullators, norators and resistors (see Table 5.9, Chapter 5) and thus, the circuit shown in Fig. 6.11 is obtained for the time after closing the switch ($t>0$), its graph being drawn in Fig. 6.12, a.

No branch of the network belongs to group 1 and 8. Branches 1, 2, 3 are to be assigned to group 2, branch 4 to group 3, branch 5 to group 4, branch 8 to group 6, and branches 9, 10, 11 to group 7. The remaining branches, containing resistors, i.e. branches 6 and 7, should be included in group 5 to obtain a tree of the graph (Fig. 6.12, b) along with the branches of group 6 and 7. The matrix of the fundamental set of loops generated by this tree is:

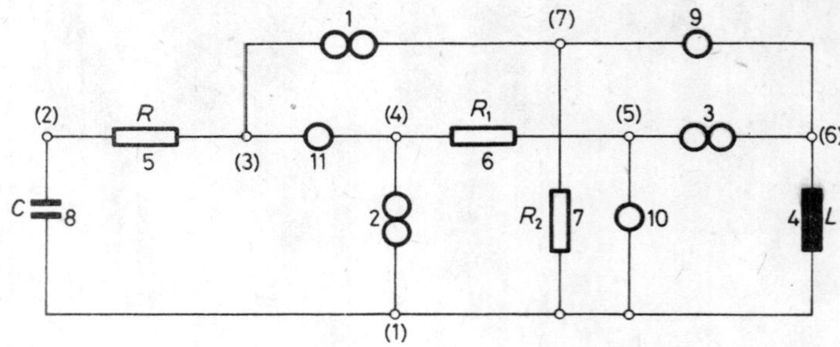

Fig. 6.11

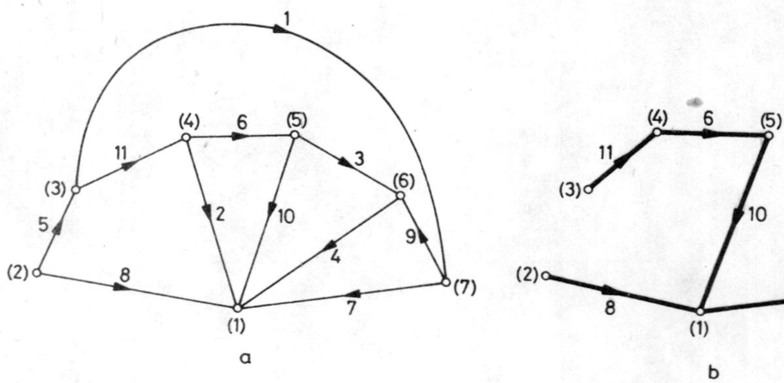

Fig. 6.12

$$
\boldsymbol{B}=\left[\begin{array}{ccc|c|c|cc|c|ccc}
1 & 0 & 0 & 0 & 0 & -1 & 1 & 0 & 0 & -1 & -1 \\
0 & 1 & 0 & 0 & 0 & -1 & 0 & 0 & 0 & -1 & 0 \\
0 & 0 & 1 & 0 & 0 & 0 & 1 & 0 & -1 & -1 & 0 \\
\hline
0 & 0 & 0 & 1 & 0 & 0 & -1 & 0 & 1 & 0 & 0 \\
\hline
0 & 0 & 0 & 0 & 1 & 1 & 0 & -1 & 0 & 1 & 1
\end{array}\right],
$$

i.e. $\boldsymbol{F}_{11}$, $\boldsymbol{F}_{12}$, $\boldsymbol{F}_{13}$, $\boldsymbol{F}_{14}$, $\boldsymbol{F}_{24}$, $\boldsymbol{F}_{34}$, $\boldsymbol{F}_{44}$ do not exist, and

$$
\boldsymbol{F}_{21}=\begin{bmatrix} -1 & 1 \\ -1 & 0 \\ 0 & 1 \end{bmatrix}, \qquad \boldsymbol{F}_{22}=\begin{bmatrix} 0 \\ 0 \\ 0 \end{bmatrix}, \qquad \boldsymbol{F}_{23}=\begin{bmatrix} 0 & -1 & -1 \\ 0 & -1 & 0 \\ -1 & -1 & 0 \end{bmatrix},
$$

$$
\boldsymbol{F}_{31}=[0 \quad -1], \qquad \boldsymbol{F}_{32}=\ 0, \qquad \boldsymbol{F}_{33}=[1 \quad 0 \quad 0],
$$

$$
\boldsymbol{F}_{41}=[1 \quad 0], \qquad \boldsymbol{F}_{42}=-1, \qquad \boldsymbol{F}_{43}=[0 \quad 1 \quad 1].
$$

Further matrices necessary for the analysis are:

$$\boldsymbol{L}=L,$$

$$\boldsymbol{G}_4=\frac{1}{R},$$

$$\boldsymbol{R}_5=\langle R_1 \;\; R_2\rangle,$$

$$\boldsymbol{C}=C.$$

According to the above, (6.88) ... (6.95) yield:

$$\boldsymbol{D}_{11}=[0 \;\; 1]\begin{bmatrix} 1 & 0 \\ -\dfrac{R_2}{R} & 1 \end{bmatrix}\begin{bmatrix} R_1 & 0 \\ 0 & R_2 \end{bmatrix}\begin{bmatrix} -1 \\ 0 \end{bmatrix}=-\frac{R_1 R_2}{R},$$

$$\boldsymbol{D}_{12}=\begin{bmatrix} \dfrac{R_1 R_2}{R} & -R_2 \end{bmatrix}\cdot\begin{bmatrix} 0 \\ 1 \end{bmatrix}\left(-\frac{1}{R}\right)=\frac{R_2}{R},$$

$$\boldsymbol{D}_{21}=-\begin{bmatrix} \dfrac{R_1}{R} & 0 \end{bmatrix}\begin{bmatrix} -1 \\ 0 \end{bmatrix}=\frac{R_1}{R},$$

$$\boldsymbol{D}_{22}=-\frac{1}{R}.$$

There are no sources in the network (no branch belongs to groups 1 and 8), thus $\boldsymbol{E}_{11}$, $\boldsymbol{E}_{12}$, $\boldsymbol{E}_{21}$, $\boldsymbol{E}_{22}$ do not exist.

The state equations of the network are

$$\begin{bmatrix} \dot{i}_4 \\ \dot{u}_8 \end{bmatrix}=\begin{bmatrix} \dfrac{1}{L} & 0 \\ 0 & \dfrac{1}{C} \end{bmatrix}\frac{1}{R}\begin{bmatrix} -R_1 R_2 & R_2 \\ R_1 & -1 \end{bmatrix}\begin{bmatrix} i_4 \\ u_8 \end{bmatrix}.$$

To solve the state equations initial conditions must be known. Before the switch closes, the inductor has no stored energy:

$$i_4(0)=0,$$

while the voltage of the capacitor is

$$u_8(0)=U_0.$$

The eigenvalues of matrix

$$\boldsymbol{D}=\begin{bmatrix} -\dfrac{R_1R_2}{RL} & \dfrac{R_2}{RL} \\ \dfrac{R_1}{RC} & -\dfrac{1}{RC} \end{bmatrix}$$

are obtained by the solution of

$$\left(\frac{R_1R_2}{RL}+\lambda\right)\left(\frac{1}{RC}+\lambda\right)-\frac{R_1R_2}{R^2LC}=0,$$

i.e.

$$\lambda_1=0,$$

$$\lambda_2=-\frac{1}{R}\left(\frac{1}{C}+\frac{R_1R_2}{L}\right).$$

The Lagrange polynomials of matrix $\boldsymbol{D}$ are:

$$\boldsymbol{L}_1(\boldsymbol{D})=\frac{1}{L+CR_1R_2}\begin{bmatrix} L & CR_2 \\ LR_1 & CR_1R_2 \end{bmatrix},$$

$$\boldsymbol{L}_2(\boldsymbol{D})=\frac{1}{L+CR_1R_2}\begin{bmatrix} CR_1R_2 & -CR_2 \\ -LR_1 & L \end{bmatrix}$$

and thus

$$e^{\boldsymbol{D}t}=\boldsymbol{L}_1e^{\lambda_1t}+\boldsymbol{L}_2e^{\lambda_2t}=\frac{1}{L+CR_1R_2}\begin{bmatrix} L+CR_1R_2e^{\lambda_2t} & CR_2(1-e^{\lambda_2t}) \\ LR_1(1-e^{\lambda_2t}) & CR_1R_2+Le^{\lambda_2t} \end{bmatrix}.$$

The solution of the state equations according to (6.108) is:

$$\begin{bmatrix} i_4 \\ u_8 \end{bmatrix}=\frac{1}{L+CR_1R_2}\begin{bmatrix} L+CR_1R_2e^{\lambda_2t} & CR_2(1-e^{\lambda_2t}) \\ LR_1(1-e^{\lambda_2t}) & CR_1R_2+Le^{\lambda_2t} \end{bmatrix}\begin{bmatrix} 0 \\ U_0 \end{bmatrix}=$$

$$=\frac{U_0}{L+CR_1R_2}\begin{bmatrix} CR_2\left(1-e^{-\frac{1}{R}\left(\frac{1}{C}+\frac{R_1R_2}{L}\right)t}\right) \\ CR_1R_2+Le^{-\frac{1}{R}\left(\frac{1}{C}+\frac{R_1R_2}{L}\right)t} \end{bmatrix} \quad \text{for } t\geq 0$$

i_8 can be obtained from u_8 since $i_8=C\dfrac{du_8}{dt}$.

CHAPTER 7

THE APPLICATION OF SIGNAL FLOW GRAPHS TO THE ANALYSIS OF LINEAR SYSTEMS

A system and its signal-flow graph

A commonly-used model of linear systems consists of a connection of basic elements (block-diagram elements). Each basic element has an input and an output, and the output signal V_j of a basic element is given by the relationship

$$V_j = W_k V_i \tag{7.1}$$

where V_i is the input signal, and W_k is the transfer function characterizing the basic element (Fig. 7.1, a). (7.1) expresses the dependence of the output signal of the basic element upon the input signal. However, signal transmission through the element is unidirectional, so that the input does not necessarily depend on the output in the way shown in (7.1), i.e. (7.1) is not in general invertible. (7.1) may be considered to represent a linear causal relationship, in which V_i is the cause and V_j is the effect. In case of signals varying sinusoidally with time, the variables in (7.1) are complex quantities, while Fourier or Laplace transforms may be used to describe signals of general time-variation. The output signal of a particular network part may in reality depend not solely upon the relevant input signals, but also on other signals present in the system. However, in our analysis, for those network parts modelled by basic elements, this dependence upon other signals either does not occur or is negligible, thus permitting the description of the relation between input and output signals by (7.1). If the inputs of several basic elements are connected, their input signals are identical (e.g. the basic elements described by transfer functions W_1 and W_3 in Fig. 7.2), and if the outputs of several basic elements are connected together, the net output is the sum of the output signals of the individual basic elements (e.g. in Fig. 7.2: $V_3 = W_2 V_2 + W_3 V_1$). Such a model may be related to the block diagram of a linear control system. Thus, for example, the model shown in Fig. 7.3, b has been derived directly from the block diagram in Fig. 7.3, a. A model consisting of basic elements can serve for the discussion of all systems described by a set of linear equations.

A directed graph can be associated with the system modelled in the above way as follows. [1, 9, 10, 32, 34, 36, 41, 48.] An edge of the graph corresponds to the transfer function of each basic element, while the two vertices of the edge are associated with the input and output signals of the basic element. The orientation of the edge is directed from the vertex corresponding to the input signal towards that

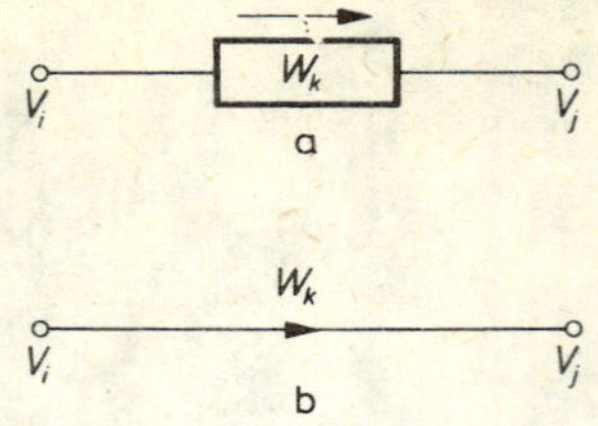

Fig. 7.1

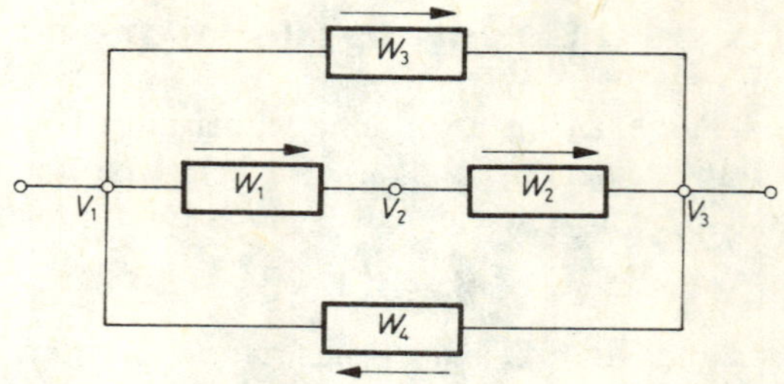

Fig. 7.2

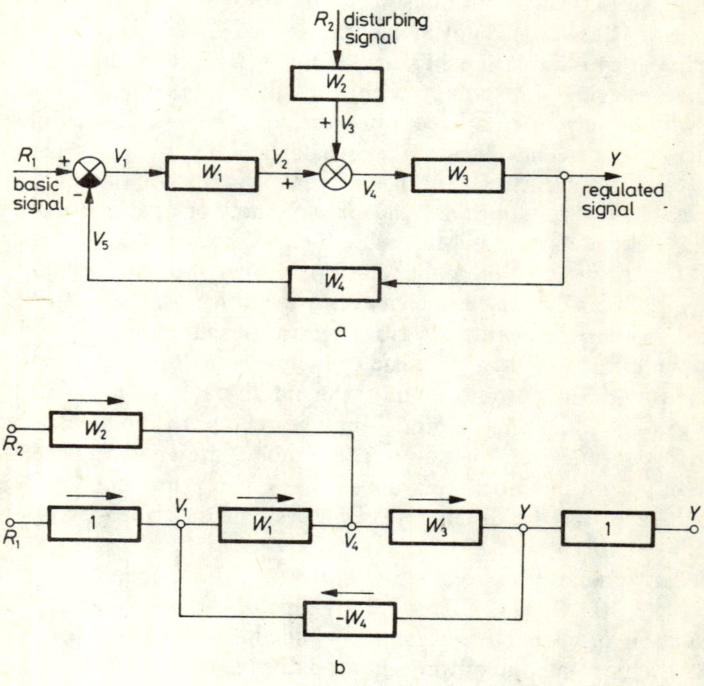

Fig. 7.3

associated with the output signal (Fig. 7.1, b). Thus, relating an edge to each basic element, a signal flow graph (Fig. 7.4, b) of the system (Fig. 7.4, a) is obtained. The input signals of the system, in contrast to the input signals of basic elements, are denoted by R_i $(i=1, 2, \ldots, n_1)$, and called excitation signals. The output signals of the system are response signals, denoted by Y_k $(k=1, 2, \ldots, n_3)$. The signals associated with the remaining vertices, the internal vertices of the signal flow graph, are denoted by V_j $(j=1, 2, \ldots, n_2)$, and are called internal signals.

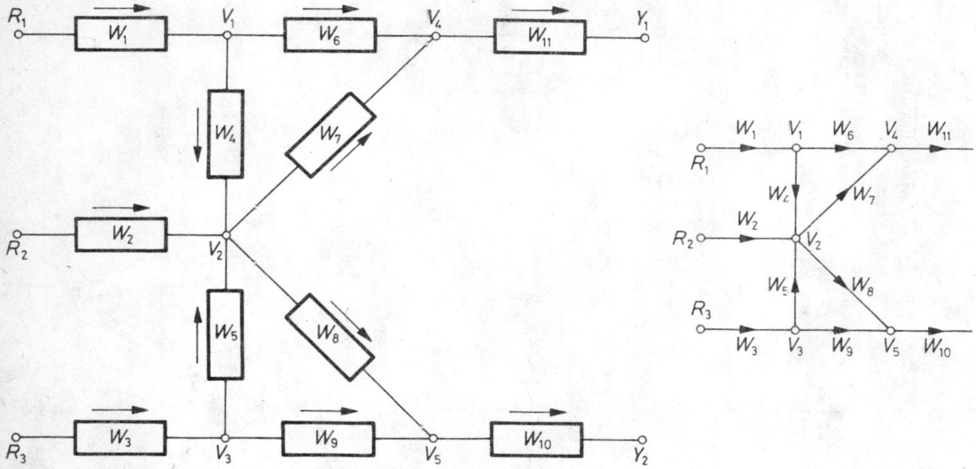

$$Y_1 = W_{11}[W_1(W_6 + W_4W_7)R_1 + W_2W_7R_2 + W_3W_5W_7R_3]$$

$$Y_2 = W_{10}[W_1W_4W_8R_1 + W_2W_8R_2 + W_3(W_9 + W_5W_8)R_3]$$

a b

Fig. 7.4

The linear, time-invariant systems which can be associated with signal flow graphs may represent not only electrical but also e.g. mechanical, thermal or combined processes. In such cases input and output signals may be voltage, current, force, pressure, displacement, temperature, heat, etc.

Signal flow graphs can be transformed or simplified, taking the relationships between the signals associated with the vertices into account. The rows of Table 7.1 show signal flow graphs which are equivalent from the point of view of the relationship between input and output signals. It is seen in row 1, that "series, unidirected" edges, i.e. those corresponding to the chain-connected basic elements of the network, can be replaced by a single edge, with the transfer function associated with the edge equal to the product of the transfer functions of the "series" edges. According to row 2, "unidirected, parallel" edges may be replaced by an edge with its transfer function being the sum of the transfer functions of the "parallel" edges. Rows 3 and 4 show examples for the relocation of an edge.

Table 7.1

Number	Equivalent signal flow graphs	Output signals	Note
1	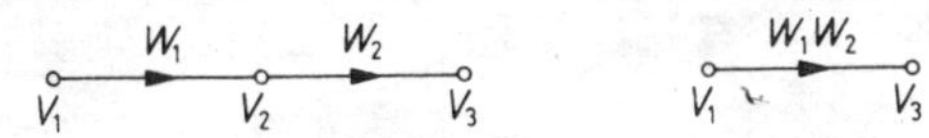	$V_3 = W_1 W_2 V_1$	Chain connection multiplication
2		$V_2 = (W_1 + W_2) V_1$	Parallel connection addition
3	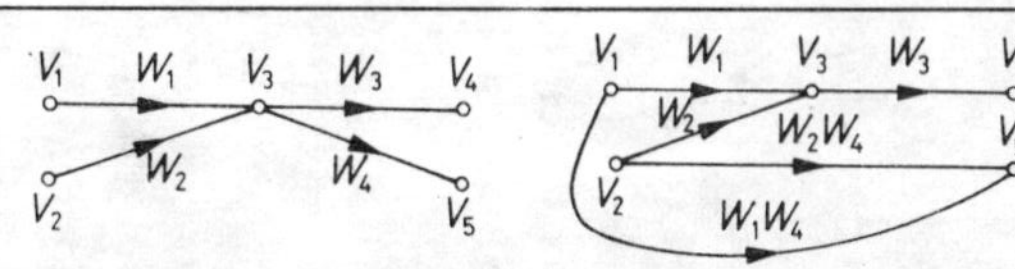	$V_4 = W_3(W_1 V_1 + W_2 V_2)$ $V_5 = W_4(W_1 V_1 + W_2 V_2)$	Replacement of connection point
4	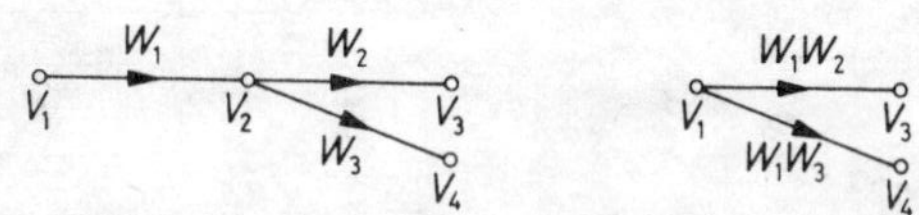	$V_3 = W_1 W_2 V_1$ $V_4 = W_1 W_3 V_1$	Replacement of connection point
5	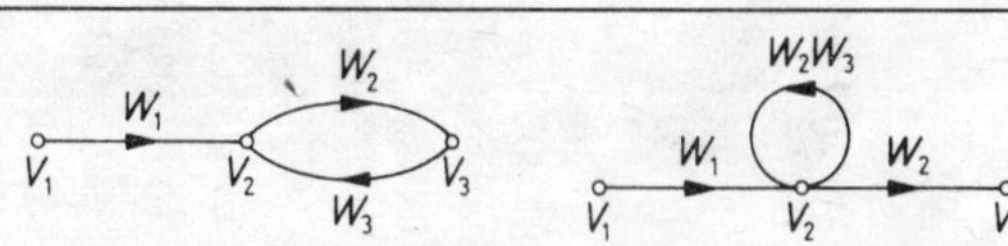	$V_3 = \dfrac{W_1 W_2}{1 - W_2 W_3} V_1$	Feedback
6	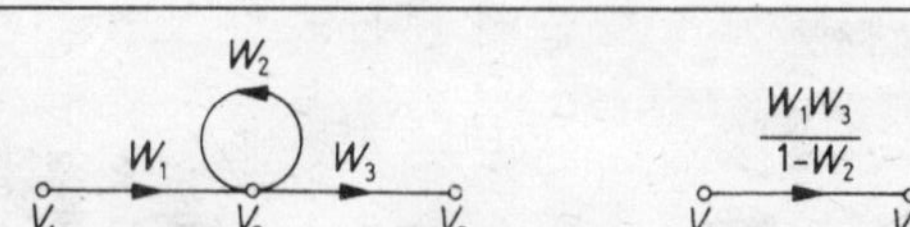	$V_3 = \dfrac{W_1 W_3}{1 - W_2} V_1$	Feedback

Signal flow graphs may include self-loops. A part of such a signal flow graph is shown in Fig. 7.5. In this case $V_3 = W_1 V_1 + W_2 V_2 + W_3 V_3$, and hence $V_3 = (W_1 V_1 + W_2 V_2)/(1 - W_3)$.

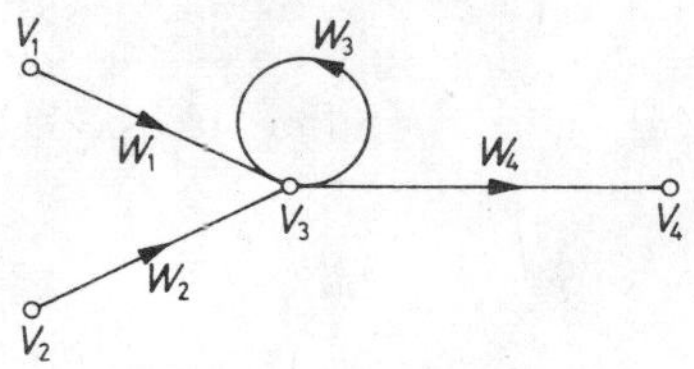

Fig. 7.5

In accordance with row 5 of Table 7.1, "counter-directed, parallel" edges, (i.e. those corresponding to feedback), may be replaced by a self-loop. Row 6 shows a graph containing a self-loop, and one without a self-loop, equivalent to it.

Signal flow graph of networks characterized by a linear set of algebraic equations or by rational transfer functions

A signal flow graph can also be associated with linear networks described by a set of linear algebraic equations, i.e. with networks consisting of impedances as well as dependent and controlled sources. The graph employed for the methods discussed in Chapters 2–6 can be determined from the connections of the networks, and with its aid the network equations may be written and the unknown voltages and currents determined from a knowledge of the impedances, source-voltages and source-currents of the network. The signal flow graph of the network, on the other hand, can only be derived after a suitable number of equations have been written for the network. Therefore, signal flow graphs are of less importance for the analysis of such networks. On the basis of the equations written, signal flow graphs may be associated with the same network in several ways.

Particular subgraphs of a signal flow graph may be derived from the equations relating to the relevant network parts. Thus, for example, the signal flow subgraphs corresponding to the relationships $U = ZI$ or $I = 1/ZU$ of an impedance Z have been drawn in Fig. 7.6, while Table 7.2 shows the signal flow subgraphs associated with a two-port characterized by its various parameters.

The number of linearly independent equations of a linear network which are necessary to determine its signal flow graph equals that required for the solution of the network analysis problem (see Chapter 2). These are the Kirchhoff equations and Ohm's law describing the relationship between branch-currents and branch-voltages. The known excitation signals and the unknown signals are associated with the vertices of the signal flow graph, i.e. the number of vertices is the sum of the number of excitation signals and unknowns. The set of equations is arranged so as to have each equation constitute an expression for one unknown with a different unknown in each equation. The vertex corresponding to the unknown selected is connected by edges with the vertices associated with the excitation signals or the unknowns appearing in the expression for the selected unknown. The edges are directed towards the vertex corresponding to the selected unknown, and the

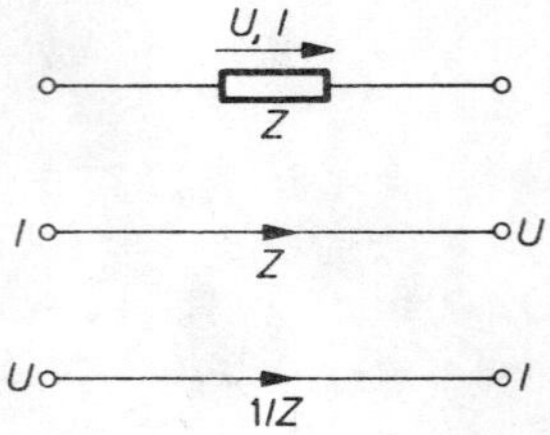

Fig. 7.6

coefficient of the quantity associated with the other vertex of the edge is assigned to the edge as its transfer function. By selecting each unknown in turn, and drawing the corresponding vertices and edges, a signal flow graph of the network is obtained.

Because of the various ways the equations may be written and the unknowns expressed, various alternative signal flow graphs may be associated with the same network.

The transfer functions of linear electrical networks consisting of resistors, inductors and capacitors are rational functions. On this basis, a signal flow graph of the network may be constructed, and, as will be shown later, this may be used to write a set of state equations of the network.

The transfer functions of passive, linear electrical networks are rational functions with the degree of the numerator not greater than that of the denominator. Therefore, only signal flow graphs of such functions will be dealt with. A model of such systems may be given, such that all transfer functions of basic elements are either constant or s^{-1}. The transfer function s^{-1} corresponds in the time domain to an integrating basic element. The construction of such models is discussed in the following paragraphs.

Three methods are customary for the construction of the signal flow graph: the direct, the iterative (also called series) and the parallel method [10, 32, 47]. These yield three different models of the system. The three methods of construction will be illustrated by the construction of the signal flow graphs of the transfer function

$$W(s)=\frac{Y(s)}{R(s)}=\frac{b_1 s+b_2}{s^2+a_1 s+a_2}; \quad b_1 \neq 0. \tag{7.2}$$

For the application of the direct method the numerator and denominator of $W(s)$ are multiplied by s^{-n}, where n is the degree of the denominator, and in our example $n=2$. The numerator and the denominator are further multiplied by a function $P(s)$,

Table 7.2

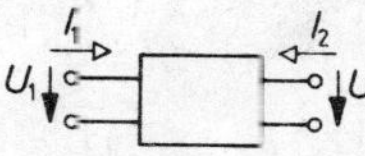

The equation characterizing the two-port	Signal flow graph
$\begin{bmatrix} U_1 \\ U_2 \end{bmatrix} = \begin{bmatrix} z_{11} & z_{12} \\ z_{21} & z_{22} \end{bmatrix} \begin{bmatrix} I_1 \\ I_2 \end{bmatrix}$	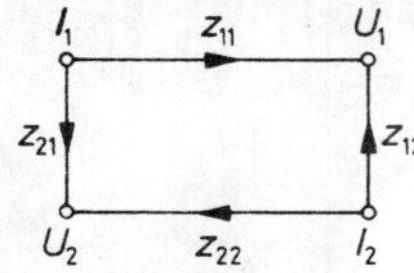
$\begin{bmatrix} I_1 \\ I_2 \end{bmatrix} = \begin{bmatrix} y_{11} & y_{12} \\ y_{21} & y_{22} \end{bmatrix} \begin{bmatrix} U_1 \\ U_2 \end{bmatrix}$	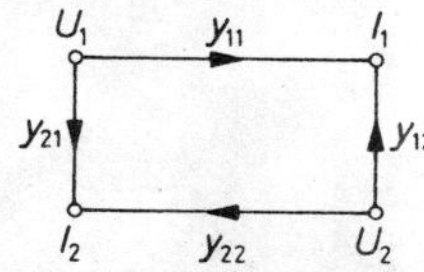
$\begin{bmatrix} U_1 \\ I_2 \end{bmatrix} = \begin{bmatrix} h_{11} & h_{12} \\ h_{21} & h_{22} \end{bmatrix} \begin{bmatrix} I_1 \\ U_2 \end{bmatrix}$	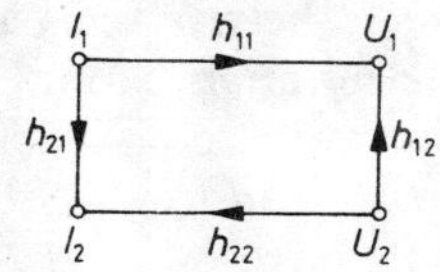
$\begin{bmatrix} I_1 \\ U_2 \end{bmatrix} = \begin{bmatrix} d_{11} & d_{12} \\ d_{21} & d_{22} \end{bmatrix} \begin{bmatrix} U_1 \\ I_2 \end{bmatrix}$	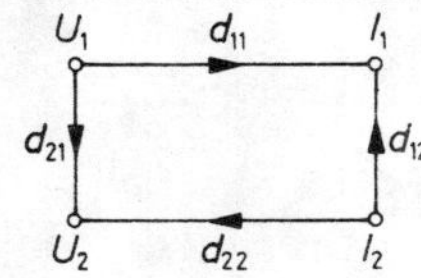
$\begin{bmatrix} U_1 \\ I_1 \end{bmatrix} = \begin{bmatrix} a_{11} & a_{12} \\ a_{21} & a_{22} \end{bmatrix} \begin{bmatrix} U_2 \\ I_2 \end{bmatrix}$	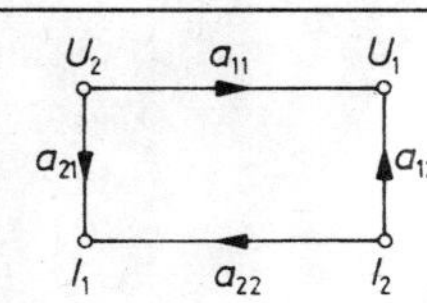

ensuring that the numerator is the Laplace transfoım of the response signal, while the denominator is that of the excitation signal:

$$\frac{Y(s)}{R(s)} = \frac{(b_1 s^{-1} + b_2 s^{-2})\, P(s)}{(1 + a_1 s^{-1} + a_2 s^{-2})\, P(s)} \tag{7.3}$$

and

$$Y(s) = b_1 s^{-1} P(s) + b_2 s^{-2} P(s) \tag{7.4}$$

$$R(s) = P(s) + a_1 s^{-1} P(s) + a_2 s^{-2} P(s)\,. \tag{7.5}$$

this may also be written in the following form:

$$P(s)=R(s)-a_1 s^{-1}P(s)-a_2 s^{-2}P(s)\,. \tag{7.6}$$

$P(s)$ will be seen to be the Laplace transform of the input signal of a chain of integrating basic elements. The signal flow graph is drawn on the basis of the relationships (7.4) and (7.6) in the manner discussed previously, associating the vertices of the signal flow graphs with the functions $R(s)$, $Y(s)$, $P(s)$, $s^{-1}P(s)$ and $s^{-2}P(s)$ (Fig. 7.7). Thus the internal vertices have been associated with the signals

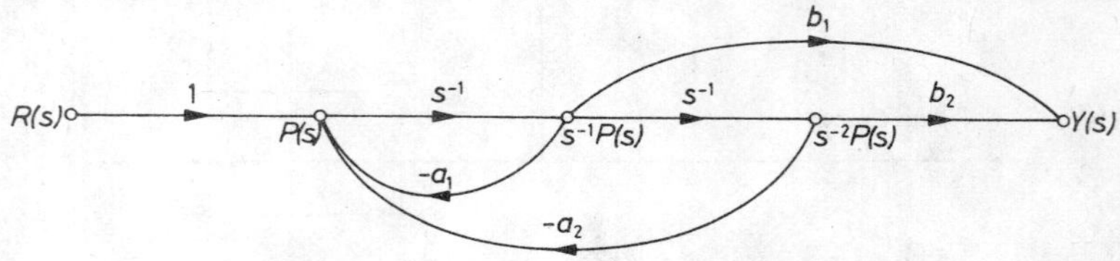

Fig. 7.7

$P(s)$, $s^{-1}P(s)$, $s^{-2}P(s)$ so as to have an edge corresponding to a basic element with transfer function s^{-1} between any two internal vertices.

In the case of the iterative method the coefficients of the highest degree term in the numerator and denominator are factored out, and the numerator and denominator are written in factorized form to obtain a product of linear fractional functions:

$$W(s)=b_1\frac{s+\dfrac{b_2}{b_1}}{(s-\beta_1)(s-\beta_2)}=\frac{b_1}{s-\beta_1}\,\frac{s+\dfrac{b_2}{b_1}}{s-\beta_2}, \tag{7.7}$$

where β_1 and β_2 are the roots of the denominator of $W(s)$. The signal flow graphs of the rational functions in (7.7) are now constructed with the aid of the direct method presented above, and then the signal flow graphs of the two rational functions are connected in "series" (Fig. 7.8). In our example the notations

$$Y_1(s)=b_1 s^{-1}P_1(s); \qquad R_1(s)=(1-\beta_1 s^{-1})\,P_1(s) \tag{7.8}$$

$$Y_2(s)=\left(1+\frac{b_2}{b_1}s^{-1}\right)P_2(s); \qquad R_2(s)=(1-\beta_2 s^{-1})\,P_2(s) \tag{7.9}$$

have been employed. Thus $R(s)=R_1(s)$ and $Y(s)=Y_2(s)$, and further $R_2(s)=Y_1(s)$.

For the parallel method the transfer function $W(s)$ is decomposed into partial fractions. If the numerator and denominator of $W(s)$ are of the same degree, it is rewritten as a sum of a constant and a proper rational function, and this proper rational function is then decomposed into partial fractions. The signal flow graphs corresponding to the partial fractions are constructed by the direct method, and the response signal is obtained by the "parallel" connection of the graphs thus derived.

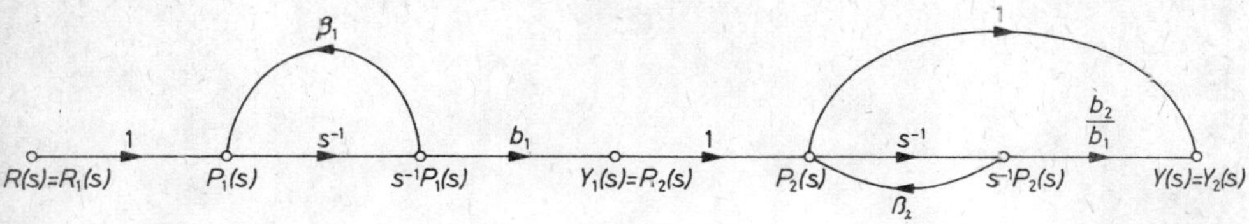

Fig. 7.8

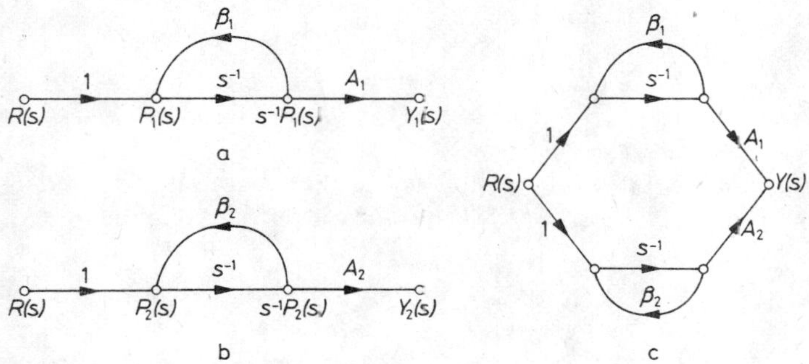

Fig. 7.9

In our example:

$$W(s) = \frac{A_1}{s-\beta_1} + \frac{A_2}{s-\beta_2} = \frac{A_1 s^{-1}}{1-\beta_1 s^{-1}} + \frac{A_2 s^{-1}}{1-\beta_2 s^{-1}}. \tag{7.10}$$

The signal flow graphs of the two fractions are separately shown in Figs 7.9, a and b, and Fig. 7.9, c, shows the parallel structure for (7.2).

The Laplace transform $Y(s)$ of the system response signal can only be written in the form

$$Y(s) = W(s)\, R(s) \tag{7.11}$$

similar to (7.2), if the excitation signal is applied to a system with no stored energy. In this case a signal flow graph of the system was seen to be constructible, in which the transfer functions associated with the edges are either s^{-1} or constants. If only one edge points towards the vertex corresponding to the signal $V_2(s)$, and the edge has the transfer function s^{-1} (Fig. 7.10, a),

$$V_2(s) = s^{-1}\, V_1(s)\,, \tag{7.12}$$

where $V_1(s)$ is the signal associated with the other vertex of the edge. Hence, using the notation

$$\mathcal{L}^{-1} V_1(s) = v_1(t), \qquad \mathcal{L}^{-1} V_2(s) = v_2(t):$$

$$v_2(t) = \int_0^t v_1(\tau)\,\mathrm{d}\tau\,, \tag{7.13}$$

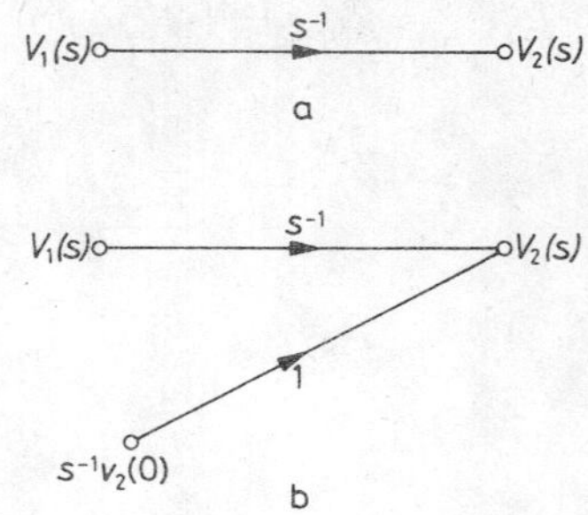

Fig. 7.10

i.e. $v_2(0)=0$. If the excitation signal acts on a system with initial stored energy, $Y(s)$ cannot be written in the form of (7.11), and the output signal of the integrating basic element is

$$v_2(t) = \int_0^t v_1(\tau)\,d\tau + v_2(0)\,, \tag{7.14}$$

where in general $v_2(0) \neq 0$. This initial condition must be taken into account in the signal flow graph as well. The Laplace transform of (7.14) is

$$V_2(s) = s^{-1}V_1(s) + s^{-1}v_2(0) \tag{7.15}$$

and hence the signal flow graph drawn in Fig. 7.10, b may be constructed. The signal flow graph shown in Fig. 7.10, a has been augmented here by associating the initial value multiplied by s^{-1} with a vertex of the graph, and adding an edge with transfer function 1 pointing towards this vertex from the vertex corresponding to $V_2(s)$.

In our further analysis such initial values multiplied by s^{-1} are treated as excitation signals.

The determination of the transfer-function matrix in a network modelled by a signal flow graph

As a result of the excitation signals acting on the input or inputs of networks consisting of basic elements, response signals appear at the output or outputs. The problem is frequently the determination of the relationship between excitation and response signals. This relationship may be described, in the case of networks consisting of linear basic elements, by a transfer-function matrix. The calculation of the transfer-function matrix will next be considered. It may be noted that Mason's formula [34] serves for the solution of similar problems. This, however, is not discussed here.

The n_1 excitation signals, n_3 response signals and n_2 "internal signals" of the system examined are arranged in column matrices **R**, **Y** and **V** respectively, in the order of the numbering of the vertices:

$$\mathbf{R}=\begin{bmatrix} R_1 \\ R_2 \\ \vdots \\ R_{n1} \end{bmatrix}; \qquad \mathbf{Y}=\begin{bmatrix} Y_1 \\ Y_2 \\ \vdots \\ Y_{n3} \end{bmatrix}; \qquad \mathbf{V}=\begin{bmatrix} V_1 \\ V_2 \\ \vdots \\ V_{n2} \end{bmatrix}. \tag{7.16}$$

The problem is to determine the transfer-function matrix $\boldsymbol{W}_0$ appearing in the formula

$$\mathbf{Y}=\boldsymbol{W}_0\mathbf{R}. \tag{7.17}$$

For the determination of transfer-function matrices of networks by means of signal flow graphs, the signal flow graph is selected so as to have those edges connected to the vertices corresponding to each excitation signal pointing away from these vertices, and those connected to the vertices associated with each response signal pointing towards the vertices. If the signal flow graph obtained as above does not satisfy this condition, additional edges with unity transfer function are connected to the relevant vertices, pointing towards the excitation signal and away from the response signals (Figs 7.11, a and b).

Suppose that the number of vertices and edges in the signal flow graph are n and b, respectively. Vertices and edges are given order numbers. The vertices associated with the Laplace transforms of excitation signals and with the initial values

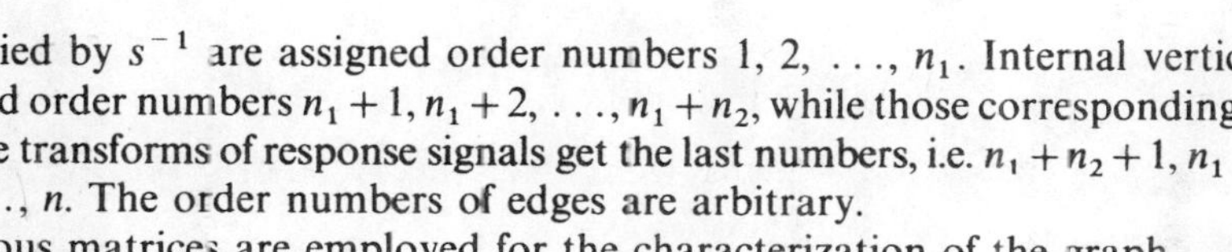

Fig. 7.11

multiplied by s^{-1} are assigned order numbers 1, 2, ..., n_1. Internal vertices are assigned order numbers $n_1+1, n_1+2, \ldots, n_1+n_2$, while those corresponding to the Laplace transforms of response signals get the last numbers, i.e. n_1+n_2+1, n_1+n_2+ $+2, \ldots, n$. The order numbers of edges are arbitrary.

Various matrices are employed for the characterization of the graph.

The matrices

$$\boldsymbol{M}_t=\frac{1}{2}(\boldsymbol{A}_{tC}+\boldsymbol{A}_t)+\boldsymbol{L} \tag{7.18}$$

$$\boldsymbol{N}_t=\frac{1}{2}(\boldsymbol{A}_{tO}+\boldsymbol{A}_t)+\boldsymbol{L} \tag{7.19}$$

are constructed from the (non-basis) incidence matrices $\boldsymbol{A}_t$ including orientations and $\boldsymbol{A}_{t0}$ disregarding orientations, and matrix $\boldsymbol{L}$ describing the incidence of self-loops with vertices (see Chapter 1). Since all the elements in the last n_3 rows of $\boldsymbol{M}_t$ and in the first n_1 rows of $\boldsymbol{N}_t$ equal zero, these rows are deleted, and the matrices thus obtained are denoted by $\boldsymbol{M}$ and $\boldsymbol{N}$, respectively.

The diagonal matrix $\boldsymbol{W}$ is formed by the transfer-functions associated with the edges, with its rows and columns corresponding to the edges:

$$\boldsymbol{W} = \langle W_1 \ \ W_2 \ \ \ldots \ \ W_b \rangle . \tag{7.20}$$

Let the column matrix of the output signals of the edges (basic elements) arranged in the order of the numbering of edges be denoted by $\mathbf{U}$. The equations corresponding to (7.1) for all basic elements may be summarized, with the aid of the matrices defined above, in

$$\mathbf{U} = \boldsymbol{W}\boldsymbol{M}^{+}\begin{bmatrix}\mathbf{R}\\ \mathbf{V}\end{bmatrix}. \tag{7.21}$$

Summing the elements of $\mathbf{U}$ corresponding to the same vertex

$$\begin{bmatrix}\mathbf{V}\\ \mathbf{Y}\end{bmatrix} = \boldsymbol{N}\boldsymbol{W}\boldsymbol{M}^{+}\begin{bmatrix}\mathbf{R}\\ \mathbf{V}\end{bmatrix} = \boldsymbol{W}_t\begin{bmatrix}\mathbf{R}\\ \mathbf{V}\end{bmatrix} \tag{7.22}$$

is obtained. Partitioning the vertex transfer matrix

$$\boldsymbol{W}_t = \boldsymbol{N}\boldsymbol{W}\boldsymbol{M}^{+} \tag{7.23}$$

appearing here, (7.22) can be written in the form

$$\begin{bmatrix}\mathbf{V}\\ \mathbf{Y}\end{bmatrix} = \begin{bmatrix}\boldsymbol{W}_{11} & \boldsymbol{W}_{12}\\ \boldsymbol{W}_{21} & \boldsymbol{W}_{22}\end{bmatrix}\begin{bmatrix}\mathbf{R}\\ \mathbf{V}\end{bmatrix}, \tag{7.24}$$

where $\boldsymbol{W}_{12}$ is a square block of order n_2. Since for the signal flow graphs selected for the calculation of the transfer matrix, $\boldsymbol{W}_{21} = \boldsymbol{0}$, (with the exception of systems consisting of one basic element), (7.24) yields

$$\mathbf{V} = \boldsymbol{W}_{11}\mathbf{R} + \boldsymbol{W}_{12}\mathbf{V} \tag{7.25}$$

$$Y = \boldsymbol{W}_{22}\mathbf{V}. \tag{7.26}$$

Expressing $\mathbf{V}$ from (7.25), and substituting it into (7.26)

$$\mathbf{Y} = \boldsymbol{W}_{22}(\boldsymbol{1} - \boldsymbol{W}_{12})^{-1}\boldsymbol{W}_{11}\mathbf{R} = \boldsymbol{W}_0\mathbf{R} \tag{7.27}$$

is obtained, i.e.

$$\boldsymbol{W}_0 = \boldsymbol{W}_{22}(\boldsymbol{1} - \boldsymbol{W}_{12})^{-1}\boldsymbol{W}_{11} \tag{7.28}$$

is the transfer matrix of the network. Thus, from a knowledge of the excitation signals, the response signals can be determined with the aid of (7.17).

It may be noted that $\boldsymbol{W}_t$ can be written directly from the signal flow graph. The columns of $\boldsymbol{W}_t$ correspond to the first $n_1 + n_2$ vertices in the order of the numbering, while its rows also correspond to the vertices with the exception of the first n_1. The elements of $\boldsymbol{W}_t$ are the transfer-functions associated with edges pointing from the vertex corresponding to the relevant column towards that corresponding to the relevant row. If no edge points from the vertex corresponding to the column towards that corresponding to the row, the relevant element of $\boldsymbol{W}_t$ is zero.

Writing the state equations based upon the signal flow graph

The state equations of linear, time-invariant networks consisting of two-terminal elements has been written in Chapter 6 from a knowledge of the network connections. The following method is presented for writing the state equations of linear systems [11] with the aid of the signal flow graph based upon the block diagram or equations of the system.

It has been seen that a signal flow graph can be associated with a system whose transfer-function is a rational function, with the transfer-functions of the edges in the signal flow graph obtained being s^{-1} or constants. The transfer-functions associated with the edges are similar in the case that the initial time-domain values of signals corresponding to the internal vertices of the signal flow graph are nonzero. Such signal flow graphs will be used for writing the state equations, i.e. one with the transfer-functions associated with the edges being s^{-1} or constants. Each variable which is associated with a vertex of the graph having an edge with transfer function s^{-1} pointing towards it, will be chosen to be the Laplace transform of a state variable. At most one further edge may point towards such vertices: one with unity transfer function, pointing away from the vertex associated with the initial value of the state variable multiplied by s^{-1}. After the selection of the state variables it is expedient to simplify the signal flow graph by combining edges so that all internal vertices connect only to edges with transfer function s^{-1}.

The signals associated with the vertices of the signal flow graph can be classified into the following four groups:

1. Laplace transforms of excitation signals, the initial values of state variables multiplied by s^{-1};
2. Laplace transforms of the first time-derivatives of state variables;
3. Laplace transforms of state variables;
4. Laplace transforms of response signals.

Let the vertices be assigned order numbers in the order of the classification, ensuring that the vertices associated with the state variables and their Laplace transforms are numbered in the same order. The column matrices formed by the signals in the 1st, 2nd, 3rd and 4th groups arranged in the order of the numbering are denoted by $\mathbf{R}(s)$, $\dot{\mathbf{X}}(s)$, $\mathbf{X}(s)$ and $\mathbf{Y}(s)$, respectively. With the aid of the vertex transfer matrix based upon the the signal flow graph:

$$\begin{bmatrix} \dot{\mathbf{X}}(s) \\ \mathbf{X}(s) \\ \mathbf{Y}(s) \end{bmatrix} = \boldsymbol{W}_t \begin{bmatrix} \mathbf{R}(s) \\ \dot{\mathbf{X}}(s) \\ \mathbf{X}(s) \end{bmatrix} = \begin{bmatrix} \boldsymbol{W}_{11} & \boldsymbol{W}_{12} & \boldsymbol{W}_{13} \\ \boldsymbol{W}_{21} & \boldsymbol{W}_{22} & \boldsymbol{W}_{23} \\ \boldsymbol{W}_{31} & \boldsymbol{W}_{32} & \boldsymbol{W}_{33} \end{bmatrix} \begin{bmatrix} \mathbf{R}(s) \\ \dot{\mathbf{X}}(s) \\ \mathbf{X}(s) \end{bmatrix}. \tag{7.29}$$

Hence, the Laplace transform of the state equations of the system is

$$\dot{\mathbf{X}}(s) = (\boldsymbol{1} - \boldsymbol{W}_{12})^{-1} \boldsymbol{W}_{13} \mathbf{X}(s) + (\boldsymbol{1} - \boldsymbol{W}_{12})^{-1} \boldsymbol{W}_{11} \mathbf{R}(s) \tag{7.30}$$

and further

$$\mathbf{Y}(s) = [\boldsymbol{W}_{32}(\boldsymbol{1} - \boldsymbol{W}_{12})^{-1} \boldsymbol{W}_{13} + \boldsymbol{W}_{33}] \mathbf{X}(s) + \boldsymbol{W}_{32}(\boldsymbol{1} - \boldsymbol{W}_{12})^{-1} \boldsymbol{W}_{11} \mathbf{R}(s) \tag{7.31}$$

is the column matrix of the Laplace transforms of the response signals where $\boldsymbol{W}_{31} = \boldsymbol{0}$ has been taken into account. By inverse Laplace transformation (7.30) yields the state equations of the system and (7.31) yields the time functions of the response signals.

Examples

1. To draw the signal flow graph of the network shown in Fig. 7.12 the following equations may be written:

$$U = Z_1 I_1 + Z_3 I_3$$

$$Z_1 I_1 + Z_5 I_5 - Z_2 I_2 = 0$$

$$Z_3 I_3 - Z_4 I_4 - Z_5 I_5 = 0$$

$$I_2 + I_5 = I_4$$

$$I_1 = I_3 + I_5 .$$

Hence

$$I_1 = \frac{U}{Z_1} - \frac{Z_3}{Z_1} I_3$$

$$I_2 = \frac{Z_1}{Z_2} I_1 + \frac{Z_5}{Z_2} I_5$$

$$I_3 = \frac{Z_4}{Z_3} I_4 + \frac{Z_5}{Z_3} I_5$$

$$I_4 = I_2 + I_5$$

$$I_5 = I_1 - I_3 .$$

Accordingly, the signal flow graph shown in Fig. 7.13 may be drawn.

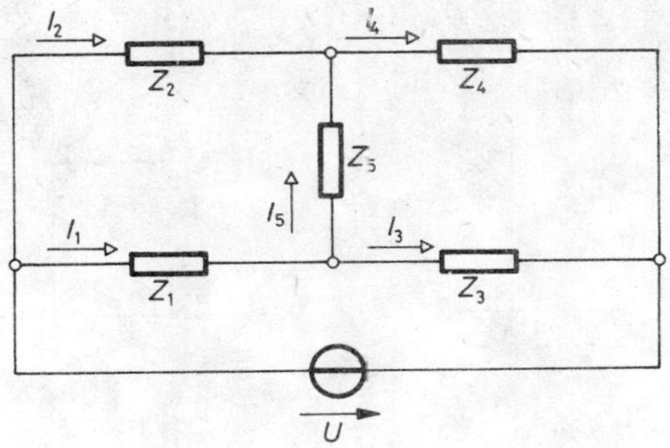

Fig. 7.12

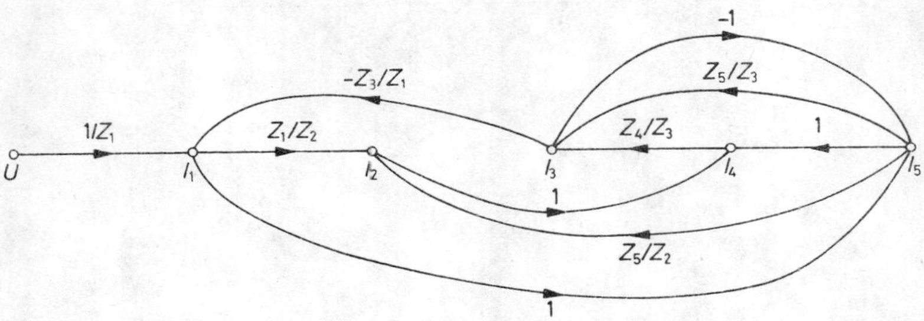

Fig. 7.13

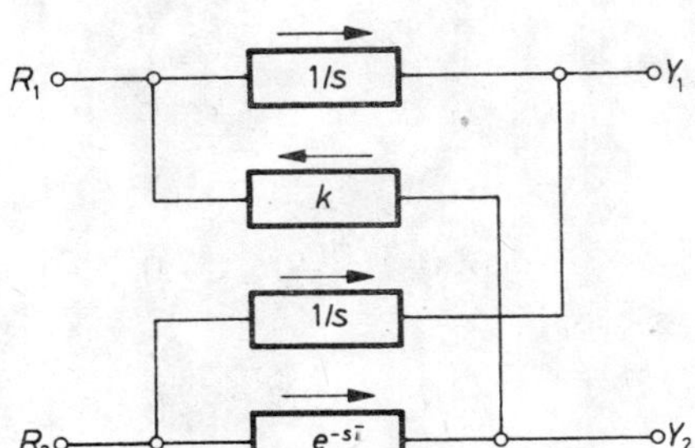

Fig. 7.14

2. The network drawn in Fig. 7.14 has no internal stored energy for $t<0$. The Laplace transforms of its input signals (Fig. 7.15) are: $R_1 = U_1/(s+\alpha)$ and $R_2 = U_2(1-e^{-sT/2})/s$. To determine the output signals Y_1 and Y_2 the signal flow graph of the network has been drawn in Fig. 7.16. The encircled numbers in the figure are the order numbers of the edges. Accordingly:

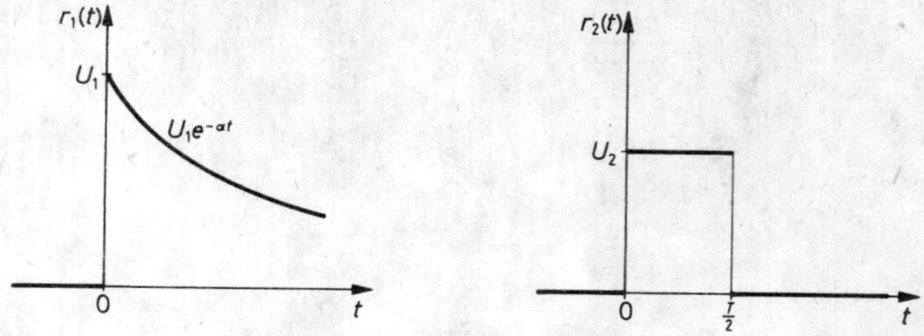

Fig. 7.15

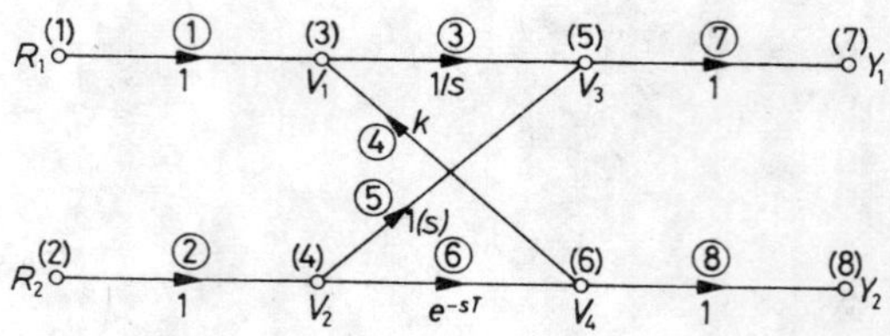

Fig. 7.16

$$\boldsymbol{M}=\begin{bmatrix} 1 & 0 & 0 & 0 & 0 & 0 & 0 & 0 \\ 0 & 1 & 0 & 0 & 0 & 0 & 0 & 0 \\ 0 & 0 & 1 & 0 & 0 & 0 & 0 & 0 \\ 0 & 0 & 0 & 0 & 1 & 1 & 0 & 0 \\ 0 & 0 & 0 & 0 & 0 & 0 & 1 & 0 \\ 0 & 0 & 0 & 1 & 0 & 0 & 0 & 1 \end{bmatrix},$$

$$\boldsymbol{N}=\begin{bmatrix} 1 & 0 & 0 & 1 & 0 & 0 & 0 & 0 \\ 0 & 1 & 0 & 0 & 0 & 0 & 0 & 0 \\ 0 & 0 & 1 & 0 & 1 & 0 & 0 & 0 \\ 0 & 0 & 0 & 0 & 0 & 1 & 0 & 0 \\ 0 & 0 & 0 & 0 & 0 & 0 & 1 & 0 \\ 0 & 0 & 0 & 0 & 0 & 0 & 0 & 1 \end{bmatrix},$$

$$\boldsymbol{W}=\langle 1 \;\; 1 \;\; 1/s \;\; k \;\; 1/s \;\; e^{-sT} \;\; 1 \;\; 1 \rangle .$$

Hence, according to (7.23):

$$\mathbf{W}_t=\left[\begin{array}{cc|cccc} 1 & 0 & 0 & 0 & 0 & \mathrm{k} \\ 0 & 1 & 0 & 0 & 0 & 0 \\ 0 & 0 & 1/s & 1/s & 0 & 0 \\ 0 & 0 & 0 & e^{-sT} & 0 & 0 \\ \hline 0 & 0 & 0 & 0 & 1 & 0 \\ 0 & 0 & 0 & 0 & 0 & 1 \end{array}\right],$$

where the partitioning in accordance with (7.24) has been indicated. Substitution into (7.27) yields:

$$\mathbf{Y}=\begin{bmatrix} Y_1 \\ Y_2 \end{bmatrix}=\begin{bmatrix} 0 & 0 & 1 & 0 \\ 0 & 0 & 0 & 1 \end{bmatrix}\begin{bmatrix} 1 & 0 & 0 & -k \\ 0 & 1 & 0 & 0 \\ -\frac{1}{s} & -\frac{1}{s} & 1 & 0 \\ 0 & -e^{-sT} & 0 & 1 \end{bmatrix}^{-1}\begin{bmatrix} 1 & 0 \\ 0 & 1 \\ 0 & 0 \\ 0 & 0 \end{bmatrix}\begin{bmatrix} R_1 \\ R_2 \end{bmatrix}=$$

$$=\begin{bmatrix} \dfrac{R_1+R_2}{s}+k\dfrac{R_2}{s}e^{-sT} \\ R_2e^{-sT} \end{bmatrix}=\begin{bmatrix} \dfrac{U_1}{s(s+\alpha)}+\dfrac{U_2}{s^2}(1-e^{-sT/2}+ke^{-sT}-ke^{-s3T/2}) \\ \dfrac{U_2}{s}(e^{-sT}-e^{-s3T/2}) \end{bmatrix},$$

$$\mathbf{y}(t)=\begin{bmatrix} y_1(t) \\ y_2(t) \end{bmatrix}=$$

$$=\begin{bmatrix} 1(t)\dfrac{U_1}{\alpha}(1-e^{-\alpha t})+U_2\left[1(t)t-1\left(t-\dfrac{T}{2}\right)\left(t-\dfrac{T}{2}\right)+\right. \\ \left.+1(t-T)k(t-T)-1\left(t-\dfrac{3T}{2}\right)k\left(t-\dfrac{3T}{2}\right)\right] U_2\left[1(t-T)-1\left(t-\dfrac{3T}{2}\right)\right] \end{bmatrix}.$$

Accordingly, the time functions of the response signals have been drawn in Fig. 7.17.

3. The transistor of the small-signal circuit model shown in Fig. 7.18 is characterized by its hybrid parameters. To determine the relationship between I_3 and U_g with the aid of a signal flow graph, the equivalent circuit drawn in Fig. 7.19 is employed, with the assumption that $h_{12}\approx 0$. The following equations can be written

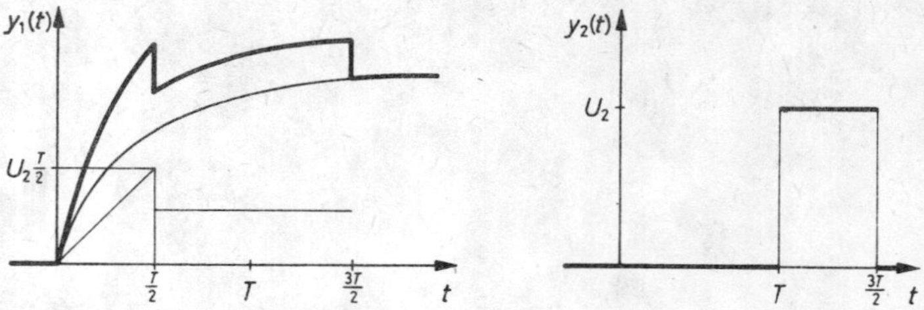

Fig. 7.17

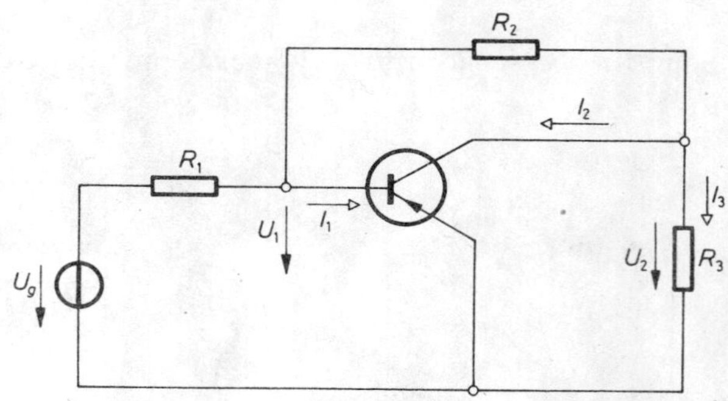

Fig. 7.18

for this:

$$U_g=(I_1+I_2+I_3)R_1+I_1h_{11}$$
$$I_1h_{11}=(I_2+I_3)R_2+I_3R_3$$
$$I_3R_3=(I_2-h_{21}I_1)/h_{22}\,.$$

Hence I_1, I_2 and I_3 may be expressed as follows:

$$I_1=I_2R_2/h_{11}+I_3(R_2+R_3)/h_{11}$$
$$I_2=U_g/R_1-I_1(R_1+h_{11})/R_1-I_3$$
$$I_3=I_2/(h_{22}R_3)-I_1h_{21}/(h_{22}R_3)\,.$$

From these equations, the signal flow graph of the network is drawn in Fig. 7.20. The matrices needed to write the relationship between the response signal I_3 and the excitation signal U_g, are:

$$\boldsymbol{M}=\begin{bmatrix} 1 & 0 & 0 & 0 & 0 & 0 & 0 & 0 \\ 0 & 0 & 1 & 0 & 1 & 0 & 0 & 0 \\ 0 & 1 & 0 & 0 & 0 & 0 & 1 & 0 \\ 0 & 0 & 0 & 1 & 0 & 1 & 0 & 1 \end{bmatrix}$$

$$N=\begin{bmatrix} 0 & 1 & 0 & 1 & 0 & 0 & 0 & 0 \\ 1 & 0 & 1 & 0 & 0 & 1 & 0 & 0 \\ 0 & 0 & 0 & 0 & 1 & 0 & 1 & 0 \\ 0 & 0 & 0 & 0 & 0 & 0 & 0 & 1 \end{bmatrix}$$

$$W=<\frac{1}{R_1} \quad \frac{R_2}{h_{11}} \quad -\frac{R_1+h_{11}}{R_1} \quad \frac{R_2+R_3}{h_{11}} \quad -\frac{h_{21}}{h_{22}R_3} \quad -1 \quad \frac{1}{h_{22}R_3} \quad 1>.$$

Hence, from (7.23):

$$W=\left[\begin{array}{c|ccc} 0 & 0 & \frac{R_2}{h_{11}} & \frac{R_2+R_3}{h_{11}} \\ \frac{1}{R_1} & -\frac{R_1+h_{11}}{R_1} & 0 & -1 \\ 0 & -\frac{h_{21}}{h_{22}R_3} & \frac{1}{h_{22}R_3} & 0 \\ \hline 0 & 0 & 0 & 1 \end{array}\right].$$

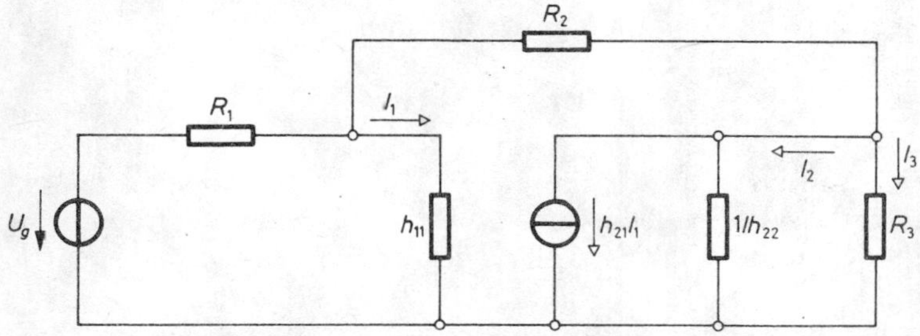

Fig. 7.19

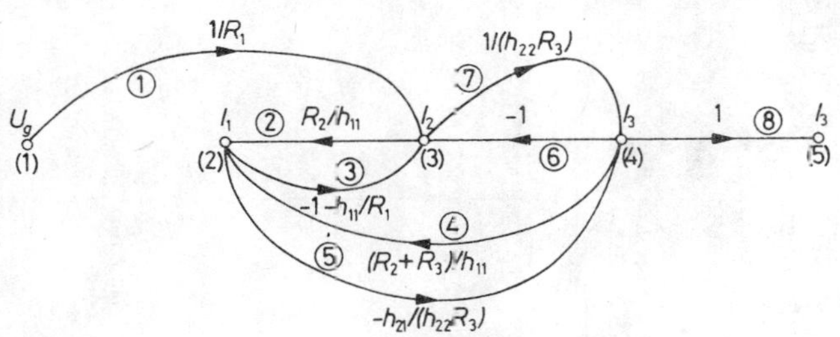

Fig. 7.20

With the partitioning according to (7.24), (7.27) yields:

$$Y=I_3=[0\ 0\ 1]\begin{bmatrix} 1 & -\dfrac{R_2}{h_{11}} & -\dfrac{R_2+R_3}{h_{11}} \\ \dfrac{R_1+h_{11}}{R_1} & 1 & 1 \\ \dfrac{h_{21}}{h_{22}R_3} & -\dfrac{1}{h_{22}R_3} & 1 \end{bmatrix}^{-1} \begin{bmatrix} 0 \\ \dfrac{1}{R_1} \\ 0 \end{bmatrix} U_g=$$

$$=\frac{1}{D}\frac{U_g}{R_1\ R_3h_{22}}\left(1-\frac{R_2h_{21}}{h_{11}}\right),$$

where

$$D=\frac{R_1R_3\ h_{11}h_{22}+(R_2+R_3+R_2\ R_3h_{22})(R_1+h_{11})+R_1(R_3h_{21}+h_{11})}{R_1R_3\ h_{11}h_{22}},$$

i.e.

$$I_3=\frac{h_{11}-R_2h_{21}}{R_1(R_3h_{11}h_{22}+R_3h_{21}+h_{11})+(R_2+R_3+R_2R_3h_{22})(R_1+h_{11})}U_g$$

is the current of the resistance R_3.

4. The signal flow graphs constructed with the aid of the direct, iterative and parallel methods based on the transfer-function

$$W(s)=\frac{b_1s+b_2}{s^2+a_1s+a}$$

have been shown in Figs 7.7, 7.8 and 7.9. In the following example, the state equations of the system and the response signal will be derived from the signal flow graph. The denominator of $W(s)$ is of 2nd degree, and is assumed to have two distinct real roots.

(a) Among the quantities associated with the vertices of the signal flow graph shown in Fig. 7.7, $X_1^{(1)}(s)=s^{-1}P(s)$ and $X_2^{(1)}(s)=s^{-2}P(s)$ are the state variables. To write the state equations, the signal flow graph is redrawn to have distinct vertices correspond to the state variables and to those multiplied by s (Fig. 7.21).

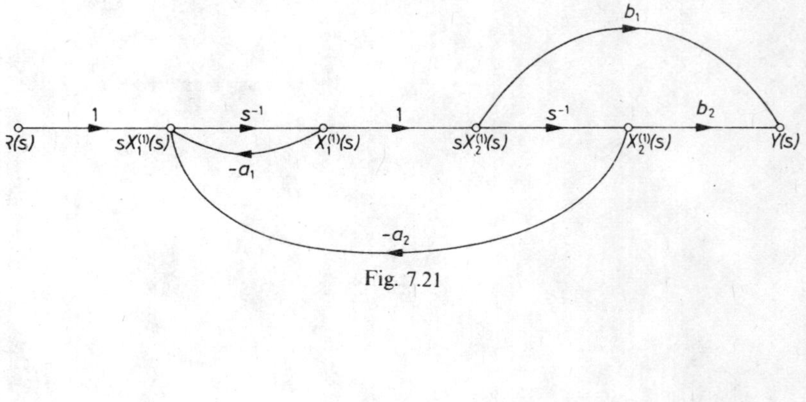

Fig. 7.21

Accordingly, the equation corresponding to (7.29) is

$$\left[\begin{array}{c} sX_1^{(1)}(s) \\ sX_2^{(1)}(s) \\ \hline X_1^{1}(s) \\ X_2^{(1)}(s) \\ \hline Y(s) \end{array}\right] = \left[\begin{array}{c|c|c|cc} 1 & 0 & 0 & -a_1 & -a_2 \\ 0 & 0 & 0 & 1 & 0 \\ \hline 0 & s^{-1} & 0 & 0 & 0 \\ 0 & 0 & s^{-1} & 0 & 0 \\ \hline 0 & 0 & b_1 & 0 & b_2 \end{array}\right] \left[\begin{array}{c} R(s) \\ \hline sX_1^{(1)}(s) \\ sX_2^{(1)}(s) \\ \hline X_1^{(1)}(s) \\ X_2^{(1)}(s) \end{array}\right]$$

The Laplace transform of the state equations, in accordance with (7.30), is:

$$\begin{bmatrix} sX_1^{(1)}(s) \\ sX_2^{(1)}(s) \end{bmatrix} = \begin{bmatrix} -a_1 & -a_2 \\ 1 & 0 \end{bmatrix} \begin{bmatrix} X_1^{(1)}(s) \\ X_2^{(1)}(s) \end{bmatrix} + \begin{bmatrix} 1 \\ 0 \end{bmatrix} R(s),$$

i.e. the state equations are

$$\begin{bmatrix} \dot{x}_1^{(1)}(t) \\ \dot{x}_2^{(1)}(t) \end{bmatrix} = \begin{bmatrix} -a_1 & -a_2 \\ 1 & 0 \end{bmatrix} \begin{bmatrix} x_1^{(1)}(t) \\ x_2^{(1)}(t) \end{bmatrix} + \begin{bmatrix} 1 \\ 0 \end{bmatrix} r(t).$$

The Laplace transform of the response signals is by (7.31)

$$Y(s) = \left\{ [0 \;\; b_1] \begin{bmatrix} -a_1 & -a_2 \\ 1 & 0 \end{bmatrix} + [0 \;\; b_2] \right\} \begin{bmatrix} X_1^{(1)}(s) \\ X_2^{(1)}(s) \end{bmatrix} +$$

$$+ [0 \;\; b_1] \begin{bmatrix} 1 \\ 0 \end{bmatrix} R(s) = [b_1 \;\; b_2] \begin{bmatrix} X_1^{(1)}(s) \\ X_2^{(1)}(s) \end{bmatrix},$$

and

$$y(t) = [b_1 \;\; b_2] \begin{bmatrix} x_1^{(1)}(t) \\ x_2^{(1)}(t) \end{bmatrix}$$

is the time function of the response signal.

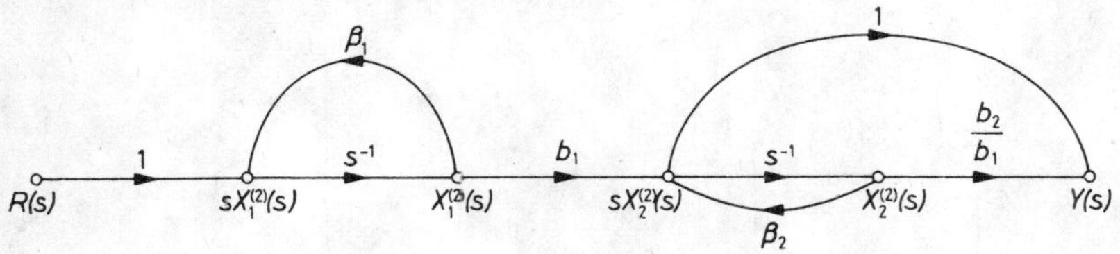

Fig. 7.22

(b) The signal flow graph of Fig. 7.8 has been redrawn for writing the state equations as shown in Fig. 7. 22, by combining edges, and the notation $X_1^{(2)}(s) = s^{-1}P_1(s)$ and $X_2^{(2)}(s) = s^{-1}P_2(s)$ has been employed. From this signal flow graph, the equation corresponding to (7.29) is

$$\begin{bmatrix} sX_1^{(2)}(s) \\ sX_2^{(2)}(s) \\ \hline X_1^{(2)}(s) \\ X_2^{(2)}(s) \\ \hline Y(s) \end{bmatrix} = \left[\begin{array}{c|cc|cc} 1 & 0 & 0 & \beta_1 & 0 \\ 0 & 0 & 0 & b_1 & \beta_2 \\ \hline 0 & s^{-1} & 0 & 0 & 0 \\ 0 & 0 & s^{-1} & 0 & 0 \\ \hline 0 & 0 & 1 & 0 & b_2/b_1 \end{array}\right] \begin{bmatrix} R(s) \\ \hline sX_1^{(2)}(s) \\ sX_2^{(2)}(s) \\ \hline X_1^{(2)}(s) \\ X_2^{(2)}(s) \end{bmatrix},$$

i.e.

$$\begin{bmatrix} sX_1^{(2)}(s) \\ sX_2^{(2)}(s) \end{bmatrix} = \begin{bmatrix} \beta_1 & 0 \\ b_1 & \beta_2 \end{bmatrix} \begin{bmatrix} X_1^{(2)}(s) \\ X_2^{(2)}(s) \end{bmatrix} + \begin{bmatrix} 1 \\ 0 \end{bmatrix} R(s),$$

and hence

$$\begin{bmatrix} \dot{x}_1^{(2)}(t) \\ \dot{x}_2^{(2)}(t) \end{bmatrix} = \begin{bmatrix} \beta_1 & 0 \\ b_1 & \beta_2 \end{bmatrix} \begin{bmatrix} x_1^{(2)}(t) \\ x_2^{(2)}(t) \end{bmatrix} + \begin{bmatrix} 1 \\ 0 \end{bmatrix} r(t)$$

are the state equations. (7.31) further yields

$$Y(s) = \left\{ [0 \;\; 1] \begin{bmatrix} \beta_1 & 0 \\ b_1 & \beta_2 \end{bmatrix} + [0 \;\; b_2/b_1] \right\} \begin{bmatrix} X_1^{(2)}(s) \\ X_2^{(2)}(s) \end{bmatrix} +$$

$$+ [0 \;\; b_2/b_1] \begin{bmatrix} 1 \\ 0 \end{bmatrix} R(s) = [b_1 \;\; b_2/b_1 + \beta_2] \begin{bmatrix} X_1^{(2)}(s) \\ X_2^{(2)}(s) \end{bmatrix}$$

and

$$y(t) = [b_1 \;\; b_2/b_1 + \beta_2] \begin{bmatrix} x_1^{(2)}(t) \\ x_2^{(2)}(t) \end{bmatrix}$$

the Laplace transform and time function of the response signal.

(c) Fig. 7.9, c is used to write the state equations, after indicating the state variables $X_1^{(3)}(s)$ and $X_2^{(3)}(s)$ (Fig. 7.23). For this, the equation corresponding to (7.29) is:

$$\begin{bmatrix} sX_1^{(3)}(s) \\ sX_2^{(3)}(s) \\ \hline X_1^{(3)}(s) \\ X_2^{(3)}(s) \\ \hline Y(s) \end{bmatrix} = \left[\begin{array}{c|cc|cc} 1 & 0 & 0 & \beta_1 & 0 \\ 1 & 0 & 0 & 0 & \beta_2 \\ \hline 0 & s^{-1} & 0 & 0 & 0 \\ 0 & 0 & s^{-1} & 0 & 0 \\ \hline 0 & 0 & 0 & A_1 & A_2 \end{array}\right] \begin{bmatrix} R(s) \\ \hline sX_1^{(3)}(s) \\ sX_2^{(3)}(s) \\ \hline X_1^{(3)}(s) \\ X_2^{(3)}(s) \end{bmatrix}.$$

Hence, according to (7.30):

$$\begin{bmatrix} sX_1^{(3)}(s) \\ sX_2^{(3)}(s) \end{bmatrix} = \begin{bmatrix} \beta_1 & 0 \\ 0 & \beta_2 \end{bmatrix} \begin{bmatrix} X_1^{(3)}(s) \\ X_2^{(3)}(s) \end{bmatrix} + \begin{bmatrix} 1 \\ 1 \end{bmatrix} R(s)$$

and its inverse Laplace transform is

$$\begin{bmatrix} \dot{x}_1^{(3)}(t) \\ \dot{x}_2^{(3)}(t) \end{bmatrix} = \begin{bmatrix} \beta_1 & 0 \\ 0 & \beta_2 \end{bmatrix} \begin{bmatrix} x_1^{(3)}(t) \\ x_2^{(3)}(t) \end{bmatrix} + \begin{bmatrix} 1 \\ 1 \end{bmatrix} r(t)$$

and further in accordance with (7.31):

$$Y(s) = [A_1 \ \ A_2] \begin{bmatrix} X_1^{(3)}(s) \\ X_2^{(3)}(s) \end{bmatrix}$$

and hence

$$y(t) = [A_1 \ \ A_2] \begin{bmatrix} x_1^{(3)}(t) \\ x_2^{(3)}(t) \end{bmatrix}$$

is the time function of the response signal.

The signal flow graphs have been constructed from the transfer function $W(s)$, and so the initial values of the state variables are zero.

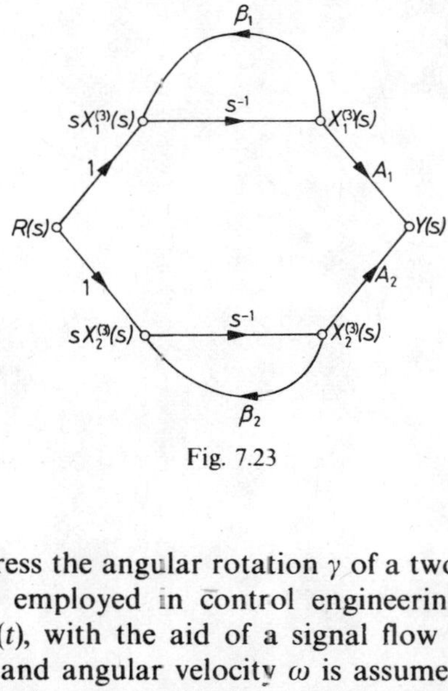

Fig. 7.23

5. Our aim is to express the angular rotation γ of a two-phase induction motor (Fig. 7.24), commonly employed in control engineering, as a function of the controlling voltage $u_1(t)$, with the aid of a signal flow graph. The relationship between the torque m and angular velocity ω is assumed to be linear. Thus the following equations may be written for the time-functions of the angular velocity, the torque and the angular rotation:

$$\frac{d\gamma}{dt} = \omega(t)$$

$$m(t) = K_1 \omega(t) + K_2 u_1(t)$$

$$\Theta \frac{d\omega}{dt} + K_3 \omega(t) = m(t)$$

$$\gamma(t=0) = \gamma_0, \qquad \omega(t=0) = \omega_0,$$

where K_1 is the slope of the linearized moment — angular frequency characteristic, K_2 is the ratio between the torque of the stationary motor and the controlling voltage, K_3 is the coefficient of viscous friction, Θ is the moment of inertia of the rotor and

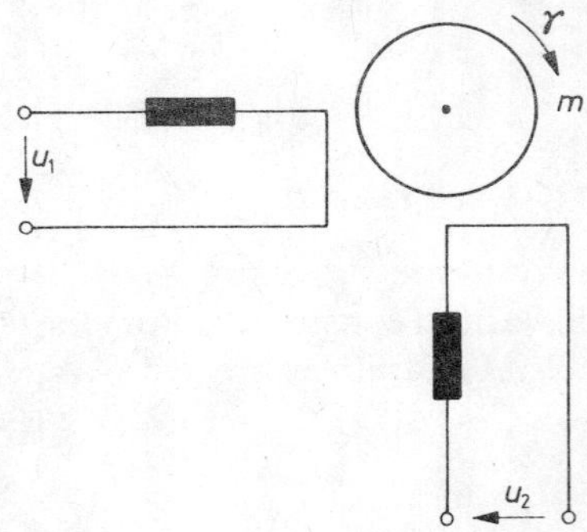

Fig. 7.24

$u_2(t)$ is the reference voltage. The Laplace transforms of the above equations are:

$$s\Gamma(s) - \gamma_0 = \Omega(s)$$

$$M(s) = K_1 \Omega(s) + K_2 U_1(s)$$

$$\Theta s \Omega(s) - \Theta \omega_0 + K_3 \Omega(s) = M(s).$$

Hence

$$M(s) = K_1 \Omega(s) + K_2 U_1(s)$$

$$\Omega(s) = \frac{M(s) + \Theta \omega_0}{s\Theta + K_3}$$

$$\Gamma(s) = \frac{1}{s}[\Omega(s) + \gamma_0].$$

The signal flow graph relating to the Laplace transforms is shown in Fig. 7.25. Regarding U_1, ω_0 and γ_0 as excitation signals, the response signal Γ can be expressed in terms of them. The matrices describing the signal flow graph are:

$$
\boldsymbol{M}=\begin{bmatrix}
1 & 0 & 0 & 0 & 0 & 0 & 0\\
0 & 1 & 0 & 0 & 0 & 0 & 0\\
0 & 0 & 1 & 0 & 0 & 0 & 0\\
0 & 0 & 0 & 1 & 0 & 1 & 0\\
0 & 0 & 0 & 0 & 1 & 0 & 0\\
0 & 0 & 0 & 0 & 0 & 0 & 1
\end{bmatrix}
$$

$$
\boldsymbol{N}=\begin{bmatrix}
0 & 1 & 0 & 0 & 1 & 0 & 0\\
1 & 0 & 0 & 1 & 0 & 0 & 0\\
0 & 0 & 1 & 0 & 0 & 1 & 0\\
0 & 0 & 0 & 0 & 0 & 0 & 1
\end{bmatrix}
$$

$$
\boldsymbol{W}=<K_2 \quad \frac{\Theta}{s\Theta+K_3} \quad \frac{1}{s} \quad K_1 \quad \frac{1}{s\Theta+K_2} \quad \frac{1}{s} \quad 1>.
$$

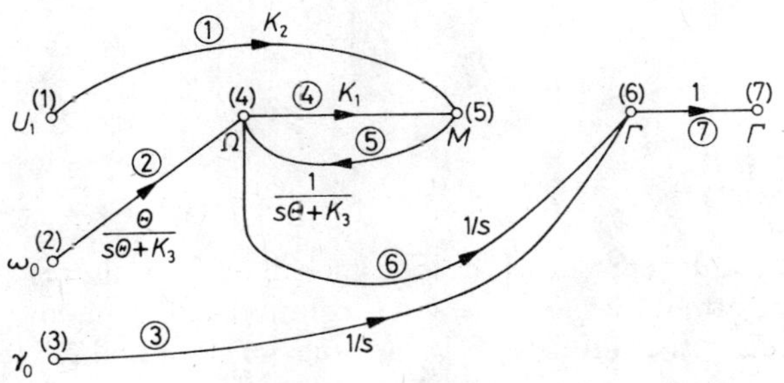

Fig. 7.25

Hence, in accordance with (7.23):

$$
\boldsymbol{W}_t=\left[\begin{array}{ccc:ccc}
0 & \frac{\Theta}{s\Theta+K_3} & 0 & 0 & \frac{1}{s\Theta+K_3} & 0\\
K_2 & 0 & 0 & K_1 & 0 & 0\\
0 & 0 & \frac{1}{s} & \frac{1}{s} & 0 & 0\\
\hdashline
0 & 0 & 0 & 0 & 0 & 1
\end{array}\right].
$$

Taking the partitioning of (7.24) into account, (7.27) yields:

$\mathbf{Y}(s)=\Gamma(s)=$

$$=[0\ 0\ 1]\begin{bmatrix}1 & -\dfrac{1}{s\Theta+K_3} & 0\\ -K_1 & 1 & 0\\ -\dfrac{1}{s} & 0 & 1\end{bmatrix}^{-1}\begin{bmatrix}0 & \dfrac{\Theta}{s\Theta+K_3} & 0\\ K_2 & 0 & 0\\ 0 & 0 & \dfrac{1}{s}\end{bmatrix}\begin{bmatrix}U_1(s)\\ \omega_0\\ \gamma_0\end{bmatrix},$$

i.e.

$$\Gamma(s)=\frac{K_2}{s(s\Theta+K_3-K_1)}U_1(s)+\frac{\Theta\omega_0}{s(s\Theta+K_3-K_1)}+\frac{\gamma_0}{s}$$

is the Laplace transform of the angular rotation. Its time-function may also be determined from this in the knowledge of U_1.

6. The following state equations may be written for a direct current motor controlled from the armature circuit (Fig. 7.26), if the field current, i_2, is constant:

$$\begin{bmatrix}\dot{i}_1\\ \dot{\omega}\\ \dot{\gamma}\end{bmatrix}=\begin{bmatrix}-\dfrac{R}{L} & 0 & 0\\ \dfrac{K_1}{\Theta} & -\dfrac{K_2}{\Theta} & 0\\ 0 & 1 & 0\end{bmatrix}\begin{bmatrix}i_1(t)\\ \omega(t)\\ \gamma(t)\end{bmatrix}+\begin{bmatrix}\dfrac{1}{L}u_1(t)\\ 0\\ 0\end{bmatrix},$$

where i_1 is the armature current, u_1 the terminal voltage, R is the resistance and L the inductance of the armature, γ is the angular rotation, ω the angular velocity, Θ the moment of inertia of the rotor, K_1 is the torque constant and K_2 the coefficient of viscous friction. Taking into account that

$$i_1(t=0)=i_0;\qquad \omega(t=0)=\omega_0;\qquad \gamma(t=0)=\gamma_0,$$

the Laplace transform of the state equations yields the following:

$$I_1(s)=\frac{U_1(s)}{R+sL}+\frac{Li_0}{R+sL}$$

$$\Omega(s)=\frac{\Theta\omega_0}{K_2+s\Theta}+\frac{K_1I_1(s)}{K_2+s\Theta}$$

$$\Gamma(s)=\frac{\Omega(s)}{s}+\frac{\gamma_0}{s}.$$

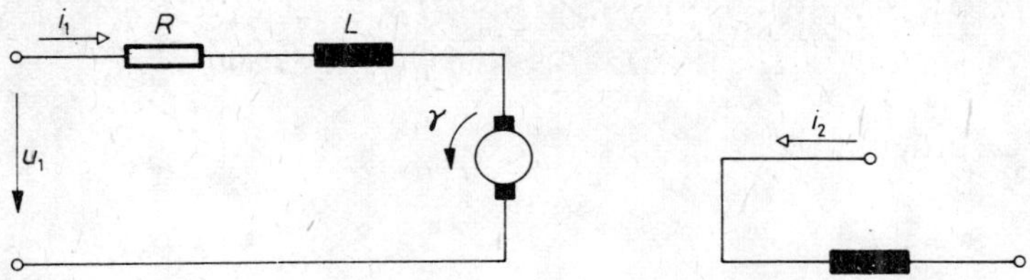

Fig. 7.26

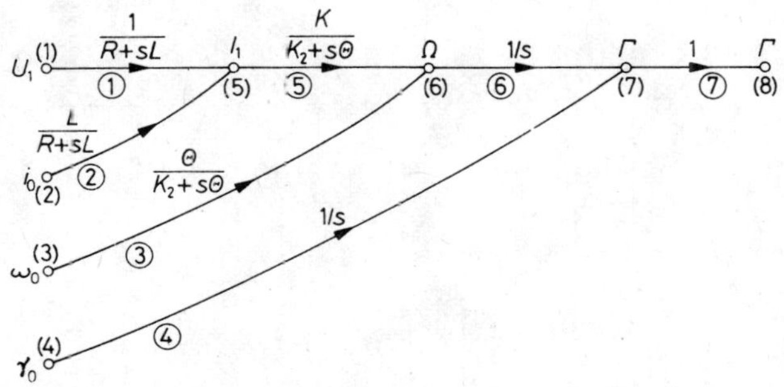

Fig. 7.27

Hence the signal flow graph shown in Fig. 7.27 may be obtained. The matrices characterizing the graph are:

$$\boldsymbol{M}=\begin{bmatrix} 1 & 0 & 0 & 0 & 0 & 0 & 0 \\ 0 & 1 & 0 & 0 & 0 & 0 & 0 \\ 0 & 0 & 1 & 0 & 0 & 0 & 0 \\ 0 & 0 & 0 & 1 & 0 & 0 & 0 \\ 0 & 0 & 0 & 0 & 1 & 0 & 0 \\ 0 & 0 & 0 & 0 & 0 & 1 & 0 \\ 0 & 0 & 0 & 0 & 0 & 0 & 1 \end{bmatrix},$$

$$\boldsymbol{N}=\begin{bmatrix} 1 & 1 & 0 & 0 & 0 & 0 & 0 \\ 0 & 0 & 1 & 0 & 1 & 0 & 0 \\ 0 & 0 & 0 & 1 & 0 & 1 & 0 \\ 0 & 0 & 0 & 0 & 0 & 0 & 1 \end{bmatrix},$$

$$W = \left\langle \frac{1}{R+sL} \quad \frac{L}{R+sL} \quad \frac{\Theta}{K_2+s\Theta} \quad \frac{1}{s} \quad \frac{K_1}{K_2+s\Theta} \quad \frac{1}{s} \quad 1 \right\rangle.$$

Hence, in accordance with (7.23):

$$W_t = \left[\begin{array}{cccc|ccc} \frac{1}{R+sL} & \frac{L}{R+sL} & 0 & 0 & 0 & 0 & 0 \\ 0 & 0 & \frac{\Theta}{K_2+s\Theta} & 0 & \frac{K_1}{K_2+s\Theta} & 0 & 0 \\ 0 & 0 & 0 & \frac{1}{s} & 0 & \frac{1}{s} & 0 \\ \hline 0 & 0 & 0 & 0 & 0 & 0 & 1 \end{array}\right]$$

and substituting into (7.27):

$$Y = \Gamma(s) =$$

$$= [0 \;\; 0 \;\; 1] \begin{bmatrix} 1 & 0 & 0 \\ -\frac{K_1}{K_2+s\Theta} & 1 & 0 \\ 0 & -\frac{1}{s} & 1 \end{bmatrix}^{-1} \begin{bmatrix} \frac{1}{R+sL} & \frac{L}{R+sL} & 0 & 0 \\ 0 & 0 & \frac{\Theta}{K_2+s\Theta} & 0 \\ 0 & 0 & 0 & \frac{1}{s} \end{bmatrix} \begin{bmatrix} U(s) \\ i_0 \\ \omega_0 \\ g_0 \\ \gamma_0 \end{bmatrix} =$$

$$= \left[\frac{K_1}{s(K_2+s\Theta)} \quad \frac{1}{s} \quad 1 \right] \begin{bmatrix} \frac{U_1(s)}{R+sL} + \frac{Li_0}{R+sL} \\ \frac{\Theta\omega_0}{K_2+s\Theta} \\ \frac{\gamma_0}{s} \end{bmatrix},$$

i.e.

$$\Gamma(s) = \frac{K_1}{s(K_2+s\Theta)(R+sL)} U_1(s) + \frac{K_1 L}{s(K_2+s\Theta)(R+sL)} i_0 +$$

$$+ \frac{\Theta}{s(K_2+s\Theta)} \omega_0 + \frac{1}{s} \gamma_0$$

is the Laplace transform of the angular rotation.

CHAPTER 8

SAMPLED-DATA AND DIGITAL SYSTEMS

The sampling process

Sampled-data and digital systems are frequently employed in control, communication and measurement engineering. For the examination of such systems, the mathematical description of the signals arising in the course of the sampling process is first presented [10, 32, 46, 47].

In the course of the *sampling process*, a signal sequence $f_1^*(t)$ is produced from a signal $f(t)$ (Fig. 8.1, a), appearing from the time $t=0$ onwards at every time interval T, for duration τ, in which the values of the signal in the time ranges $nT \leq t \leq nT+\tau$ $(n=0, 1, 2, \ldots)$ equal those assumed by the signal $f(t)$ in the same time ranges, and are otherwise zero (Fig. 8.1, b). In the following, $f(t)$ is called a *continuous* signal, while $f_1^*(t)$ is a *sampled* or *pulse amplitude modulated signal.* The sampling process is carried out by a sampling element. In the block diagrams of systems, the sampling element is denoted as shown in Fig. 8.2. The sampling element may be regarded as a switch, closing at instants $0, T, 2T, 3T, \ldots, nT, \ldots$, staying closed for a duration τ, and then opening. The continuous signal $f(t)$ (Fig. 8.1, a) is applied to the input of the sampling element, and the sampled signal $f_1^*(t)$ (Fig. 8.1, b) is its output. The effect of the sampling element can be described with the aid of the carrier signal $1^*(t)$ (Fig. 8.3). $1^*(t)$ may be expressed in terms of the unit step function $1(t)$ as follows:

$$1^*(t) = \sum_{n=0}^{\infty} [1(t-nT) - 1(t-nT-\tau)] . \tag{8.1}$$

$1^*(t)$ can also be written with the aid of the Dirac-impulse $\delta(t, \tau)$ occurring at the instant t, with duration τ, and defined by

$$\delta(t, \tau) = \begin{cases} 1/\tau & 0 < t < \tau \\ 0 & t < 0, \qquad t \geq \tau \end{cases} \tag{8.2}$$

(see Fig. 8.4). The "area" under such an impulse is 1. Hence

$$1^*(t) = \tau \sum_{n=0}^{\infty} \delta(t-nT, \tau). \tag{8.3}$$

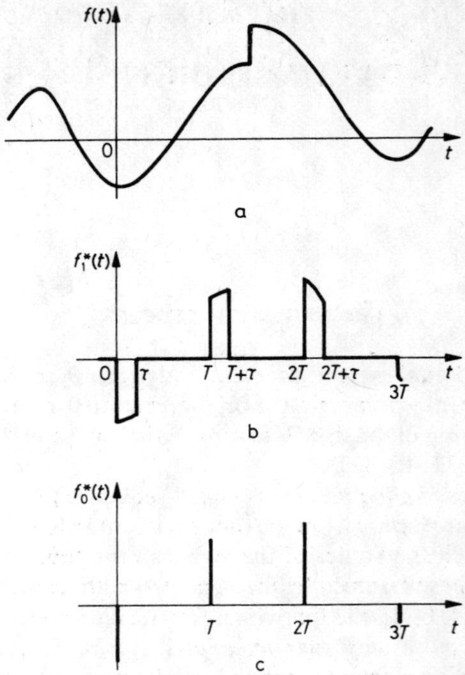

Fig. 8.1

The modulated signal $f_1^*(t)$ (Fig. 8.1, b), appearing after the switch is

$$f_1^*(t) = f(t)\,1^*(t) = f(t) \sum_{n=0}^{\infty} [1(t-nT) - 1(t-nT-\tau)] =$$

$$= \tau f(t) \sum_{n=0}^{\infty} \delta(t-nT, \tau) \tag{8.4}$$

$f_1^*(t)$ is the signal appearing at the output of the sampling element in the case of so-called physical sampling. If $\tau \ll T$ and $\tau \ll T_{0\min}$, where $T_{0\min}$ is the smallest time constant of the network to which the sampled signal is applied, a "mathematical" model of the sampled signal may be employed instead of $f_1^*(t)$, as follows. For $\tau \to 0$, $f_1^*(t) \to f_0^*(t)$ (Fig. 8.1, c) and thus the sampled signal may be described by $f_0^*(t)$. For the analysis, however, the use of the signal $f^*(t) = \lim\limits_{\tau \to 0} f_0^*(t)/\tau$ is appropriate:

$$f^*(t) = \lim_{\tau \to 0} \frac{1}{\tau} f(t)\,1^*(t) =$$

$$= \lim_{\tau \to 0} \frac{1}{\tau} f(t) \sum_{n=0}^{\infty} [1(t-nT) - 1(t-nT-\tau)] =$$

$$= f(t) \sum_{n=0}^{\infty} \delta(t-nT), \quad (8.5)$$

where

$$\delta(t-nT) = \lim_{\tau \to 0} \delta(t-nT, \tau) = \begin{cases} 0, & \text{for} \quad t \neq nT \\ \infty, & \text{for} \quad t = nT \end{cases} \quad (8.6)$$

and

$$\int_{-\infty}^{+\infty} \delta(t-nT)\, \mathrm{d}t = 1 . \quad (8.7)$$

$\delta(t-nT)$ is called the *Dirac delta*. In the following discussion, $f^*(t)$ is used for the description of the sampled signal.

It should be noted that the spectrum of a continuous bandlimited signal can be reconstructed from the spectrum of the sampled signal (Shannon's sampling theorem).

Sampled-data systems utilize the values of the sampled signals at the output of their basic elements instead of those of the continuous signals. If the input signal of a

Table 8.1

$f(t)$	$f(nT)$	$f^*(t)$	$F(s)$	$F^*(s)$	$F(z)$
$\delta(t)$	$\delta(nT) = \begin{cases} \infty, & \text{if } n=0 \\ 0, & \text{if } n \neq 0 \end{cases}$	$\delta(t)$	1	1	1
$\delta(t-kT)$	$\delta(nT-kT) = \begin{cases} \infty, & \text{if } n=k \\ 0, & \text{if } n \neq k \end{cases}$	$\delta(t-kT)$	e^{-skT}	e^{-skT}	z^{-k}
$1(t)$	1	$\sum_{n=0}^{\infty} \delta(t-nT)$	$\frac{1}{s}$	$\frac{1}{1-e^{-sT}}$	$\frac{z}{z-1}$
t	nT	$\sum_{n=0}^{\infty} nT\, \delta(t-nT)$	$\frac{1}{s^2}$	$\frac{Te^{-sT}}{(1-e^{-sT})^2}$	$\frac{Tz}{(z-1)^2}$
$e^{-\alpha t}$	$e^{-\alpha nT}$	$\sum_{n=0}^{\infty} e^{-\alpha nT}\, \delta(t-nT)$	$\frac{1}{s+\alpha}$	$\frac{1}{1-e^{-(\alpha+s)T}}$	$\frac{z}{z-e^{-\alpha T}}$
$a^{t/T}$	a^n	$\sum_{n=0}^{\infty} a^n\, \delta(t-nT)$	$\frac{1}{s-\frac{1}{T}\ln a}$	$\frac{1}{1-a\,e^{-sT}}$	$\frac{z}{z-a}$

basic element in the network is a sampled (discrete-time) signal, the output signal of the basic element is, with the exception of some special digital basic elements, a continuous signal. In the case of *discrete-time basic elements*, the output signal is sampled with the same period T, synchronously with the input signal. A non-discrete-time network part together with a sampling element, connected as shown in

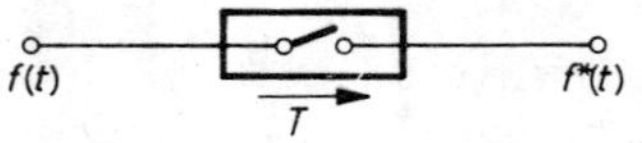

Fig. 8.2

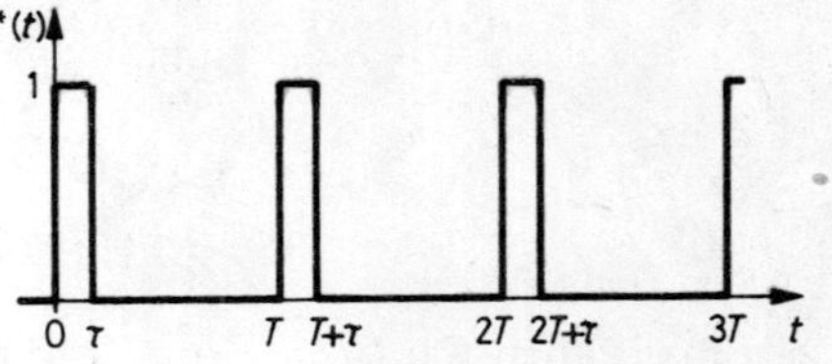

Fig. 8.3

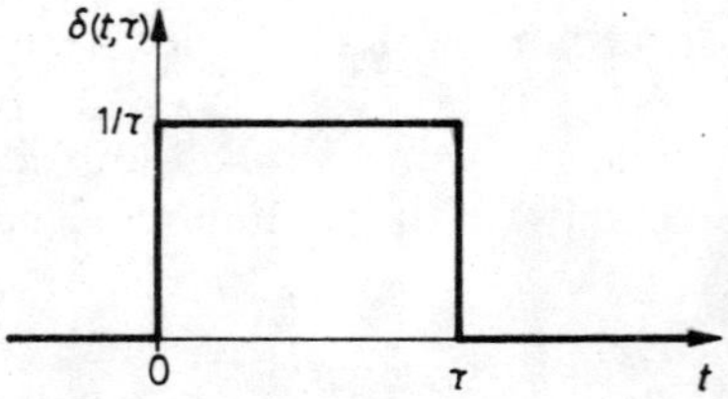

Fig. 8.4

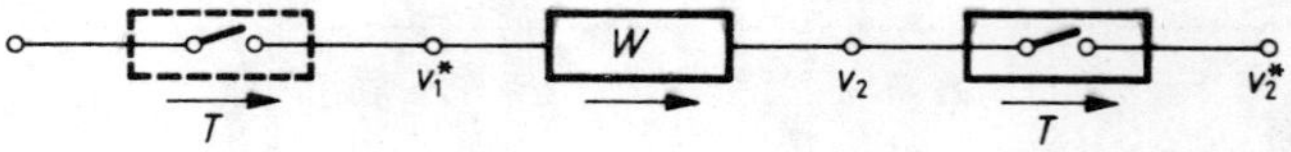

Fig. 8.5

Fig. 8.5, can be regarded as a discrete time basic element. If a system is constructed of discrete-time basic elements only, and thus all of its signals are discrete-time, it is called a *discrete-time system*. If the signals formed by sequences of pulses with varying amplitudes in a discrete-time system are encoded into sequences of constant-amplitude pulses (e.g. by analogue-to-digital conversion to binary code), and these sequences are processed digitally to yield response signals which are then decoded (e.g. by digital-to-analogue conversion) into pulse-amplitude modulated signals, the system is called a *digital system*. If both discrete-time and continuous signals occur among the input and output signals of basic elements, the system is termed a *sampled-data system*. The use of these terms in the literature is not always consistent, but it is necessary to distinguish between them for the following treatment, because of the different methods of analysis.

The discrete Laplace transformation and z-transformation

Since the value of $f^*(t)$ is zero with the exception of the sampling instants, according to (8.5):

$$f^*(t)=\sum_{n=0}^{\infty} f(t)\delta(t-nT)=\sum_{n=0}^{\infty} f(nT)\delta(t-nT). \tag{8.8}$$

If the right and left limits of $f(t)$ at $t=nT$ are not equal, $f(nT)$ means by agreement the right limit value $f(nT+0)$. $f^*(t)$ is seen to be reconstructible from a knowledge of $f(nT)$ ($n=0, 1, 2, \ldots$) according to (8.8). Let us form the Laplace transform of $f^*(t)$ on the basis of (8.8):

$$\mathcal{L}f^*(t)=F^*(s)=\sum_{n=0}^{\infty} f(nT)e^{-snT}. \tag{8.9}$$

The Laplace transform of the sampled signal $f^*(t)$ is called the *discrete Laplace transform.*

The z-transformation [10, 32, 46, 47] is obtained from the discrete Laplace transform by the substitution

$$z=e^{sT} \qquad s=\frac{1}{T}\ln z \tag{8.10}$$

i.e.

$$\mathcal{Z}f(t)=F(z)=[F^*(s)]|_{s=\frac{1}{T}\ln z}=\sum_{n=0}^{\infty} f(nT)z^{-n} \tag{8.11}$$

is the z-transform of $f(t)$. A necessary and sufficient condition of the existence of the z-transform is that $M>0$, $\alpha>0$ and $n_0>0$ can be found, such that

$$|f(nT)|\le Me^{\alpha nT} \qquad \text{if} \quad n>n_0. \tag{8.12}$$

Some theorems concerning the z-transformation are presented below:

1. *Theorem of linearity.* The z-transformation is a linear operation, hence sums and proportions remain unchanged:

$$\mathcal{Z}[af_1(t)\pm bf_2(t)]=aF_1(z)\pm bF_2(z) \tag{8.13}$$

where $\mathcal{Z}f_1(t)=F_1(z)$, $\mathcal{Z}f_2(t)=F_2(z)$ and a and b are constants.

2. *Translation theorems.* The function $f(t-kT)$ ($k=0, 1, 2, \ldots$) is obtained from $f(t)$ by translation in the time domain by kT. If T is the sampling period, and sampling starts at $t=0$, the z-transform of $1(t)f(t-kT)$ is used in our calculations:

$$\mathcal{Z}1(t)f(t-kT)=z^{-k}F(z)+z^{-k}\sum_{n=1}^{k} f(-nT)z^{n} \tag{8.14}$$

where $\mathcal{Z}f(t)=F(z)$.

If the starting time of sampling is $t=kT$, the formula

$$\mathscr{Z}1(t-kT)f(t-kT)=z^{-k}F(z) \tag{8.15}$$

can be used. The translation by $-kT$ of $f(t)$ in the time domain is given by $f(t+kT)$:

$$\mathscr{Z}1(t)f(t+kT)=z^kF(z)-z^k\sum_{n=0}^{k-1}f(nT)z^{-n} \tag{8.16}$$

The starting time of sampling here, as shown by the factor $1(t)$, is $t=0$.

3. *Theorem of scale changing.* Multiplication by $e^{\alpha t}$ in the time domain corresponds to a change of scale in the z-domain, i.e.

$$\mathscr{Z}[e^{\alpha t}f(t)]=F(e^{-\alpha T}z), \qquad \alpha \text{ is real}. \tag{8.17}$$

4. According to the *initial value theorem:*

$$\lim_{t\to 0} f^*(t)=\lim_{z\to\infty} F(z) \tag{8.18}$$

if the limit exists.

5. According to the *final value theorem:*

$$\lim_{t\to\infty} f^*(t)=\lim_{z\to 1}(1-z^{-1})F(z) \tag{8.19}$$

if $\mathscr{Z}f(t)=F(z)$ has no pole on or outside the unit circle in the z-domain.

(8.11) yields the z-transform in the form of an infinite series. In many cases, the z-transform can be written in closed form as well. E.g. the z-transform of the function

$$f(t)=1(t)e^{-\alpha t} \qquad \alpha\geq 0$$

sampled with period T may be determined as follows. On the one hand, the translation theorem (8.16) yields:

$$\mathscr{Z}1(t)e^{-\alpha(t+T)}=zF(z)-z\,,$$

on the other hand:

$$\mathscr{Z}1(t)e^{-\alpha(t+T)}=e^{-\alpha T}\mathscr{Z}1(t)e^{-\alpha t}=e^{-\alpha T}F(z)\,.$$

From the two equations:

$$zF(z)-z=e^{-\alpha T}F(z)$$

i.e.

$$F(z)=\frac{z}{z-e^{-\alpha T}}$$

is the z-transform sought. If $\alpha=0$, $f(t)=1(t)$, and

$$\mathscr{Z}1(t)=\frac{z}{z-1}\,.$$

The z-transforms of a few functions have been written in Table 8.1.

For the inverse z-transformation, the integral formula

$$f(nT) = \frac{1}{2\pi j} \oint_{\Gamma} F(z)\, z^{n-1}\, dz \qquad n=0, 1, 2, \ldots \tag{8.20}$$

may be used, based upon the formula for the inverse Laplace transformation, where Γ is a circle on the complex plane with center at the origin, and with radius r, with all the poles of $F(z)$ being inside Γ. The integral may be evaluated with the aid of the residue theorem. The inverse z-transform yields the values of the signal at the instants of sampling only. Knowing $f(nT)$ ($n=0, 1, 2, \ldots$), $f^*(t)$ may be written in accordance with (8.8).

The inverse transform may be determined by expansion into power series as follows. The function $F(z)$ is written as a power series of the variable z^{-1}:

$$F(z) = c_0 + c_1 z^{-1} + c_2 z^{-2} + \ldots + c_n z^{-n} + \ldots \tag{8.21}$$

Upon substituting $z = e^{sT}$, forming the inverse Laplace transforms of the terms,

$$f^*(t) = c_0 \delta(t) + c_1 \delta(t-T) + c_2 \delta(t-2T) + \ldots + c_n \delta(t-nT) + \ldots \tag{8.22}$$

is obtained.

In practical cases, $F(z)$ is frequently a rational function, i.e.

$$F(z) = \frac{b_0 + b_1 z^{-1} + b_2 z^{-2} + \ldots + b_m z^{-m}}{1 + a_1 z^{-1} + a_2 z^{-2} + \ldots + a_p z^{-p}} =$$

$$= \frac{1}{z^{m-p}} \frac{b_0 z^m + b_1 z^{m-1} + b_2 z^{m-2} + \ldots + b_m}{z^p + a_1 z^{p-1} + a_2 z^{p-2} + \ldots + a_p}. \tag{8.23}$$

Dividing the numerator by the denominator, a form similar to (8.21) is obtained, and the z-transform can be written in accordance with (8.22). E.g. in the case of

$$F(z) = \frac{2z^3 - 3z^2 + z + 1}{z^3 - 3z^2 + 2z} = \frac{1}{z} \frac{2z^3 - 3z^2 + z + 1}{z^2 - 3z + 2}$$

$m=3$, $p=2$. To determine the inverse transform of $F(z)$, the numerator is divided by the denominator, to yield

$$F(z) = 2 + 3z^{-1} + 6z^{-2} + 13z^{-3} + \ldots$$

and thus

$$f(0) = 2, \qquad f(T) = 3, \qquad f(2T) = 6, \qquad f(3T) = 13, \ \ldots$$

is the inverse transform.

The transfer-function of a discrete-time basic element

The discrete-time basic element drawn in Fig. 8.5 consists of a network part characterized by a weighting function $w(t)$ and a sampling element with period T. The input signal of the basic element is $v_1^*(t)$ sampled by period T. The sampling element operates synchronously with the one producing $v_1^*(t)$, denoted by dotted line in Fig. 8.5.

If the input signal of the basic element is $\delta(t)$, the output signal $v_2(t)$ of the network part is $w(t)$. Thus $\mathscr{L}w(t) = W(s)$ is the transfer-function of the network part.

The input signal of the basic element is

$$v_1^*(t) = \sum_{n=0}^{\infty} v_1(t)\,\delta(t-nT) = \sum_{n=0}^{\infty} v_1(nT)\,\delta(t-nT). \tag{8.24}$$

Its discrete Laplace transform is

$$V_1^*(s) = \sum_{n=0}^{\infty} v_1(nT)\,e^{-nsT}, \tag{8.25}$$

while its z-transform is

$$V_1(z) = \sum_{n=0}^{\infty} v_1(nT)\,z^{-n}. \tag{8.26}$$

$v_2(t)$ is due to the impulse sequence (8.24). If $v_1(nT)\,\delta(t-nT)$ is the input signal, $v_2(t) = 1(t-nT)\,v_1(nT)\,w(t-nT)$. Thus, as a result of the input signal in (8.24):

$$v_2(t) = \sum_{n=0}^{\infty} 1(t-nT)\,v_1(nT)\,w(t-nT) \tag{8.27}$$

The n-th term of the sum is zero if $t<nT$, i.e. the summation need only be carried out for $n = \left[\frac{t}{T}\right]$ terms to obtain $v_2(t)$ for $t<nT$. $\mathscr{L}v_2(t) = V_2(s)$ is obtained from (8.25) upon multiplication by the Laplace transform of the weighting function, i.e.

$$V_2(s) = V_1^*(s)\,W(s) = W(s)\sum_{n=0}^{\infty} v_1(nT)\,e^{-nsT}. \tag{8.28}$$

The time function of the signal at the output is

$$v_2^*(t) = \sum_{m=0}^{\infty} v_2(t)\,\delta(t-mT) =$$

$$= \sum_{m=0}^{\infty}\sum_{n=0}^{\infty} 1(t-nT)\,v_1(nT)\,w(t-nT)\,\delta(t-mT). \tag{8.29}$$

It suffices to carry out the first summation from $m=n$, since in the case of $m<n$, the impulse, the effect of which is being examined, has not yet appeared. Thus

$$v_2^*(t) = \sum_{m=n}^{\infty}\sum_{n=0}^{\infty} v_1(nT)\,w(mT-nT)\,\delta(t-mT). \tag{8.30}$$

Its discrete Laplace transform is

$$V_2^*(s) = \sum_{m=n}^{\infty} \sum_{n=0}^{\infty} v_1(nT)\, w(mT-nT)\, e^{-msT}. \tag{8.31}$$

Let us introduce the notation

$$m-n=p, \qquad m=p+n. \tag{8.32}$$

Hence

$$V_2^*(s) = \sum_{p=0}^{\infty} \sum_{n=0}^{\infty} v_1(nT)\, w(pT)\, e^{-psT}\, e^{-nsT}, \tag{8.33}$$

i.e.

$$V_2^*(s) = \sum_{p=0}^{\infty} w(pT)\, e^{-psT} \sum_{n=0}^{\infty} v_1(nT)\, e^{-nsT}, \tag{8.34}$$

i.e. according to the definition (8.9) of the discrete Laplace transform,

$$V_2^*(s) = W^*(s)\, V_1^*(s). \tag{8.35}$$

Upon substituting $z=e^{sT}$, z-transforms are obtained, thus

$$V_2(z) = W(z)\, V_1(z). \tag{8.36}$$

On the basis of the above considerations, the discrete Laplace transform and z-transform of the output signal in the system drawn in Fig. 8.5 can be determined. The function

$$W(z) = \frac{V_2(z)}{V_1(z)} = \sum_{p=0}^{\infty} w(pT)\, z^{-p} \tag{8.37}$$

is called the *pulse transfer-function* or *z-transfer function*.

Signal flow graphs and state equations of digital systems

In the following section, digital systems as defined previously will be dealt with. In digital systems (for example in digital computers) the conversion of pulse sequences with varying amplitudes into coded sequences of constant amplitude pulses can be achieved only with certain quantization errors. These errors are not taken into account in our analysis.

The excitation signal $r(t)$ and response signal $y(t)$ of digital systems are considered to be sampled signals as described by (8.5). The relationship between the z-transforms of $r(t)$ and $y(t)$ is given by

$$W(z) = \frac{Y(z)}{R(z)} = \frac{b_0 z^m + b_1 z^{m-1} + b_2 z^{m-2} + \ldots + b_m}{z^p + a_1 z^{p-1} + a_2 z^{p-2} + \ldots + a_p}. \tag{8.38}$$

$$p \geq m.$$

Hence

$$Y(z)\,(z^p+a_1z^{p-1}+a_2z^{p-2}+\ldots+a_p)=$$
$$=R(z)\,(b_0z^m+b_1z^{m-1}+b_2z^{m-2}+\ldots+b_m)\,. \qquad (8.39)$$

Three elementary operations appear in this formula: multiplication by a constant, addition and translation by $\pm T$ in the time domain. The following holds for their z-transforms in accordance with (8.13), (8.14) and (8.16):

$$\mathscr{Z}[av(t)]=a\mathscr{Z}v(t)=aV(z) \qquad (8.40)$$

$$\mathscr{Z}[v_1(t)+v_2(t)]=\mathscr{Z}v_1(t)+\mathscr{Z}v_2(t)=V_1(z)+V_2(z) \qquad (8.41)$$

If $v_1(t)=v_2(t+T)$ or $v_2(t)=v_1(t-T)$, their z-transforms are:

$$V_1(z)=zV_2(z)-zv_2(0) \qquad (8.42)$$

and

$$V_2(z)=z^{-1}V_1(z)+v_2(0)\,, \qquad (8.43)$$

i.e. the determination of the z-transform necessitates a knowledge of $v_2(0)=v_1(-T)$. The signal flow graphs corresponding to the three elementary operations have been drawn in Figs 8.6, a, b, c in accordance with (8.40), (8.41) and (8.43).

According to (8.39) the relationship

$$y^*(t+pT)+a_1y^*(t+pT-T)+a_2y^*(t+pT-2T)+\ldots+a_py^*(t)=$$
$$=b_0r^*(t+mT)+b_1r^*(t+mT-T)+\ldots+b_mr^*(t) \qquad (8.44)$$

may be written for the time functions of the excitation and response signals. This is a difference equation of order p, the solution of which necessitates in general the specification of p initial conditions.

Accordingly, the digital system under examination may be regarded a system similar to those discussed in Chapter 7, with the z-transfer functions of the basic

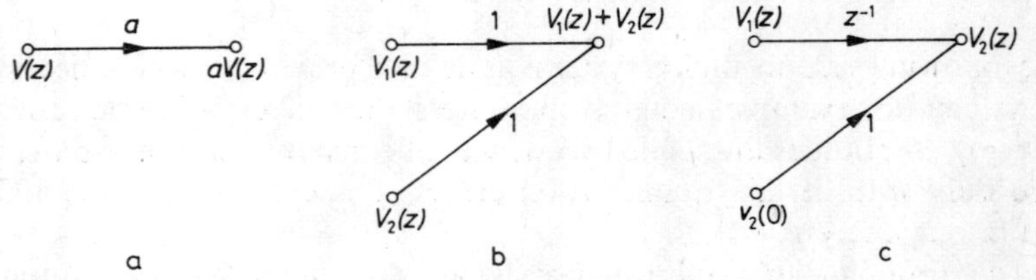

Fig. 8.6

elements being constant or z^{-1}. Naturally, the equations written here relate, in contrast to those in chapter 7, to z-transforms. These may be rewritten for discrete Laplace transforms upon substituting $z=e^{sT}$.

Equations (8.38) and (8.39) may be employed for the construction of a signal flow graph of the digital system. The signal flow graph is obtained, similarly to the procedure presented in chapter 7, by the direct, iterative or parallel method. The variables associated with the vertices here are z-transforms of signals, and the transfer functions corresponding to the edges are z^{-1} or constants. An example for the direct method is shown in Fig. 8.9.

The diagonal matrix

$$\boldsymbol{W}(z) = \langle W_1(z) \quad W_2(z) \quad \ldots \quad W_b(z) \rangle \tag{8.45}$$

may be formed from the z-transfer functions associated with the edges of the signal flow graph, similarly to (7.20). Its rows and columns correspond to the edges of the signal flow graph. To characterize the graph, the matrices

$$\boldsymbol{M}_t = \frac{1}{2}(\boldsymbol{A}_{t0} + \boldsymbol{A}_t) + \boldsymbol{L} \tag{8.46}$$

$$\boldsymbol{N}_t = \frac{1}{2}(\boldsymbol{A}_{t0} + \boldsymbol{A}_t) + \boldsymbol{L} \tag{8.47}$$

are formed from the non-basis incidence matrices $\boldsymbol{A}_t$ including orientations and $\boldsymbol{A}_{t0}$ disregarding orientations as well as from matrix $\boldsymbol{L}$ characterizing the incidence of self-loops, if any, with vertices. Since the elements in the rows corresponding to response signals in $\boldsymbol{M}_t$ and in those corresponding to excitation signals in $\boldsymbol{N}_t$ are all zero, these rows are deleted from $\boldsymbol{M}_t$ and $\boldsymbol{N}_t$ to yield the matrices $\boldsymbol{M}$ and $\boldsymbol{N}$. These are employed to write the vertex transfer matrix

$$\boldsymbol{W}_t = \boldsymbol{N}\boldsymbol{W}\boldsymbol{M}^+ \tag{8.48}$$

used in our calculations. $\boldsymbol{W}_t$ has been seen to be capable of also being written directly from the signal flow graph.

On the basis of the signal flow graph state equations of the system may be written, which, in the case of digital systems, are a set of difference equations [10, 11, 32, 46, 48]. According to (8.44), the transfer functions of the edges in the signal flow graph, constructed by one of the three methods mentioned, are z^{-1} or constants. Multiplication by z^{-1} means a delay by T in the time domain. If the output signal of the basic element corresponding to such an edge is $x^*(t)$, its input signal is $x^*(t+T)$. Since the value of $x^*(t+T)$ is zero for $t<0$, and thus also for $t=-T$, $x^*(0)=0$, i.e. the initial value of $x^*(t)$ is zero. If at $t=0$, i.e. at the switching on of the excitation signal $r^*(t)$ of the system nonzero signals are present in the system, even in the case $r^*(0)=0$ the initial values of the output signals $x^*(t)$ of the edges with transfer function z^{-1} are not in general zero. In such cases, a further edge with transfer function 1 points towards the vertex associated with $\mathscr{Z}x^*(t)$ in the signal flow graph as well as the edge with transfer function z^{-1}. The other vertex of this edge is associated with $x(0)$, i.e. with the initial value of $x^*(t)$ (Fig. 8.7). Such signal flow graphs, augmented to take the initial values into account, are shown e.g. in Fig. 8.11 (iterative method) and Fig. 8.12 (parallel method).

To write the state equations, the z-transforms of the state variables are selected from the signals associated with the vertices of the signal flow graph of the system.

They are those signals which correspond to vertices having an edge with transfer function z^{-1} pointing towards them. In the case of the use of the method discussed, at most one other edge may point towards each such vertex. This other edge must have a unity transfer function and points away from the vertex associated with the initial value of the state variable. After the selection of the state variables, it is expedient to simplify the signal flow graph, by drawing edges together, so that the

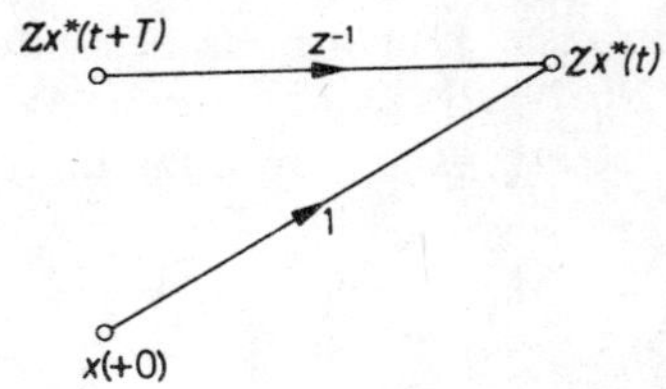

Fig. 8.7

vertices connected to edges with transfer function z^{-1} become the only internal vertices.

The signals associated with the vertices of the signal flow graph may be classified into the following four groups:

1. the z-transforms of excitation signals $r_i^*(t)$ $(i=1, 2, \ldots, n_1)$, the initial values $x_j(0)$ of state variables $(j=1, 2, \ldots, n_2)$;
2. the z-transforms of $x_j^*(t+T)$;
3. the z-transforms of state variables $x_j^*(t)$;
4. the z-transforms of response signals $y_k^*(t)$ $(k=1, 2, \ldots, n_3)$.

Let the vertices be assigned order numbers in the order of the classification, ensuring that in the 1st group order numbers are first given to the vertices corresponding to the excitation signals and then to those associated with initial values, and further, that the signals in the 1st, 2nd and 3rd groups are arranged in the same order of the index j. The column matrices of the signals in the 1st, 2nd, 3rd and 4th groups are denoted by $\mathbf{R}(z)$, $\mathbf{X}_T(z)$, $\mathbf{X}(z)$ and $\mathbf{Y}(z)$, respectively. It is noted, that according to (8.16)

$$\mathbf{X}_T(z)=\mathscr{Z}\mathbf{x}^*(t+T)=z\mathbf{X}(z)-z\mathbf{x}(0)\,, \tag{8.49}$$

where $\mathbf{x}(0)$ is the column matrix formed by the initial values of the state variables.

On the basis of the signal flow graph the vertex transfer matrix $\boldsymbol{W}_t$, and hence the relationship

$$\begin{bmatrix} \mathbf{X}_T(z) \\ \mathbf{X}(z) \\ \mathbf{Y}(z) \end{bmatrix} = \boldsymbol{W}_t \begin{bmatrix} \mathbf{R}(z) \\ \mathbf{X}_T(z) \\ \mathbf{X}(z) \end{bmatrix} = \begin{bmatrix} \boldsymbol{W}_{11} & \boldsymbol{W}_{12} & \boldsymbol{W}_{13} \\ \boldsymbol{W}_{21} & \boldsymbol{W}_{22} & \boldsymbol{W}_{23} \\ \boldsymbol{W}_{31} & \boldsymbol{W}_{32} & \boldsymbol{W}_{33} \end{bmatrix} \begin{bmatrix} \mathbf{R}(z) \\ \mathbf{X}_T(z) \\ \mathbf{X}(z) \end{bmatrix} \tag{8.50}$$

between the signals associated with the vertices may be written. Hence

$$\mathbf{X}_T(z)=(\boldsymbol{1}-\boldsymbol{W}_{12})^{-1}\,\boldsymbol{W}_{13}\mathbf{X}(z)+(\boldsymbol{1}-\boldsymbol{W}_{12})^{-1}\,\boldsymbol{W}_{11}\mathbf{R}(z) \tag{8.51}$$

is the z-transform of the state equations of the system, while

$$\mathbf{Y}(z)=[\boldsymbol{W}_{32}(\boldsymbol{1}-\boldsymbol{W}_{12})^{-1}\,\boldsymbol{W}_{13}+\boldsymbol{W}_{33}]\,\mathbf{X}(z)+$$
$$+[\boldsymbol{W}_{32}(\boldsymbol{1}-\boldsymbol{W}_{12})^{-1}\,\boldsymbol{W}_{11}+\boldsymbol{W}_{31}]\,\mathbf{R}(z) \qquad (8.52)$$

is that of the response signals. Since among the blocks of $\boldsymbol{W}_t$, $\boldsymbol{W}_{21}$, $\boldsymbol{W}_{22}$ and $\boldsymbol{W}_{23}$ may only have z^{-1} elements, the remaining elements being constant, (8.51) yields

$$\mathbf{x}^*(t+T)=(\boldsymbol{1}-\boldsymbol{W}_{12})^{-1}\,\boldsymbol{W}_{13}\,\mathbf{x}^*(t)+(\boldsymbol{1}-\boldsymbol{W}_{12})^{-1}\,\boldsymbol{W}_{11}\,\mathbf{r}^*(t) \qquad (8.53)$$

the state equations of the system, and from (8.52):

$$\mathbf{y}^*(t)=[\boldsymbol{W}_{32}(\boldsymbol{1}-\boldsymbol{W}_{12})^{-1}\,\boldsymbol{W}_{13}+\boldsymbol{W}_{33}]\,\mathbf{x}^*(t)+$$
$$+[\boldsymbol{W}_{32}(\boldsymbol{1}-\boldsymbol{W}_{12})^{-1}\,\boldsymbol{W}_{11}+\boldsymbol{W}_{31}]\,\mathbf{r}^*(t) \qquad (8.54)$$

is the column matrix of the time functions of response signals.

Signal flow graphs and the calculation of response signals in sampled-data systems

In sampled-data systems, in contrast to digital systems, both discrete-time and continuous signals may be present. As a consequence, the calculation of response signals is often difficult. The transfer-function of the system cannot always be given, i.e. the transform of the response signal can not be written as the product of the transfer-function and the transform of the excitation signal. Certain known methods [2, 10, 34] aim at the construction of a signal flow graph, suitable for the application of Mason's formula, to determine the response signal. The application of these methods is tedious, especially in case of several excitation and response signals, and the methods are not general enough. The following method has a relatively wide applicability. The signals occurring in the system are taken into account with their Laplace transforms in the course of the calculations, since this is suitable for the description of both continuous and discrete-time signals. The system is considered to be built up of basic elements. If a network part characterized by transfer function $W_n(s)$ is followed by a sampling element, and its input signal is discrete-time it is expedient to simplify the block diagram of the system to have the network part and the following sampling element represented by one basic element with transfer function $W_n^*(s)$. Sampling processes throughout the system must occur at the same periods, synchronously.

A signal flow graph may be associated with the system, similar to that discussed in Chapter 7, with the following differences. Distinct vertices are associated with the input and output of a sampling element, but no edge represents the sampling element itself. The vertices in the signal flow graph, corresponding to the output signals of sampling elements, and to the excitation and response signals of the system, are incident with one edge only. The transfer function associated with the edge incident with a vertex corresponding to a response signal is $W(s)=1$, or some other constant. If the construction yields a signal flow graph not meeting these requirements, it is augmented so that it does meet them.

The excitation signals of the system are $R_1(s)$, $R_2(s)$, . . ., its response signals are $Y_1(s)$, $Y_2(s)$, ... Among the signals associated with the vertices, those to be sampled are denoted by $P_1(s)$, $P_2(s)$, . . ., the respective sampled signals are $P_1^*(s)$, $P_2^*(s)$, . . ., while the remaining signals are $V_1(s)$, $V_2(s)$, ... Then vertices are given order numbers as follows. The first order numbers are assigned to the vertices corresponding to the excitation signals of the system, the order numbers following are assigned to the vertices corresponding to the output signals of the sampling elements in some arbitrary order of the sampling elements. The next order numbers are assigned to the vertices associated with the input signals of the sampling elements in accordance with the above order of the sampling elements. The following order numbers correspond to the remaining internal signals, and finally follow the order numbers of the vertices associated with the response signals of the system.

The vertex transfer matrix $\boldsymbol{W}_t$ of the network is now written, as discussed in Chapter 7 (see 7.23). The column matrix of internal signals is arranged in three blocks. **P***, **P** and **V** denote the column matrices of the sampled signals, the signals to be sampled and the remaining internal signals, respectively, arranged in the order of the numbering of the vertices. Thus the relationship

$$\begin{bmatrix} \mathbf{P} \\ \mathbf{V} \\ \mathbf{Y} \end{bmatrix} = \boldsymbol{W}_t \begin{bmatrix} \mathbf{R} \\ \mathbf{P}^* \\ \mathbf{P} \\ \mathbf{V} \end{bmatrix} \tag{8.55}$$

can be written between the signals. The matrix $\boldsymbol{W}_t$ is partitioned in accordance with the blocks of the column matrices, regarding the nonsampled signals **P** and **V** as one block. I.e.

$$\begin{bmatrix} \mathbf{P} \\ \mathbf{V} \\ \hline \mathbf{Y} \end{bmatrix} = \begin{bmatrix} \boldsymbol{W}_{11} & \boldsymbol{W}_{12} & \boldsymbol{W}_{13} \\ \boldsymbol{0} & \boldsymbol{W}_{22} & \boldsymbol{W}_{23} \end{bmatrix} \begin{bmatrix} \mathbf{R} \\ \mathbf{P}^* \\ \hline \mathbf{P} \\ \mathbf{V} \end{bmatrix}, \tag{8.56}$$

where $\boldsymbol{W}_{21}=\boldsymbol{0}$ follows from the requirements stated for the signal flow graph, with the exception of a system of one basic element. From (8.56):

$$\begin{bmatrix} \mathbf{P} \\ \mathbf{V} \end{bmatrix} = \boldsymbol{W}_{11}\mathbf{R} + \boldsymbol{W}_{12}\mathbf{P}^* + \boldsymbol{W}_{13}\begin{bmatrix} \mathbf{P} \\ \mathbf{V} \end{bmatrix} \tag{8.57}$$

and

$$\mathbf{Y} = \boldsymbol{W}_{22}\mathbf{P}^* + \boldsymbol{W}_{23}\begin{bmatrix}\mathbf{P}\\ \mathbf{V}\end{bmatrix}. \tag{8.58}$$

The elements of $\boldsymbol{W}_{22}$ and $\boldsymbol{W}_{23}$ are independent of z. From (8.57):

$$\begin{bmatrix}\mathbf{P}\\ \mathbf{V}\end{bmatrix} = [\boldsymbol{1} - \boldsymbol{W}_{13}]^{-1}\,\boldsymbol{W}_{11}\mathbf{R} - [\boldsymbol{1} - \boldsymbol{W}_{13}]^{-1}\,\boldsymbol{W}_{12}\mathbf{P}^*. \tag{8.59}$$

This may alternatively be written as

$$\begin{bmatrix}\mathbf{P}\\ \mathbf{V}\end{bmatrix} = [\boldsymbol{1} - \boldsymbol{W}_{13}]^{-1}\,\boldsymbol{W}_{11}\mathbf{R} + [\boldsymbol{1} - \boldsymbol{W}_{13}]^{-1}[\boldsymbol{W}_{12}\ \ \boldsymbol{0}]\begin{bmatrix}\mathbf{P}\\ \mathbf{V}\end{bmatrix}^*. \tag{8.60}$$

Let us form its sampled function:

$$\begin{bmatrix}\mathbf{P}\\ \mathbf{V}\end{bmatrix}^* = \{[\boldsymbol{1} - \boldsymbol{W}_{13}]^{-1}\,\boldsymbol{W}_{11}\mathbf{R}\}^* + \{[\boldsymbol{1} - \boldsymbol{W}_{13}]^{-1}[\boldsymbol{W}_{12}\ \ \boldsymbol{0}]\}^*\begin{bmatrix}\mathbf{P}\\ \mathbf{V}\end{bmatrix}^* \tag{8.61}$$

hence

$$\begin{bmatrix}\mathbf{P}\\ \mathbf{V}\end{bmatrix}^* = [\boldsymbol{1} - \{[\boldsymbol{1} - \boldsymbol{W}_{13}]^{-1}[\boldsymbol{W}_{12}\ \ \boldsymbol{0}]\}^*]^{-1}\{[\boldsymbol{1} - \boldsymbol{W}_{13}]^{-1}\,\boldsymbol{W}_{11}\mathbf{R}\}^*. \tag{8.62}$$

If it is sufficient to determine $\mathbf{Y}^*$, the sampled functions of response signals are formed in (8.58), and (8.62) is substituted here:

$$\mathbf{Y}^* = [\boldsymbol{W}_{22}\ \ \boldsymbol{0}]^*\begin{bmatrix}\mathbf{P}\\ \mathbf{V}\end{bmatrix}^* + \boldsymbol{W}_{23}\begin{bmatrix}\mathbf{P}\\ \mathbf{V}\end{bmatrix}^* = \{[\boldsymbol{W}_{22}^*\ \ \boldsymbol{0}] +$$
$$+ \boldsymbol{W}_{23}\}[\boldsymbol{1} - \{[\boldsymbol{1} - \boldsymbol{W}_{13}]^{-1}[\boldsymbol{W}_{12}\ \ \boldsymbol{0}]\}^*]^{-1}\{[\boldsymbol{1} - \boldsymbol{W}_{13}]^{-1}\,\boldsymbol{W}_{11}\mathbf{R}\}^*. \tag{8.63}$$

To determine $\mathbf{Y}$, (8.60) is substituted into (8.58):

$$\mathbf{Y} = \{[\boldsymbol{W}_{22}\ \ \boldsymbol{0}] + \boldsymbol{W}_{23}[\boldsymbol{1} - \boldsymbol{W}_{13}]^{-1}[\boldsymbol{W}_{12}\ \ \boldsymbol{0}]\}\begin{bmatrix}\mathbf{P}\\ \mathbf{V}\end{bmatrix}^* +$$
$$+ \boldsymbol{W}_{23}[\boldsymbol{1} - \boldsymbol{W}_{13}]^{-1}\,\boldsymbol{W}_{11}\mathbf{R}. \tag{8.64}$$

The employment of (8.62) yields

$$\mathbf{Y} = \{[\boldsymbol{W}_{22}\ \ \boldsymbol{0}] + \boldsymbol{W}_{23}[\boldsymbol{1} - \boldsymbol{W}_{13}]^{-1}[\boldsymbol{W}_{12}\ \ \boldsymbol{0}]\}[\boldsymbol{1} -$$
$$- \{[\boldsymbol{1} - \boldsymbol{W}_{13}]^{-1}[\boldsymbol{W}_{12}\ \ \boldsymbol{0}]\}^*]^{-1}\{[\boldsymbol{1} - \boldsymbol{W}_{13}]^{-1}\,\boldsymbol{W}_{11}\mathbf{R}\}^* +$$
$$+ \boldsymbol{W}_{23}[\boldsymbol{1} - \boldsymbol{W}_{13}]^{-1}\,\boldsymbol{W}_{11}\mathbf{R}, \tag{8.65}$$

the Laplace transform of the response signals.

Formula (8.52) has been seen to permit the determination of the z-transform of the response signal in a digital system and hence the time function of the response signal, knowing the z-transforms of the state variables. A solution of a different nature for this problem is given by (8.65). This latter equation is applicable not only for digital but also for sampled-data systems, i.e. for systems in which both discrete-time and continuous signals are present. The Laplace and discrete Laplace transformations have been used for the calculation.

Examples

1. The signal flow graph of the system characterized by transfer function

$$W(z) = \frac{Y(z)}{R(z)} = \frac{b_0 + b_1 z^{-1} + b_2 z^{-2}}{1 + a_1 z^{-1} + a_2 z^{-2}}$$

will be now constructed. From the transfer function:

$$Y(z) = b_0 P(z) + b_1 z^{-1} P(z) + b_2 z^{-2} P(z)\,,$$

$$P(z) = R(z) - a_1 z^{-1} P(z) - a_2 z^{-2} P(z)\,.$$

The vertices of the signal flow graph are associated with the functions $R(z)$, $P(z)$, $z^{-1} P(z)$, $z^{-2} P(z)$, $Y(z)$, and the former two equations permit the construction of the signal flow graph shown in Fig. 8.8.

To write the state equations, two state variables are selected, in accordance with the fact that the denominator of $W(z)$ is of second order, and under the assumption that it has two distinct real roots. The two state variables are $X_1(z) = z^{-1} P(z)$ and

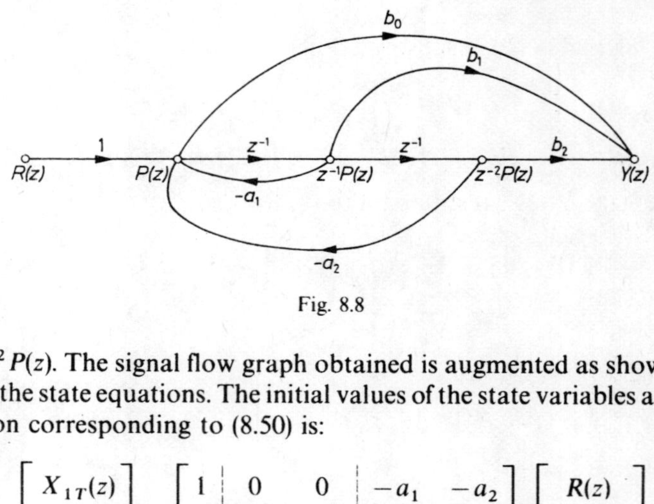

Fig. 8.8

$X_2(z) = z^{-2} P(z)$. The signal flow graph obtained is augmented as shown in Fig. 8.9 for writing the state equations. The initial values of the state variables are zero. Thus the equation corresponding to (8.50) is:

$$\begin{bmatrix} X_{1T}(z) \\ X_{2T}(z) \\ \hdashline X_1(z) \\ X_2(z) \\ \hdashline Y(z) \end{bmatrix} = \begin{bmatrix} 1 & 0 & 0 & -a_1 & -a_2 \\ 0 & 0 & 0 & 1 & 0 \\ 0 & z^{-1} & 0 & 0 & 0 \\ 0 & 0 & z^{-1} & 0 & 0 \\ 0 & b_0 & 0 & b_1 & b_2 \end{bmatrix} \begin{bmatrix} R(z) \\ \hdashline X_{1T}(z) \\ X_{2T}(z) \\ \hdashline X_1(z) \\ X_2(z) \end{bmatrix}.$$

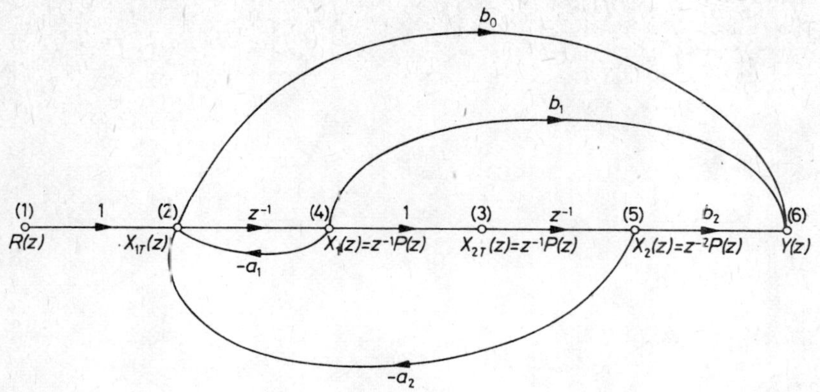

Fig. 8.9

Hence, on the basis of (8.53):

$$\begin{bmatrix} x_1^*(t+T) \\ x_2^*(t+T) \end{bmatrix} = \begin{bmatrix} -a_1 & -a_2 \\ 1 & 0 \end{bmatrix} \begin{bmatrix} x_1^*(t) \\ x_2^*(t) \end{bmatrix} + \begin{bmatrix} 1 \\ 0 \end{bmatrix} r^*(t)$$

are the state equations of the system, and according to (8.54)

$$y^*(t) = \left\{ [b_0 \ 0] \begin{bmatrix} -a_1 & -a_2 \\ 1 & 0 \end{bmatrix} + [b_1 \ b_2] \right\} \begin{bmatrix} x_1^*(t) \\ x_2^*(t) \end{bmatrix} + b_0 r^*(t) =$$

$$= [-a_1 b_0 + b_1 \quad -a_2 b_0 + b_2] \begin{bmatrix} x_1^*(t) \\ x_2^*(t) \end{bmatrix} + b_0 r^*(t)$$

is the time function of the response signal.

2. A signal flow graph of the system drawn in Fig. 8.10 is shown in Fig. 8.11, taking into account, that $x_1(0)$ and $x_2(0)$ are given. Note that the graph shown in the figure is the signal flow graph obtained by the iterative method from the transfer function $W(z)$ discussed in example 1. From the equation

$$\begin{bmatrix} X_{1T}(z) \\ X_{2T}(z) \\ \hline X_1(z) \\ X_2(z) \\ \hline Y(z) \end{bmatrix} = \left[\begin{array}{ccc|cc|cc} 1 & 0 & 0 & 0 & 0 & \beta_1 & 0 \\ 0 & 0 & 0 & b_0 & 0 & -\alpha_1 & \beta_2 \\ \hline 0 & 1 & 0 & z^{-1} & 0 & 0 & 0 \\ 0 & 0 & 1 & 0 & z^{-1} & 0 & 0 \\ \hline 0 & 0 & 0 & 0 & 1 & 0 & -\alpha_2 \end{array}\right] \begin{bmatrix} R(z) \\ x_1(0) \\ x_2(0) \\ \hline X_{1T}(z) \\ X_{2T}(z) \\ \hline X_1(z) \\ X_2(z) \end{bmatrix}$$

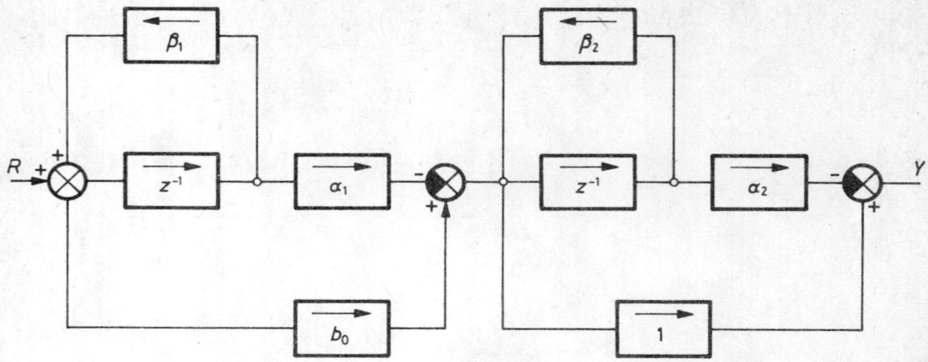

Fig. 8.10

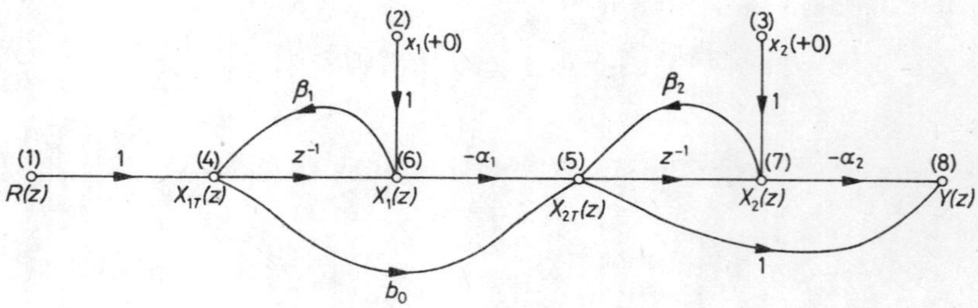

Fig. 8.11

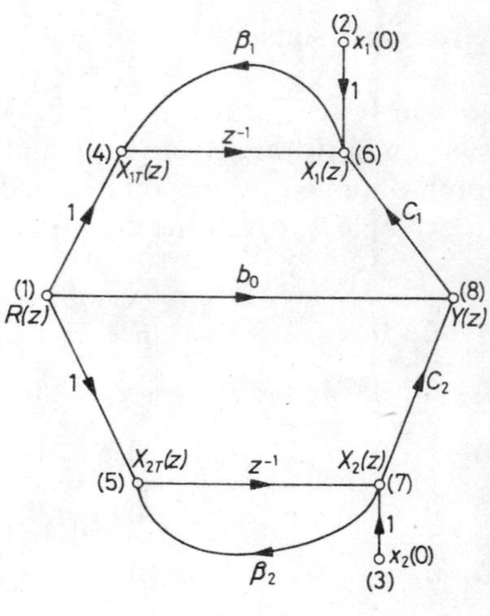

Fig. 8.12

corresponding to (8.50), the state equations of the network have been written based upon (8.53):

$$\begin{bmatrix} x_1^*(t+T) \\ x_2^*(t+T) \end{bmatrix} = \begin{bmatrix} \beta_1 & 0 \\ b_0\beta_1-\alpha_1 & \beta_2 \end{bmatrix} \begin{bmatrix} x_1^*(t) \\ x_2^*(t) \end{bmatrix} + \begin{bmatrix} 1 & 0 & 0 \\ b_0 & 0 & 0 \end{bmatrix} \begin{bmatrix} r^*(t) \\ x_1(0) \\ x_2(0) \end{bmatrix},$$

and further, according to (8.54):

$$y^*(t) = [b_0\beta_1-\alpha_1 \quad \beta_2-\alpha_2] \begin{bmatrix} x_1^*(t) \\ x_2(t) \end{bmatrix} + b_0 r^*(t)$$

is the time function of the response signal.

3. The signal flow graph of a system is shown in Fig. 8.12. On the basis of this, the equation corresponding to (8.50) is:

$$\begin{bmatrix} X_{1T}(z) \\ X_{2T}(z) \\ \hline X_1(z) \\ X_2(z) \\ \hline Y(z) \end{bmatrix} = \left[\begin{array}{ccc|cc|cc} 1 & 0 & 0 & 0 & 0 & \beta_1 & 0 \\ 1 & 0 & 0 & 0 & 0 & 0 & \beta_2 \\ \hline 0 & 1 & 0 & z^{-1} & 0 & 0 & 0 \\ 0 & 0 & 1 & 0 & z^{-1} & 0 & 0 \\ \hline b_0 & 0 & 0 & 0 & 0 & C_1 & C_2 \end{array}\right] \begin{bmatrix} R(z) \\ x_1(0) \\ x_2(0) \\ \hline X_{1T}(z) \\ X_{2T}(z) \\ \hline X_1(z) \\ X_2(z) \end{bmatrix}.$$

Hence, according to (8.54) the time function of the response signal is

$$y^*(t) = [C_1 \quad C_2] \begin{bmatrix} x_1^*(t) \\ x_2(t) \end{bmatrix} + b_0 r^*(t).$$

4. The block diagram of a sampled-data system is shown in Fig. 8.13. The signal flow graph constructed as explained has been drawn in Fig. 8.14. On the basis of this the vertex admittance matrix of the system can be written:

$$\mathbf{W}_t = \begin{array}{c} \\ (4) \\ (5) \\ (6) \\ (7) \end{array} \begin{array}{c} \begin{array}{cccccc} (1) & (2) & (3) & (4) & (5) & (6) \end{array} \\ \left[\begin{array}{c|cc|ccc} 1 & 0 & 0 & 0 & -1 & 0 \\ 0 & 0 & 0 & 0 & 0 & W_2 \\ 0 & W_1 & W_3 & 0 & 0 & 0 \\ \hline 0 & 0 & 0 & 0 & 1 & 0 \end{array}\right] \end{array}.$$

The correspondence between the vertices and the rows and columns of the matrix has been indicated. The partitioning of the matrix has been carried out in accordance with the classification of the signals associated with the vertices. Thus

$$\boldsymbol{W}_{11}=\begin{bmatrix}1\\0\\0\end{bmatrix}; \qquad \boldsymbol{W}_{12}=\begin{bmatrix}0 & 0\\0 & 0\\W_1 & W_3\end{bmatrix}; \qquad \boldsymbol{W}_{13}=\begin{bmatrix}0 & -1 & 0\\0 & 0 & W_2\\0 & 0 & 0\end{bmatrix};$$

$$\boldsymbol{W}_{22}=\boldsymbol{0}; \qquad \boldsymbol{W}_{23}=[0\ 1\ 0].$$

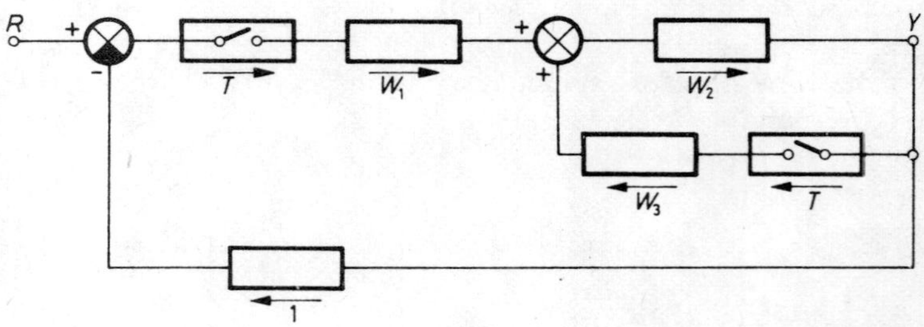

Fig. 8.13

According to (8.62):

$$\begin{bmatrix}P_1\\P_2\\V\end{bmatrix}^* = \frac{R^*}{1+(W_1W_2)^*-(W_2W_3)^*}\begin{bmatrix}1-(W_2W_3)^*\\(W_1W_2)^*\\W_1^*+(W_1W_2)^*W_3^*-W_1^*(W_2W_3)^*\end{bmatrix}$$

and from (8.60):

$$\begin{bmatrix}P_1\\P_2\\V\end{bmatrix}=\begin{bmatrix}R\\0\\0\end{bmatrix}+$$

$$+\begin{bmatrix}-W_1W_2 & -W_2W_3 & 0\\W_1W_2 & W_2W_3 & 0\\W_1 & W_3 & 0\end{bmatrix}\begin{bmatrix}1-(W_2W_3)^*\\(W_1W_2)^*\\W_1^*+(W_1W_2)^*W_3^*-W_1^*(W_2W_3)^*\end{bmatrix}\times$$

$$\times\frac{R^*}{1+(W_1W_2)^*-(W_2W_3)^*}.$$

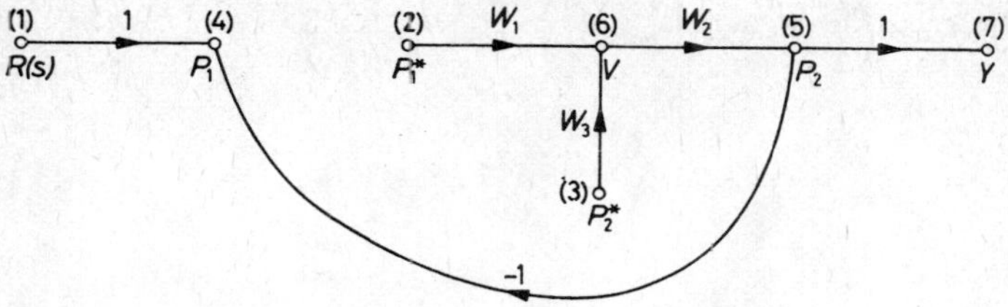

Fig. 8.14

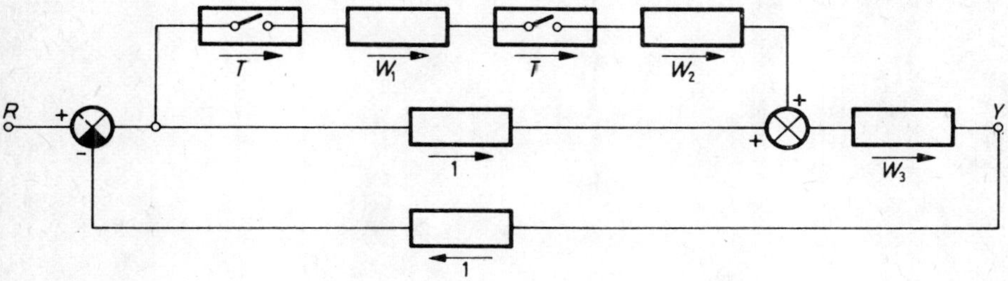

Fig. 8.15

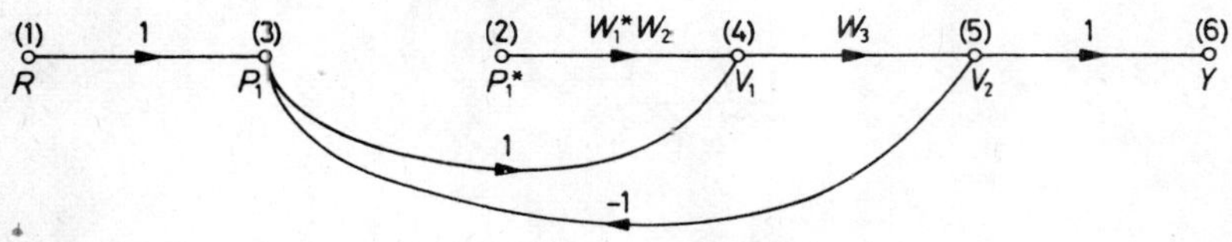

Fig. 8.16

The response signal in accordance with (8.65) is

$$Y = \frac{W_1 W_2 - W_1 W_2 (W_2 W_3)^* + (W_1 W_2)^* W_2 W_3}{1 + (W_1 W_2)^* - (W_2 W_3)^*} R^* .$$

The coefficient of R^* is the transfer function of the system.

The calculation is somewhat simpler if the sampled function of the response signal is determined. Then according to (8.63), the function sought is

$$Y^* = \frac{(W_1 W_2)^*}{1 + (W_1 W_2)^* - (W_2 W_3)^*} R^* .$$

5. The block diagram of a sampled-data system is shown in Fig. 8.15. Sampling elements are here connected before and after the network part of transfer function W_1. These together may be regarded as a basic element with transfer function W_1^*, connected in cascade with the basic element of transfer function W_2. In the signal flow graph shown in Fig. 8.16, one edge has been associated with these, of transfer function $W_1^* W_2$.

From the signal flow graph

$$\boldsymbol{W}_t=\left[\begin{array}{c|c|ccc} 1 & 0 & 0 & 0 & -1 \\ 0 & W_1^* W_2 & 1 & 0 & 0 \\ 0 & 0 & 0 & W_3 & 0 \\ \hline 0 & 0 & 0 & 0 & 1 \end{array}\right]$$

is the vertex transfer-matrix of the system, yielding

$$\boldsymbol{W}_{11}=\begin{bmatrix} 1 \\ 0 \\ 0 \end{bmatrix}; \qquad \boldsymbol{W}_{12}=\begin{bmatrix} 0 \\ W_1^* W_2 \\ 0 \end{bmatrix}; \qquad \boldsymbol{W}_{13}=\begin{bmatrix} 0 & 0 & -1 \\ 1 & 0 & 0 \\ 0 & W_3 & 0 \end{bmatrix};$$

$$\boldsymbol{W}_{22}=\boldsymbol{0}; \qquad \boldsymbol{W}_{23}=[0\ 0\ 1].$$

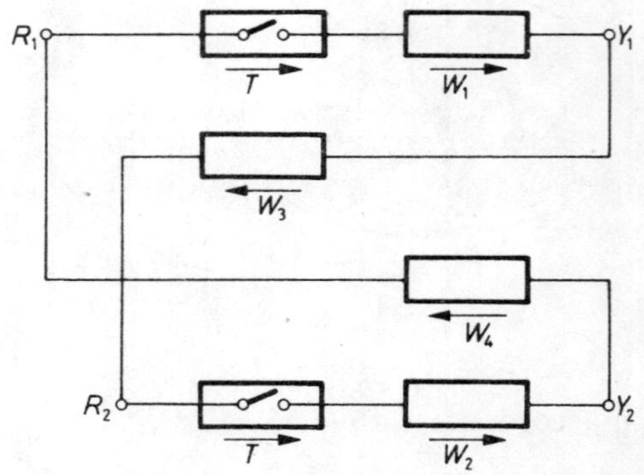

Fig. 8.17

Substituting these into (8.65),

$$Y=\frac{W_3}{1+W_3}R+\frac{1}{1+W_3}\frac{W_1^* W_2 W_3}{1+W_1^*\left(\dfrac{W_2 W_3}{1+W_3}\right)^*}\left(\frac{R}{1+W_3}\right)^*$$

is obtained for the response signal, which, as it is seen, cannot be written as a product of the excitation signal and a transfer function.

6. The block diagram of a sampled-data system with two excitation and two response signals is shown in Fig. 8.17. The signal flow graph associated with the

system has been drawn in Fig. 8.18. Hence, the vertex transfer matrix is

$$\mathbf{W}_t = \left[\begin{array}{cc|cc|cccc} 1 & 0 & 0 & 0 & 0 & 0 & 0 & W_4 \\ 0 & 1 & 0 & 0 & 0 & 0 & W_3 & 0 \\ 0 & 0 & W_1 & 0 & 0 & 0 & 0 & 0 \\ 0 & 0 & 0 & W_2 & 0 & 0 & 0 & 0 \\ \hline 0 & 0 & 0 & 0 & 0 & 0 & 1 & 0 \\ 0 & 0 & 0 & 0 & 0 & 0 & 0 & 1 \end{array}\right],$$

i.e.

$$\mathbf{W}_{11} = \begin{bmatrix} 1 & 0 \\ 0 & 1 \\ 0 & 0 \\ 0 & 0 \end{bmatrix}; \quad \mathbf{W}_{12} = \begin{bmatrix} 0 & 0 \\ 0 & 0 \\ W_1 & 0 \\ 0 & W_2 \end{bmatrix}; \quad \mathbf{W}_{13} = \begin{bmatrix} 0 & 0 & 0 & W_4 \\ 0 & 0 & W_3 & 0 \\ 0 & 0 & 0 & 0 \\ 0 & 0 & 0 & 0 \end{bmatrix};$$

$$\mathbf{W}_{22} = \begin{bmatrix} 0 & 0 \\ 0 & 0 \end{bmatrix}; \quad \mathbf{W}_{23} = \begin{bmatrix} 0 & 0 & 1 & 0 \\ 0 & 0 & 0 & 1 \end{bmatrix}.$$

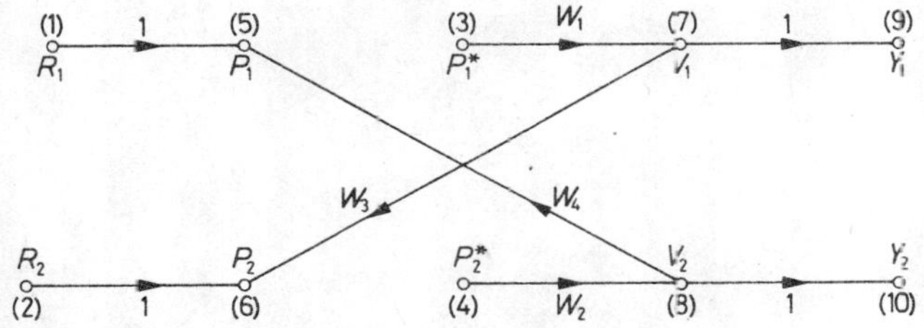

Fig. 8.18

Upon substituting in (8.62), and thereafter into (8.60) the following is obtained:

$$\begin{bmatrix} P_1 \\ P_2 \\ V_1 \\ V_2 \end{bmatrix}^* = \frac{1}{1-(W_1 W_3)^* (W_2 W_4)^*} \begin{bmatrix} 1 & (W_2 W_4)^* \\ (W_1 W_3)^* & 1 \\ W_1^* & W_1^*(W_2 W_4)^* \\ W_2^*(W_1 W_3)^* & W_2^* \end{bmatrix} \begin{bmatrix} R_1 \\ R_2 \end{bmatrix}^*$$

$$\begin{bmatrix} P_1 \\ P_2 \\ V_1 \\ V_2 \end{bmatrix} = \frac{1}{1-(W_1 W_3)^* (W_2 W_4)^*} \begin{bmatrix} W_2 W_4 (W_1 W_3)^* & W_2 W_4 \\ W_1 W_3 & W_1 W_3 (W_2 W_4)^* \\ W_1 & W_1 (W_2 W_4)^* \\ W_2 (W_1 W_3)^* & W_2 \end{bmatrix} \begin{bmatrix} R_1 \\ R_2 \end{bmatrix}^* +$$

$$+ \begin{bmatrix} 1 & 0 \\ 0 & 1 \\ 0 & 0 \\ 0 & 0 \end{bmatrix} \begin{bmatrix} R_1 \\ R_2 \end{bmatrix}.$$

Further, (8.65) yields the column matrix of the response signals:

$$\mathbf{Y} = \begin{bmatrix} Y_1 \\ Y_2 \end{bmatrix} = \frac{1}{1-(W_1 W_3)^* (W_2 W_4)^*} \begin{bmatrix} W_1 & W_1 (W_2 W_4)^* \\ W_2 (W_1 W_3)^* & W_2 \end{bmatrix} \begin{bmatrix} R_1 \\ R_2 \end{bmatrix}^*.$$

REFERENCES

[1] ABRAHAMS, J. R. and COVERLEY, G. P.: *Signal Flow Analysis.* Pergamon, Oxford, 1965.

[2] ASH, R., KIM, W. H. and KRANC, G. M.: A general flow graph technique for the solution of multiloop sampled systems. *Trans. ASME* (1960), 360–370.

[3] BALABANIAN, N. and BICKART, Th. A.: *Electrical Network Theory.* Wiley, New York, 1969.

[4] BERGE, C.: *Graphs and Hypergraphs.* North-Holland, Amsterdam, 1973.

[5] BLACKWELL, W. A. and GRIGSBY, L. L.: Systems of one-port linear components containing through-across type drives. *J. Franklin Inst.*, 278 (1964), 237–247.

[6] BONDY, J. A. and MURTY, U. S. R.: *Graph Theory with Applications.* Elsevier, New York, 1976.

[7] BRUTON, L. T.: *RC-active Circuits.* Prentice-Hall, Englewood Cliffs, N. J., 1980.

[8] CEYHUN, Y. and GIRESUNLU, E.: State equations for network containing nullators and norators. *Proc. 3rd Int. Symposium on Network Theory, Split, Yugoslavia* Sept. 1–5, 1975, 113–120.

[9] CHEN, W. K.: *Applied Graph Theory.* North-Holland, Amsterdam, 1976.

[10] CSÁKI, F.: *Modern Control Theories, Nonlinear, Optimal and Adaptive Systems.* Akadémiai Kiadó, Budapest, 1972.

[11] CSÁKI, F.: *State-space methods for control systems.* Akadémiai Kiadó, Budapest, 1976.

[12] CVETKOVIC, D. M.: *Spectra of Graphs.* VEB Deutscher Verlag der Wissenschaften, Berlin, 1980.

[13] DAVIES, A. C.: Matrix analysis of networks containing nullators and norators. *Electron. Lett.*, 2 (No. 2), (1966), 48–49.

[14] DAVIES, A. C.: Nullator–norator equivalent networks for controlled sources. *Proc. IEEE* (1967), 722–723.

[15] DEO, N.: *Graph Theory with Applications to Engineering and Computer Science.* Prentice-Hall, Englewood Cliffs, N. J., 1974.

[16] DESOER, Ch. A. and KUH, E. S.: *Basic Circuit Theory.* Wiley, New York, 1969.

[17] EULER, L.: Solutio problematis ad geometriam situs pertinantis. *Acad. Petropolitanae*, 8 (1736), 128–140.

[18] FODOR, GY., SIMONYI, K. and VÁGÓ, I.: *Elméleti villamosságtan példatár* (Collection of Examples in Theoretical Electrotechniques. In Hungarian). Tankönyvkiadó, Budapest, 1980.

[19] FODOR, GY.: *Laplace Transforms in Engineering.* Akadémiai Kiadó, Budapest, 1965.

[20] FODOR, GY.: The analysis of linear networks containing two-ports and coupled two poles. *Period. Polytech. Electr. Eng.*, 17 (1973), 321–332.

[21] FODOR, GY. and VÁGÓ, I.: Setting the operating point of electronic circuits with the aid of graph theory. *2nd Conference on Electronic Circuits, Prague, 1976*, 167–170.

[22] FU, Y.: Realization of circuit matrices. *IEEE Trans. Circuit Theory*, CT–12 (1965), 604–607.

[23] GALANI, Z. and SZENTIRMAI, G.: D. C. operation of three-transistor gyrators. *IEEE Trans. Circuit Theory*, (1971), 738–739.

[24] GUILLEMIN, E. A.: *Communication Networks.* Wiley, New York, 1936.

[25] GUILLEMIN, E. A.: *Introductory Circuit Theory.* Wiley, New York, 1953.

[26] GUILLEMIN, E. A.: *Theory of Linear Physical Systems.* Wiley, New York, 1963.

[27] HARARY, F.: *Graph Theory.* Addison–Wesley, Reading, 1969.

[28] HARRIS, B.: *Graph Theory and its Application.* Academic Press, New York, 1970.
[29] HOLLÓS, E.: Two-port models containing nullators and norators. *Period. Politech. Electr. Eng.* (1981), No. 3, 167–170.
[30] KIRCHHOFF, G.: Über die Auflösung der Gleichungen, auf welche man bei der Untersuchung der linearen Vertheilung galvanischer Ströme geführt wird. *Poggendorf Ann. Phys.*, 72 (1847), 497–508.
[31] KÖNIG, D.: *Theorie der endlichen und unendlichen Graphen.* Chelsea, New York, 1950.
[32] KUO, B. C.: *Automatic Control Systems.* Prentice-Hall, Englewood Cliffs, N. J.,
[33] MARSHALL, C.: *Applied Graph Theory Interscience.* Wiley, New York, 1971.
[34] MASON, S. J.: Feedback theory – Further properties of signal flow graphs. *Proc. IRE*, 44 (1956), 920–926.
[35] MAYEDA, W.: A proof of Tutte's realizability condition. *IEEE Trans.*, CT–17 (1970), 506–511.
[36] MAYEDA, W.: *Graph Theory.* Wiley, New York,
[37] MITRA, S. K.: *Analysis and Synthesis of Linear Active Networks.* Wiley, New York, 1969.
[38] NOLTEMEIER, H.: *Graphentheorie mit Algorithmen und Anwendungen.* Gruyter, Berlin, 1976.
[39] PICARD, C. F.: *Graphs and Questionnaires.* North-Holland, Amsterdam, 1980.
[40] ROGERS, F. E.: *Topology and Matrices in the Solution of Networks.* Iliffe, London, 1965.
[41] SALZER, J. M.: Signal flow reductions in sampled data systems. *1957 IRE Wescon Convention Record*, Part 4, 166–170.
[42] SESHU, S. and REED, M. B.: *Linear Graphs and Electrical Networks.* Addison–Wesley, London, 1961.
[43] SIGORSKI, V. P. and PETRENKO, A. I.: *Algoritmi analiza elektronni skhem* (Algorithms for the Analysis of Electrical Networks. In Russian). Technika, Kiev, 1970.
[44] SIGORSKI, V. P. and PETRENKO, A. I.: *Osnovi teorii elektronni skhem* (The Foundation of Electrical Network Theory. In Russian). Vishcha Shkola, Kiev, 1971.
[45] TINHOFER, G.: *Methoden der angewandten Graphentheorie.* Springer, Vienna, 1976.
[46] TOU, J. T.: *Digital and Sampled-data Control Systems.* McGraw-Hill, New York, 1959.
[47] TOU, J. T.: *Modern Control Theory.* McGraw-Hill, New York, 1964.
[48] TOU, J. T.: *Optimum Design of Digital Control Systems.* Academic Press, New York, 1963.
[49] TUTTE, W. T.: From matrices to graphs. *Can. J. Math.*, 16 (1964), 108–127.
[50] VÁGÓ, I.: Over-ground power transmission line systems. *Period. Polytech. Electr. Eng.*, 9 (1965), 67–96.
[51] VÁGÓ, I.: The calculation of transmission networks with the aid of graph theory. *Period. Polytech. Electr. Eng.*, 13 (1969), 155–168.
[52] VÁGÓ, I.: The calculation of stationary state transmission line systems by topological methods. *Period. Polytech. Electr. Eng.*, 13 (1969), 325–340.
[53] VÁGÓ, I.: Determination of the characteristic matrices of networks consisting of transmission lines. *Period. Polytech. Electr. Eng.*, 15 (1971), 381–389.
[54] VÁGÓ, I. and BÁRDI, I.: Calculation of networks containing ideal generators by using the topological matrices. *Period. Polytech. Electr. Eng.*, 17 (1973), 293–300.
[55] VÁGÓ, I. and HOLLÓS, E.: Two-port models with nullators and norators. *Period. Polytech. Electr. Eng.*, 17 (1973), 301–309.
[56] VÁGÓ, I.: Calculation of network models containing nullators and norators. *Period. Polytech. Electr. Eng.*, 17 (1973), 311–319.
[57] VÁGÓ, I.: Analyse des linearen Netzen mit gesteuerten Generatoren. *XX. Internationales Wissenschaftliches Kolloquium Ilmenau*, 1975, Heft 2, 43–45.
[58] VÁGÓ, I.: Analysis of steady state linear networks containing controlled generators. *Period. Polytech. Electr. Eng.*, 20 (1976), 129–139.
[59] VÁGÓ, I.: State equations for linear network models containing nullators and norators. *Period. Polytech. Electr. Eng.*, 20 (1976), 399–409.
[60] VÁGÓ, I.: State equation for linear networks. *Period. Polytech. Electr. Eng.*, 20 (1976), 411–416.
[61] VÁGÓ, I.: Calculation of impedance characterizing the 750 kV transmission line. *Period. Polytech. Electr. Eng.*, 24 (1980), 191–199.
[62] WEINBERG, L.: Kirchhoff's third and fourth laws. *Trans. IRE*, CT–5 (1958), 8–30.

SUBJECT INDEX